OXFORD MONOGRAPHS ON GEOLOGY AND GEOPHYSICS

The Precambrian–Cambrian Boundary

Edited by

J. W. COWIE

Department of Geology
University of Bristol, UK

M. D. BRASIER

Department of Earth Sciences
University of Oxford, UK

CLARENDON PRESS · OXFORD
1989

Oxford University Press, Walton Street, Oxford OX2 6DP
Oxford New York Toronto
Delhi Bombay Calcutta Madras Karachi
Petaling Jaya Singapore Hong Kong Tokyo
Nairobi Dar es Salaam Cape Town
Melbourne Auckland
and associated companies in
Berlin Ibadan

Oxford is a trade mark of Oxford University Press

Published in the United States
by Oxford University Press, New York

British Library Cataloguing in Publication Data
The Precambrian–Cambrian boundary
1. Pre-Cambrian strata   2. Cambrian
strata
I. Cowie, J. W.   II. Brasier, M. D.
III. Series
551.7'1
ISBN 0-19-854481-2

Library of Congress Cataloging in Publication Data
The Precambrian–Cambrian boundary/edited by J. W. Cowie, M. D.
Brasier.
p. cm. Includes index.
1. Geology, Stratigraphic—Pre-Cambrian.   2. Geology,
Stratigraphic—Cambrian.   3. Paleontology—Pre-Cambrian.
4. Paleontology—Cambrian.   I. Cowie, J. W.   II. Brasier, M. D.
QE653.P7322 1989
551.7' 1—dc19   88-17865   CIP
ISBN 0-19-854481-2

Typeset by Cotswold Typesetting Limited, Gloucester, UK
Printed in Great Britain by
The Alden Press, Oxford

Preface

This book is intended to serve as a brief review of the state of research on the Precambrian–Cambrian boundary problem in the later 1980s; a problem which has been examined with increasing intensity since the nineteenth century. It is couched as a statement by some English geologists in the hope of an impartial stance, but the authors are solely responsible for their chapters and do not represent in any formal way the views of British geologists (if there are such homogeneous views—and we know of no evidence for them) or the views of the Working Group on the Precambrian–Cambrian (P€–€) boundary (IUGS) as a whole, although the authors in this volume are all members of that Working Group. At the time of writing there is no consensus or majority view evident as to the recommended placing of the P€–€ boundary either in stratigraphical level or geographical position for the Global Stratotype Section and Point (GSSP).

The chapters of this book are on topics selected because they are considered to be of current (and probably future) interest, but no attempt has been made to be comprehensive on this global subject. A really complete review would probably be a very large volume (or a multi-volume monograph set). It is hoped that it will appeal to specialists in the problem and also to general geologists in academic and resource fields as a state-of-the-art entrée to further reading.

Bristol   J.W.C.
Oxford   M.D.B.
July 1988

Contents

The plates for Chapter 7 fall between pages 134 and 135

Contributors

M. D. Brasier

Department of Earth Sciences, University of Oxford, Parks Road, Oxford OX1 3PR, UK

S. Conway Morris

Department of Earth Sciences, University of Cambridge, Downing Street, Cambridge CB2 3EQ, UK

J. W. Cowie

Department of Geology, University of Bristol, Bristol BS8 1RJ, UK

T. P. Crimes

Department of Geological Sciences, University of Liverpool, Brownlow Street, Liverpool L69 3BX, UK

W. B. Harland

Department of Earth Sciences, University of Cambridge, Downing Street, Cambridge CB2 3EQ, UK

PART I
REGIONAL OVERVIEWS

1

Introduction

J. W. Cowie

It can be argued as a principle that the stratigraphical level of the Global Stratotype Section and Point (GSSP) should be decided first, and then after that attention is directed to the selection of the best geographical position for the GSSP. Many geologists, including non-stratigraphers, are looking for a practical outcome to the problem, however, and pragmatism sees the need for compromise in these man-made decisions. Thus, a simultaneous search for both a stratigraphical level and a geographical position may be the way to get a useful, practical outcome which is guided but not dictated by certain stratigraphical principles decided a priori. As discussed at some length in *Guidelines and Statutes of the International Commission on Stratigraphy (ICS)* a Global Stratotype Section and Point (GSSP) is 'a unique and specific point in a specific sequence of rock strata in a unique and specific geographical location' (Cowie *et al.* 1986, p. 5) and is 'an actual point in rock and is not an abstract concept' and therefore cannot be diachronic unlike all other practical, real methods for defining a 'unique time signal for the world geological stratigraphic time scale'.

Changes found in the geological record moving from Precambrian into Cambrian rocks (however defined) are numerous and can include many esoteric and speculative factors. A simple selection was made by Rozanov (1984). A slightly modified form is:

(1) decrease in dolomite accumulation;
(2) sharp drop in stromatolite formation and change of morphology;
(3) first widespread appearance of red biogenic limestones;
(4) global accumulation of large phosphorite deposits (USSR, Mongolia, China, and elsewhere);
(5) first rich assemblages of skeletalized fossils;
(6) considerable changes in the morphology and incoming of inferred complex biological 'programming' seen in trace fossils.

The stratigraphical column can be divided into Eons. One scheme widely used is:

Phanerozoic
Proterozoic
Archaean.

Many geologists, probably a majority, take the Cambrian Period as being the earliest period of the Phanerozoic Eon and this, conveniently perhaps, places the Proterozoic and Archaean Eons in the informal but useful term 'Precambrian'. An alternative view would take the earliest period of the Phanerozoic as being the Sinian/Vendian/Ediacarian putative Period: this removes the end of the 'Precambrian' from coincidence with the end of the 'Proterozoic' Eon as so defined. Here we assume that the Precambrian–Cambrian chronostratigraphical boundary is at the same time as the Proterozoic–Phanerozoic chronostratigraphical boundary. It is hoped that further discussion can take place and perhaps an international decision can be furthered or even achieved at meetings in Washington DC, USA, during the 28th International Geological Congress in July 1989.

It will be apparent in this volume that in many Precambrian–Cambrian (P€–€) sequences the earliest skeletal fossils appear in facies which lack trilobites; the earliest skeletal fossils have been used to delimit sub-trilobite units of the earliest Cambrian and latest Precambrian: correlation (dependent on pragmatic necessity) is on stratigraphical position and the absence of trilobites, rather than on the presence of a distinctive assemblage. The first appearance of arthropod trace fossils (often assigned to the activities of trilobites) is usually stratigraphically well below the first trilobite skeletalized occurrences. This 'anomaly' is usually explained away by ecological-environmental arguments. The occurrence or non-occurrence of trilobites with the same genera (and possibly species) of 'small shelly fossils' (SSF), from differing levels in the early Cambrian beds has caused some unease in assessing the biostratigraphical value of earliest skeletal fossils for global correlation. See pp. 120–2 of this volume for a full discussion of this complex problem.

Since the Second World War discussion on the best placing of the boundary between Precambrian and Cambrian rocks has progressed far from the suggestion by H. E. Wheeler in 1947 that the boundary should be based on the first occurrence of trilobites. Earlier criteria were based on the classical idea that the base of the Cambrian System should be placed at the unconformity widely found between unmetamorphosed sediments and metamorphosed basement rocks (gneisses, schists, and igneous bodies). A later view was to place the PЄ–Є boundary at the first long hiatus (disconformity or unconformity) below beds with 'archaeocyathids, brachiopods, hyolithids and trilobites'.

Progress was accelerated by two meetings:

(1) a symposium on the Cambrian System and the problem of its base at the International Geological Congress (IGC) in Mexico in 1956;
(2) an international colloquium in Paris on 'Les relations entre Précambrien et Cambrien, problèmes des séries intermédiaires' in 1957.

Other forward steps were the establishment of the Subcommission on Cambrian Stratigraphy at the 1960 IGC in Copenhagen (at the suggestion of M. F. Glaessner) followed by the appointment of C. J. Stubblefield as its first President (Chairman) at the 1964 IGC in New Delhi (with, later, J. W. Cowie as Secretary). Stubblefield selected the Precambrian–Cambrian boundary problem as the prime research effort for the Cambrian Subcommission. Cowie became Chairman of the newly formed PЄ–Є Boundary Working Group at the Montreal IGC in 1972 when a symposium on the topic was held, effectively postponed since the Prague IGC in 1968.

Over one hundred members from twenty countries have been, or are, members of the PЄ–Є Boundary Working Group (PЄ–ЄBWG) which is directly responsible to the International Commission on Stratigraphy (ICS) of the International Union of Geological Sciences (IUGS). From 1974 to 1987 the same Working Group was Project 29 of the International Geological Correlation Programme (a joint programme of the United Nations Educational and Scientific Organization and IUGS).

From 1972 to 1987 a series of meetings was organized to examine and discuss Precambrian–Cambrian boundary sections. Plenary sessions and workshops were held in Montreal (Canada, 1972), Paris (France, 1974), Moscow (USSR, 1975), Leningrad (USSR, 1976), Sydney (Australia, 1976), Beijing (China, 1978), Paris (1980), Golden (Colorado, USA, 1981), Kunming (China, 1982), Bristol (UK, 1983), Moscow (1984), Uppsala (Sweden, 1986), and St. John's, Newfoundland (Canada, 1987).

Field meetings were held on fifteen occasions, involving both examinations of sections and discussions

leading to subsequent research with local geologists. The following areas have been visited: East Siberia (1973 and 1981), Normandy and Brittany (1974), Ural Mts (1975), Georgia, USSR (1975), Anti-Atlas Mts of Morocco (1975 and 1976), Flinders Ranges in South Australia (1976), Iberian Peninsula of Spain and Portugal (1976), Central and South China (1978, 1982, and 1987), eastern Newfoundland (1979 and 1987), Mackenzie Mts, Canada (1979), Nevada–California (1981), Wales and England (1983), and South Sweden (1986). In plenary sessions at Paris (1974), Cambridge (1978), Bristol (1983), and Moscow (1984) a series of important decisions were reached.

In Paris, the Working Group agreed that its primary task was the choice of a stratotype boundary point, with a secondary task to consider stratigraphical divisions above and below the boundary. Any succession selected for the boundary point must be marine, and as continuous and monofacial as possible.

'Monofacial' was included in order to exclude a situation where a radical and long-lasting change of facies at, or near, a putative boundary level coincided with the incoming of abundant skeletalized fossils. The incoming could then have been due to ecological–environmental change rather than evolution of the fossil biota. It was agreed that the main method of guidance in selection should be biostratigraphy, but all possible methods of correlation should be enlisted. It was also agreed at this time that the 'Ediacara' type fauna should be considered as Precambrian, and the 'olenellid–fallotaspid' trilobite faunas as Cambrian.

At the Cambridge meeting, the Working Group agreed to be guided primarily by the criterion that the Precambrian–Cambrian boundary should be approximately located in the chronostratigraphical scale near the evolutionary changes signalled in the rocks by the appearance of diverse fossils with hard parts. This decision was superseded in Bristol (1983) by one to place the boundary stratotype as close as practicable to the lowest known appearance of diverse shelly fossils with a good potential for correlation. The Cambridge meeting also emphasized that the boundary stratotype point is defined and placed in the rock, where it will be a reference point that should remain fixed, despite the possibility of fresh discoveries in the rocks stratigraphically below or above. It is worth noting that at this meeting, a minority view held that 'in selecting a boundary stratotype point, one should ensure that there is a maximum potential for correlation, even if this implies a higher stratigraphic position so that there obtains a Precambrian shelly fauna'.

At the 1983 Bristol meeting, candidates for the Global Stratotype Section and Point were discussed in some detail, and three were selected for further consideration: Ulakhan–Sulugur on the Aldan River in east Siberia, USSR; on the Burin Peninsula, of eastern Newfoundland, Canada; and at Meishucun in Yunnan Province, southern China. The 32 members of the Working Group present (out of a total membership of 100) voted by 19 to 13 to put the Ulakhan–Sulugur stratotype section to a postal ballot of all the Voting Members of the Group. The result of this postal vote was a rejection of the Siberian candidate by 9 against, 7 in favour, and 3 abstentions.

By December 1983, there was still no satisfactory standard available for the Newfoundland candidate, so it was decided to put the Chinese section to a postal ballot. This resulted in a clear majority (nearly 80 per cent of votes) for placing the Global Stratotype Section and Point at Meishucun in south China. This decision was submitted in June 1984 for discussion at the Moscow International Geological Congress in August 1984, to be followed by a postal ballot of Voting Members of ICS, with the result submitted for ratification by the IUGS Executive in 1985.

After discussions in Moscow, the Working Group Chairman/Project 29 Leader added a 'rider' to the June 1984 submission for presentation to the full Commission meeting in Moscow in August 1984. He stated that although the Chinese candidate is 'on the table and will stay there unless removed by postal ballot of the Voting Members of the Working Group or a Commission decision, . . . there is still an urgent need for more accurate and detailed *correlation* with other areas'.

In Uppsala, Sweden, in May 1986, specialists from the Cambrian Subcommission and the PЄ–ЄB WG (and others) had an intensive work session on the taxonomy and biostratigraphy of the earliest skeletal fossils and made progress in attempting to resolve, by studying each others' collections, current problems of taxonomy and biostratigraphy at this stratigraphical level.

The opinion has been expressed in the PЄ–ЄB WG that although there is general agreement with the principle of placing the boundary '. . . as close as practical to the first appearance of abundant shelly fossils . . .', marked provincialism of the earliest skeletal fossils and their virtual restriction to carbonate facies have hampered global correlation in the boundary interval. In August of 1987, fifty geologists from ten countries met in eastern Newfoundland for a conference and field trip on the Precambrian–Cambrian boundary at St. John's and in the Avalon, Bonavista, and Burin Peninsulas as part of the PЄ–ЄB WG series of meetings.

One major theme of the meeting was the prospect for global correlation using trace fossils. Trace fossils are especially common in siliciclastic facies, in which shelly fossils typically are rare and poorly preserved. Correlation in siliciclastic facies is critical, as these deposits comprise nearly 70 per cent of exposed rocks in the boundary interval. In his keynote address, T. P. Crimes (UK) outlined three globally correlatable trace-fossil zones that occur below the lowest trilobites. In talks on the Wernecke Mountains of north-western Canada and the Chapel Island Formation of eastern Newfoundland, G. M. Narbonne (Canada) and his co-workers emphasized the fact that trace fossils are cosmopolitan, and the same genera and species can be used to zone siliciclastic strata world-wide. Other talks by J. Paczesna (Poland), G. Kumar (India), and Jiang Zhiwen (China) pointed out the biostratigraphical potential of trace fossils. However, careful morphological analysis, rigorous taxonomy, and detailed documentation of the stratigraphical ranges of ichnotaxa are needed for more sections, particularly in Australia and the Russian Platform, to test further the correlations.

The conference was followed by a seven-day field trip to examine the Upper Proterozioc–Lower Cambrian strata of eastern Newfoundland. Interest was focused on the newly proposed Precambrian–Cambrian boundary stratotype candidate (Narbonne *et al.* 1987). The boundary is placed at the abrupt appearance of complex Phanerozoic-type trace fossils (*Phycodes pedum* Zone), which approximately corresponds with the first appearance of simple small shelly fossils (*Sabellidites cambriensis* interval) in the section. Some concerns were expressed about the relative scarcity of small shelly fossils in the immediate vicinity of the boundary, and the distinct possibility that the Acadian Orogeny may have reset the palaeomagnetic and geochronometric signatures of the section. These were offset by the accessibility of the section, the apparent absence of disconformities or marked facies changes in the boundary interval, the presence of a distinctly Precambrian fossil assemblage in the underlying strata (*Harlaniella podolica* Zone), and the excellent prospects for global correlation. At the end of the field trip, seven of the ten voting members of the Working Group present (and 73 per cent of all field-trip participants) voted that the section and horizon were suitable for a boundary stratotype. Further deliberations will be held in the International Geological Congress at Washington DC in 1988 before a postal ballot of the full voting membership is held. See also Chapter 11 of this volume: 'Concluding remarks'.

2

South-eastern Newfoundland and adjacent areas (Avalon Zone)

S. Conway Morris

2.1. INTRODUCTION

The search for an acceptable stratotype for the Precambrian–Cambrian boundary (hereafter referred to as simply the boundary) is only one project within a major programme designed to identify type sections for all the principal chronostratigraphical boundaries within the Phanerozoic. Definition of the boundary, however, has a unique importance because of the development of organisms with hard skeletal parts. Their appearance provides for the first time in Earth history a fossil record that is adequate both for detailed evolutionary and biostratigraphical investigations. The evolution of skeletal parts is only one facet of a series of major adaptive radiations in the metazoans that ushered in the Phanerozoic, the eon of 'visible life'. The factors behind this dramatic evolutionary event are still the subject of vigorous debate, although the role of extrinsic controls such as ocean chemistry and continental configurations would seem to be of some importance (Conway Morris 1987). The detailed examination of the various boundary sections around the world is providing abundant new data on these problems, and the purpose of this chapter is to review the present knowledge of the Precambrian–Cambrian transition in Newfoundland and adjacent maritime areas as far south as Massachusetts and Rhode Island (Fig. 2.1), collectively referred to as the Avalon Zone. Some of the relative merits of this area of the world, especially in comparison with existing stratotype candidates in eastern Siberia (Aldan River) and southern China (Meishucun, near Kunming), are discussed below.

2.2. THE AVALON ZONE

With the acceptance of plate tectonics, the study of fossil distributions received a renewed impetus, that in many cases helped to formulate hypotheses on the former distribution of continents and now-vanished oceans. It had been long recognized (e.g. Marcou 1889, 1890; Matthew 1892) that the early Palaeozoic faunas, especially those of the Cambrian, in south-east Newfoundland and a maritime strip that extended south-west to encompass Nova Scotia and sections of eastern New Brunswick, Massachusetts, and Rhode Island (Fig. 2.1) had a striking similarity to equivalent faunas in England and Wales. The classic paper by Wilson (1966) sought to explain these distributions by the existence of an ocean, now generally referred to as Iapetus or the 'proto-Atlantic', although in this context the remarkably prescient palaeocontinental reconstruction by Choubert (1935) also deserves special attention. In its simplest form this hypothesis proposed that the location of this ocean with respect to the major continental cratons of North America (Laurentian craton) and Europe (Balto-Scandinavian craton and the supposedly adjacent area of Avalonia) approximated to the position of the present-day North Atlantic. However, the suture of closure of Iapetus did not coincide exactly with the line of opening of the North Atlantic. Thus it was that some marginal areas of the cratons were transposed to a position which is anomalous in terms of Cambrian biogeography. Hence, south-east Newfoundland and adjacent maritime areas were detached from the region that now comprises England and Wales, and were joined to the Laurentian craton, which during the Cambrian housed a markedly different faunal assemblage that flourished on the far side of Iapetus (see Conway Morris & Rushton, in press).

However, this model involving a jigsaw-like joining and separation of large cratons is realized now to be too simplistic, and has been substantially modified by the identification of a series of exotic terranes. These had their origins off-board from the Laurentian craton, and were accreted at various times before the Carboniferous. Of particular relevance to this chapter is the recognition of the Avalon Zone (e.g. Williams & Hatcher 1982, 1983; see also Rast & Skehan 1983), which Keppie (1985) has argued is a composite structure composed of perhaps nine individual terranes. Each terrane is defined by Keppie (1985, p. 1218) 'as an area characterized by an internal continuity of geology, including stratigraphy, faunal provinces, structure, metamorphism, igneous petrology, metallogeny, and palaeomagnetic record, that is distinct from that of neighbouring terranes and cannot be explained by facies changes', while the 'Terrane boundaries are generally faults or melanges representing sutures, now largely cryptic, with complex structural histories often involving transcurrent motions.'

For convenience, the Avalon Composite Terrane or Superterrane is hereafter referred to as the Avalon Zone, because by boundary times the various terranes appear to have amalgamated into a single unit. The evidence for this lies in the faunal, lithological, and stratigraphical similarities of the Cambrian sequences of the different terranes (Keppie 1985; see also Williams & Hatcher 1983). Keppie (1985) notes that some of these similarities could be a result of still-separated terranes lying in the same facies belt (but see Krogh *et al.* 1988), but at the moment it seems best to assume that fusion into a superterrane is reflected by the late Precambrian Cadomian (or Avalonian) Orogeny, and was complete by the Lower Cambrian. However, the precise times of collision are not well constrained, and it seems doubtful whether the adjoining of the terranes was synchronous in terms of geological time.

Even greater uncertainty surrounds the relative positions of the various terranes to each other or the major cratons during the Precambrian and prior to amalgamation. Similarities to the general tectonic and sedimentary sequences of north-west Africa, especially Morocco and Algeria, have led a number of authors (e.g. Schenk 1971; Hughes 1972; Strong 1979; Rast 1980; Piqué 1981; King 1982) to argue for the proximity of the Avalon Zone to this margin of Gondwanaland. By implication this would include Armorica, which on palaeomagnetic evidence appears to have formed part of the Gondwana supercontinent during the Cambrian, possibly having converged during the late Precambrian to produce the Cadomian orogeny (Hagstrum *et al.* 1980; Perigo *et al.* 1983). A juxtaposition is also favoured, given Hagstrum *et al.*'s (1980 see also Perigo *et al.* 1983; Rast & Skehan 1983) inclusion, on palaeomagnetic evidence, of the Anglo-Welsh area within Armorica, and the faunal similarities between England and Wales and the Avalon Zone.

With respect to the Avalon Zone, however, few palaeomagnetic data are available, although there is some evidence for its proximity to Amorica and the north African edge of Gondwanaland (Johnson & Van der Voo 1985; Seguin 1985). Armorica appears to have been situated near the equator in the late Precambrian (Perroud 1985), and the palaeomagnetic results of

Seguin (1985) and Seguin & Lortie (1986) also suggest that the late Precambrian Harbour Main Group and associated intrusives (Fig. 2.8) of south-east Newfoundland were located at relatively low latitudes (c. 20–30°). Palaeomagnetic measurements of the Marystown Group, which may correlate with the Harbour Main Group (Fig. 2.8), also gave a comparable palaeolatitude of about 35°, although Irving & Strong (1985) stressed that the date of magnetization was not firmly established. If such a location receives further support, this may indicate a period of drift into more temperate climes by the Lower Cambrian, as indicated both by the sedimentary facies and relatively low faunal diversity. This is supported by palaeomagnetic measurements within the Bourinot Group of Cape Breton Islands, indicating that by the early Middle Cambrian (or possibly even the Lower Cambrian, see below) this part of the Avalon Zone lay at about $49° \text{ S} \pm 11°$ (Johnson & Van der Voo 1985). Moreover, palaeomagnetic measurements on the Fourchu Volcanic Group (and an associated gabbro) gave a palaeolatitude of about 45° S (Johnson & Van der Voo 1986). These authors stressed that the age range of this determination could straddle some time between the late Precambrian and early Silurian, but tentatively suggested an early Cambrian age on the assumption that magnetization had been imposed shortly after the Avalonian Orogeny. In this interval, Hagstrum *et al.* (1980, see also Perroud 1985) also show Armorica and northern Africa (Gondwanaland) to lie at high latitudes.

Rao *et al.* (1981, 1986), however, presented somewhat contradictory pole positions on the basis of palaeomagnetic measurements on a number of Cambrian granites (see below for discussion of radiometric dating) and sediments (Morrison River and MacCodrum Formations, Fig. 2.8) on Cape Breton Island. These palaeomagnetic data were not easy to interpret (see also Johnson & Van der Voo 1985, 1986), and in the case of the palaeomagnetic analysis of the sediments, Rao *et al.* (1986) were unable to establish the time of rock magnetization with any certainty. However, their inference that, in the Cambrian, Cape Breton had a low-latitude position (7° and 17° on the basis of granites and sediments respectively) is neither in close agreement with measurements from the adjacent Bourinot Group (Johnson & Van der Voo 1985) nor in keeping with the general lithofacies and biofacies of the Avalon Zone.

Elsewhere in the Avalon Zone palaeomagnetic measurements on boundary rocks typically yield poles that are consistent with Devonian–Carboniferous positions, and were evidently imposed during later orogenic events. For example, palaeomagnetic measurements of the Lower Cambrian Ratcliffe Brook Formation in New

Brunswick (Fig. 2.8) indicate that the signature was imposed after Carboniferous folding, and gives a pole position consistent with the later Carboniferous position of North America (Spariosu *et al.* 1983; see also Rao *et al.* 1986).

The possible links between the Avalon Zone and Gondwanaland (including Armorica) require further examination, especially with respect to the degree of biotic similarity (e.g. absence of Lower Cambrian archaeocyathans in the Avalon Zone as against their presence in Normandy and Morocco; see Conway Morris & Rushton in press) and palaeogeographical position as inferred from palaeomagnetic work. However, it seems generally agreed that the Laurentian craton was remote from the Avalon Zone (Hagstrum *et al.* 1980; Johnson & Van der Voo 1985, but see Seguin 1985), and owes its present proximity to the closure of an ocean.

The suggestions that the Avalon Zone has a composite nature may require some reassessment of the various models that have sought to explain its tectonostratigraphical evolution as a single unit. Attempts to incorporate the history of the Avalon Zone, or parts thereof, into the broad framework of plate tectonics have led to two major and discrepant models. Many of these discussions have revolved around data from south-east Newfoundland, although other areas of the Avalon Zone have also received attention. One model views the development of the Late Proterozoic sequence as reflecting proximity to a consuming plate margin, such as an ensialic island arc (e.g. Hughes & Brückner 1971; Hughes 1972; Rast *et al.* 1976; Rast & Skehan 1983). An echo of this suggestion is found also in Thorpe *et al.* (1984) who postulate an island-arc setting for the late Precambrian sequences of England and Wales, which have a number of striking similarities (e.g. Rast *et al.* 1976) with certain parts of the Avalon Zone, especially south-east Newfoundland. The alternative model appeals to data taken to interpret the Avalon Zone as having lain in a tensional regime until at least the Middle Cambrian (Greenough & Papezik 1985, 1986), experiencing repeated episodes of ensialic rifting or block faulting. On occasion, rifting may have led to creation of oceanic crust and short-lived basins (e.g. Papezik 1970, 1972; Strong *et al.* 1978; Strong 1979; Hiscott 1981).

In conclusion, although the Avalon Zone may have its origins in a series of disparate terranes, some of which may have originated from or close to the major cratons such as Gondwanaland, by the Late Proterozoic the sedimentary and faunal evolution of the region can be taken as a unit that, for the purposes of this chapter, was effectively independent of other craton blocks and microcontinents. Following its probable amalgamation

in the latest Precambrian (Keppie 1985), the Avalon Zone encompassed a wide area that now includes southeast Newfoundland and submarine extensions to the east (e.g. Lilly 1966; Pelletier 1971; Haworth & Lefort 1979; King *et al.* 1985; King *et al.* 1986) and south (Miller 1987), much of the Canadian Maritimes, and more limited areas of the eastern continental margin of the United States. This unit may be treated as a microcontinent that was at least in part ensialic. Middle Proterozoic gneisses and sediments (e.g. Green Head Group in New Brunswick and George River Group in Cape Breton Island; see Rast *et al.* 1976; Strong 1979) floor some regions. In south-east Newfoundland the identification of a Proterozoic basement (Grey River Metamorphic Complex; see O'Brien *et al.* 1983) remains equivocal, but the discovery of detrital garnets in uppermost Proterozoic sediments (Signal Hill Group, Fig. 2.8) of the Avalon Peninsula argues for relative proximity of acidic crust (Papezik 1973), and recent geophysical work (King *et al.* 1986) also indicates a basement complex offshore.

The contacts between the Avalon Zone and other components of the Appalachian Fold Belt are entirely tectonic, being defined by a series of major faults whose distribution is summarized by O'Brien *et al.* (1983). Moving southwards these tectonic junctions with the Appalachian orogen to the west include the Dover and Hermitage Faults (Newfoundland), Aspy Fault (Cape Breton Island), and Wheaton Brook Fault (New Brunswick), the latter fault now being regarded as a more likely boundary than the Belleisle Fault that lies nearby to the east (McCutcheon 1981; see Currie 1986 for a discussion of the possible continuation to the immediate south-west). The time of collision and incorporation within the Appalachian Fold Belt may be constrained by features such as the age of the granites that jointly intrude the Avalon and adjacent zones, and the timing of metamorphism (e.g. Dallmeyer *et al.* 1981). Together these indicate docking of the Avalon Zone no later than early Devonian (Williams & Hatcher 1982, 1983).

2.3. PRECAMBRIAN–CAMBRIAN BOUNDARY SEQUENCES IN THE AVALON ZONE

The geology of the Avalon Zone, and more specifically a number of the boundary sections, has received attention for over a century. Early names that are linked with this classic area include the prolific G. F. Matthew, as well as C. D. Walcott, J. P. Howley, G. Van Ingen, and B. F. Howell. Of particular note was the establishment by

Matthew (1888*a, b*; see also 1899*a*) of the Etcheminian (variously spelt Eteminian, Etchiminian) Series, that he regarded as a poorly fossiliferous sequence separate and distinct from the Cambrian as then recognized. The history of the concept of an Etcheminian Series was discussed briefly by Hayes & Howell (1937). Confusion has been engendered by Matthew applying this term to faunas, such as those from the Bourinot Group of Cape Breton Island (see below), that subsequently transpired to be of mid-Cambrian age (Hutchinson 1952). Moreover, Walcott (1900) gave a cogent rebuttal of the notion that the Etcheminian was separate from the Cambrian, which Matthew (1908, p. 132) ultimately more-or-less accepted. Nevertheless, as originally coined and generally understood, the Etcheminian is a useful and valid term that may be used to encompass the pre- or non-trilobite sequences in the Avalon Zone. Significantly, Matthew stressed the small size of many of the fossils, the relative abundance of tubicolous taxa, and absence (or rarity) of trilobites. As explained below, trilobites appear relatively late in comparison with many areas of the world. Therefore, the Etcheminian cannot be equated simply with the standard Tommotian as recognized in Siberia, but as discussed in more detail

below, it would appear to be equivalent also to the Lower Atdabanian of the Siberian sequences. Given that discussion on the pre- or non-trilobite succession with its abundant small shelly fossils has dominated the debate of where best to place the boundary stratotype, it is quite remarkable that Matthew's prescience has not received adequate recognition by later workers.

Within the Avalon Zone, by far the most important stratigraphical sequence that spans the Precambrian–Cambrian boundary is in south-east Newfoundland (Fig. 2.1). Here, composite sections on the Avalon and Burin Peninsulas range from Late Proterozoic sequences with a tillite (Gaskiers Formation), succeeded by volcaniclastics with a diverse Ediacaran fauna through to sediments yielding assemblages of small shelly fossils and trace fossils, and finally trilobites of the *Callavia* Zone *s.l.* (Fig. 2.8). Accordingly, the greater part of this chapter will be devoted to a review of this important sequence. However, sections relevant to the boundary are recognizable in various areas of Cape Breton Island, New Brunswick, eastern Massachusetts (Fig. 2.1), and Rhode Island. These latter sections tend to be thinner, less complete, and may be interrupted by unconformities, but correlations between these sections

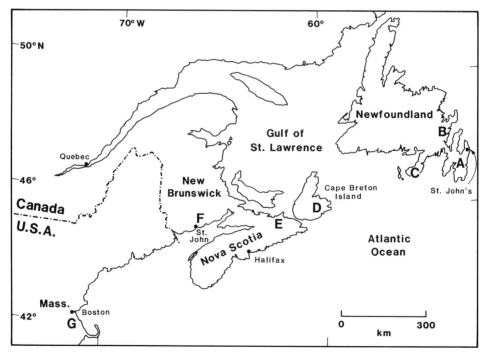

Fig. 2.1. Principal sites of Precambrian–Cambrian boundary sequences in the Avalon Zone. A, Avalon Peninsula; B, Trinity Bay and Bonavista; C, Burin Peninsula; D, south Cape Breton Island; E, Antigonish Highlands, Nova Scotia; F, southern New Brunswick; G, Massachusetts.

and south-east Newfoundland are moderately secure and allow a more regional picture to be developed.

2.3.1. South-east Newfoundland

Avalon Peninsula and Bonavista

Reviews of the stratigraphical sequences of the latest Proterozoic and early Cambrian age in the Avalon Peninsula and Bonavista area (Fig. 2.2) are excellently covered by a variety of workers (e.g. Rose 1952; Hutchinson 1953, 1962; Jenness 1963; McCartney 1967, 1969; Brückner 1969; Poole *et al.* 1970; Hughes & Brückner 1971; North 1971; Rast *et al.* 1976; Williams & King 1979; King 1980, 1982; O'Brien *et al.* 1983; Landing *et al.* 1988). This region may be divided conveniently into three broad zones (A, Isthmus of Avalon and an area west of Trinity Bay that includes Bonavista, B, Trinity Bay to Conception Bay, and C, Conception Bay to Trepassey Bay, Fig. 2.2), the Precambrian intercorrelations (Fig. 2.8) of which are discussed by Hofmann *et al.* (1979) and King (1980).

Much of the Avalon Peninsula and Bonavista region is covered by a very thick sequence of late Precambrian rocks that overall have a north-east–south-west strike, and are generally folded and cut by faults, as well as locally suffering intense tectonic deformation. However, for the most part the grade of metamorphism is low, and sedimentary and igneous structures are well preserved. The episodes of diastrophism, tectonism, and igneous intrusion within the late Precambrian have been termed the Avalonian Orogeny (Lilly 1966; Rodgers 1967; Anderson *et al.* 1975; see also Hughes 1970, 1972; Rast

& Skehan 1983). This somewhat ill-defined episode is widely regarded as synchronous and possibly the same as the Cadomian Orogeny that is recognized in Armorica (Keppie 1985), and possibly may reflect the collision of this region with Gondwanaland (Hagstrum *et al.* 1980).

The oldest of these Precambrian rocks presently exposed belong to the predominantly volcanic sequences of the Harbour Main Group (e.g. Rose 1952; McCartney 1967, 1969; Papezik 1970, 1972, 1980; Hughes & Brückner 1971; Nixon & Papezik 1979), and their assumed correlatives to the west in the form of the Love Cove Group (e.g. Jenness 1963). The igneous rocks of the Harbour Main Group (Fig. 2.8) range from mafic to felsic. Rock types include basalts, ignimbrites, tuffaceous beds, and rhyolites, as well as some interbedded sediments. Major faults define three main blocks, which Papezik (1980) suggests represent different levels within a volcanic complex. The western block may represent the lowest part of this complex, and the volcanics here are mildly alkaline to tholeiitic with mafic flows of intraplate type. The eastern block includes pillow lavas and subaqueous pyroclastics, intruded by rhyolite domes, while the central block of felsic to mafic pyroclastics is intruded by the Holyrood Granite (e.g. Rose 1952; McCartney 1967; Hughes & Brückner 1971). This is a high-level plutonic body largely composed of a pink granite, together with marginal bodies of quartz monzonite and quartz-hornblende gabbro. This intrusive body is believed to be comagmatic with the volcanics. The role of the Holyrood Granite in radiometric determinations is returned to below. The Love

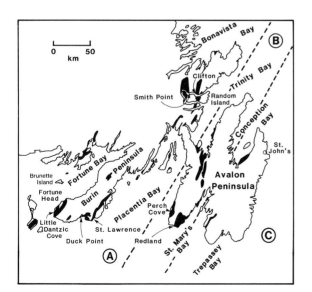

Fig. 2.2. Cambrian outcrops in south-east Newfoundland (after Hiscott 1982), and demarcation of the major tectonostratigraphical zones A–C (after Hofmann *et al.* 1979).

Cove Group (Fig. 2.8) is a schistose assemblage of predominantly intermediate to acidic volcanics and associated sedimentary rocks. Although correlated with the Harbour Main Group and giving roughly comparable radiometric ages (see below), lithological comparisons are not close (Jenness 1963). The Love Cove Group is intruded by apparently comagmatic bodies such as the Swift Current Granite.

In part contemporaneous with the Harbour Main Group is a thick succession (3–5 km) of mostly flysch-like sediments, often volcaniclastic and turbiditic, that comprise the Conception Group (e.g. Rose 1952; McCartney 1969; Williams & King 1979; King 1980; Gardiner & Hiscott 1984). Near the base of this group are the glaciogenic tillites of the Gaskier Formation (Fig. 2.8) (Brückner & Anderson 1971; Anderson & King 1981), which presumably are roughly correlative with the various Varangian tillites elsewhere in the North Atlantic region (Brückner & Anderson 1971; Williams & King 1979) and possibly beyond (Anderson 1972), although as King (1980) notes, this equivalence remains to be firmly established.

Although medusoid-like elements occur as low as the Drook Formation (Anderson 1978; Hofmann et al. 1979; King 1980), the principal development of the Ediacaran biota is found in the Mistaken Point Formation (Fig. 2.8) (Anderson 1972, 1978; Anderson & Conway Morris 1982). The fauna (Fig. 2.4 A–G) is best known from spectacular bedding-planes strewn with a diverse assemblage of soft-bodied medusoids and other cnidarian-like creatures at Mistaken Point, near Trepassey. However, many other localities around the Avalon Peninsula (Anderson 1978, his Fig. 1) indicate that this assemblage was locally abundant, and typically owes its preservation to burial under volcanic ash. Medusoids have also been found in the Hibbs Hole Formation (Fig. 2.8), which is a correlative unit to the Mistaken Point Formation, exposed further to the north-east (King 1980). However, no Ediacaran metazoans have been discovered in the Connecting Point Group (Fig. 2.8), which is believed to be more-or-less equivalent to the Conception Group (but see Landing et al. 1988).

Despite Ediacaran biotas having a world-wide distribution (Glaessner 1984), those from south-east Newfoundland appear to be largely endemic, apart from some striking similarities with the soft-bodied assemblages from Charnwood Forest, England (e.g. King 1980). Such similarities are consistent with a shared location on a single terrane, and in addition are supported by sedimentological comparisons. The distinctiveness of these Avalonian Ediacaran biotas may in part reflect biogeographical isolation, but could also be

explicable in terms of facies differences between these apparently deeper-water faunas and the shallow-water conditions that appear to have characterized the majority of other Ediacaran biotas.

The overlying St John's Group and Signal Hill Group (Fig. 2.8) represent a molassic sequence of 8–9 km thickness. Convincing correlations with the equally thick Hodgewater Group to the west are reviewed by McCartney (1969) and King (1980), but the exact equivalents within the Musgravetown Group (Fig. 2.8) are for the most part more speculative. In particular, no equivalent to the Bull Arm Formation, a heavily altered spilite–keratophyre volcanic association that includes abundant subaerial flows and pyroclastics (McCartney 1967; Hughes & Malpas 1971) is recognizable to the east.

The St John's Group is characterized by black shales and silty sandstones that King (1980) interpreted as representing initial deltaic progradation into a marine basin and concomitant marine shoaling. Prodelta deposits have yielded numerous medusoid-like impressions, referred to as *Aspidella terranovica*. They are especially abundant in the Fermeuse Formation (Fig. 2.8). The history of research into these problematic structures was reviewed by Hofmann (1971), who concluded that they were of inorganic origin. King (1980), however, re-iterated the suggestion that at least some of the larger discs could represent medusoid body fossils. Fig. 2.3 presents a frequency histogram of the diameters of a prolific assemblage of these medusoid-like structures on a bedding-plane near Ferryland Head (Fig. 2.4 H, I). Although the great majority have a diameter of less than 12 mm, the wide size range, combined with a strongly skewed distribution, may not represent a biological population. Moreover, the occurrence of a centrally located plug of sandstone, especially in the larger discs, also suggests that these medusoid-like structures are more likely to be inorganic.

The succeeding Signal Hill Group (Fig. 2.8) shows a continuation of delta advance southwards and ultimately a sequence of sands and conglomerates deposited under broadly fluvial conditions with alluvial fans (King 1980). Units within this group yield the enigmatic structure referred to as *Arumberia* (Bland 1984), which has also been noted in the Musgravetown Group. The organic nature of *Arumberia*, which was described originally from Central Australia by Glaessner & Walter (1975) was supported by Bland (1984). He drew attention to its occurrence in roughly coeval rocks from many parts of the world, arguing that it might have some potential for correlation near the Precambrian–Cambrian boundary.

The Precambrian sequences of the Avalon Peninsula have also been sampled for organic-walled microfossils

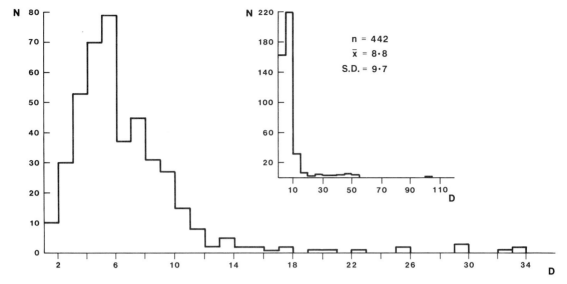

Fig. 2.3. Size frequency distribution of diameters of medusoid-like ?pseudofossil *Aspidella terranovica* from a single bedding plane (Fig. 2.4) within the Fermeuse Formation (St John's Group), near Ferrylands, Avalon Peninsula. Left histogram: size distribution of all specimens with a diameter of less than 36 mm, accounting for 96 per cent of the total sample. Inset histogram to the right: entire population of 442 individuals. N = number of specimens, D = diameter in mm.

by Hofmann *et al.* (1979). Their results were disappointing, in that the microfloras recovered were sparse, poorly preserved, relatively undiagnostic in terms of age, and appeared to hold little potential for biostratigraphical correlation. Anderson *et al.* (1982; see also Timofeyev *et al.* 1980) reported additional microfossil occurrences from the St John's Group, including *Bavlinella* from the Fermeuse Formation and Renews Head Formation (Fig. 2.8). The occurrences of this genus were taken to be indicative of a Vendian age for this stratigraphical interval, but the inference that the underlying Conception Group falls within the Upper Riphean (see Timofeyev *et al.* 1980) is considered to be less likely. Thus, although Anderson *et al.* (1982) noted that *Bavlinella* was rare in the lowest unit (Trepassey Formation) of the St John's Group, it occurs also in the Gaskiers Tillite of the Conception Group (Knoll *in* Hofmann *et al.* 1979; Timofeyev *et al.* 1980). The biostratigraphical utility of *Bavlinella*, which ranges from latest Riphean to Cambrian, and its preference for glaciogenic sediments, is reviewed by Knoll & Swett (1985). Together with radiometric determinations of the underlying Harbour Main Group, and associated intrusions, and their assumed equivalents (see below), these observations suggest that the sedimentary sequences above the Harbour Main Group are Vendian in age.

In the eastern and central zones of the Avalon Peninsula the overlying Cambrian platformal sequence of the Adeyton Group, the oldest unit of which comprises the Random Formation, rests unconformably on the Signal Hill Group and its assumed correlative in the Hodgewater Group (the Snows Pond Formation, Fig. 2.8). However, to the north-west the top of the Musgravetown Group (Crown Hill Formation) appears in places to be 'conformable, if not actually gradational' with the overlying Random Formation (McCartney 1967, p. 99). In the absence of faunal evidence, secure correlations are impossible, but it might be speculated whether the upper units of the Musgravetown Group could be equivalent to the Chapel Island Formation (Fig. 2.8) exposed on the Burin Peninsula (see below).

Across much of the Avalon Peninsula, therefore, the absence of continuous sedimentation prevents it from being considered for boundary stratotypes, and even in areas of Bonavista where the succession may be conformable, the lack of biostratigraphical control precludes serious consideration. However, the overlying Lower Cambrian succession is an important sequence, and allows for broad correlations across the Avalon Zone. The Random Formation (Fig 2.8) is reviewed by various workers (e.g. McCartney 1967, 1969; Anderson 1981), and aspects of the sedimentology of this siliciclastic unit have been addressed by Hiscott (1982). Five facies are recognized, representing environments within a tidally active, shallow sea. Locally, Hiscott (1982) found some evidence for possible fluvial deposition, but for the most part deposition appears to have taken in a coastal embayment, which, if not a broad continental shelf, was at least part of wide, shallow platform. Recognizable facies include shallow subtidal sand banks, intertidal mud-flats, and open shelf facies with evidence for storm deposition. Moreover, the maturity of the sands argues for a source area which was at least andesitic in general nature, and is consistent with a stable tectonic setting of an ensialic platform of considerable dimensions.

Apart from a variety of trace fossils in Random Formation, which are discussed below with the Burin occurrences, no other metazoan remains have been found. However, although organic-walled microfossils were recovered from the Random Formation of the Bonavista area by Nautiyal (1976) and Hofmann et al. (1979), in common with the assemblages recovered from the underlying Precambrian sequences (see above) they are poorly preserved and of limited biostratigraphical utility.

Overlying the Random Formation is the Bonavista Formation (Fig. 2.8), typically comprising red and green shales and interbedded impure limestone, that are often nodular (e.g. Walcott 1900; McCartney 1967, 1969; see also Benus & Landing 1984; Landing & Benus 1988). The basal beds are usually rich in gravels, and may have prominent stromatolitic horizons. The first shelly fossils in the Avalon Peninsula and Bonavista area occur in this unit, although comparisons with the Burin sections demonstrate that these occurrences, which are discussed below together with those of the Burin Peninsula, are relatively high in the earliest shelly fossil assemblage. In addition to the remains of skeletal metazoans, remains of fossil algae and other microfossils have been noted in the Bonavista Formation. Edhorn & Anderson (1977) recorded *Girvanella*, *Epiphyton*, and *Renalcis* from limestones at Bacon Cove, Conception Bay (Fig. 2.2), but Edhorn's documentation (Edhorn 1977) of cropping in these floras needs to be re-examined.

The succeeding Smith Point Limestone (Fig. 2.8) is a prominent red unit, that has received wide palaeontological attention (e.g. G. F. Matthew 1899b; Walcott 1890, 1900; Hutchinson 1962), especially with the first appearance of trilobites of the *Callavia* Zone *s.l.* in the uppermost beds (see also Landing & Benus 1988). Bengtson & Fletcher (1983) point out that this appearance may be facies-controlled or diachronous. The example they give to support this notion is a comparison between the trilobite faunules at Smith Point: Trinity Bay and Redland: St Mary's Bay. At the former locality

the first faunule contains *Callavia broeggeri*, *Serrodiscus bellimarginatus*, and *Strenuella strenua*, whereas at Redland this occurs above a faunule yielding *Comluella* spp. (Bengtson & Fletcher 1983, p. 532). The complex sedimentology of the Smith Point Limestone shows regional variation, but includes stromatolitic and oncolitic horizons with associated iron–manganese mineralization, hardgrounds, and shelly horizons (Landing & Benus 1984).

The overlying Brigus Formation (Fig. 2.8) also lies in the *Callavia* Zone *s.l.*, but extends upwards into *Protolenus* Zone *s.l.* The lithology is typically that of red and green shales, interspersed with thin, often nodular, carbonate horizons (McCartney 1967, 1969). The base of the Middle Cambrian is marked by the Chamberlains Brook Formation, and is succeeded by such celebrated units as the Manuels River Formation, that although of great palaeontological importance are beyond the scope of this discussion.

Burin Peninsula

While many aspects of Lower Cambrian palaeontology and biostratigraphy are admirably displayed on the Avalon Peninsula and Bonavista area, it is to the west on the Burin Peninsula (Fig. 2.2) that attention must be focused, because of the availability of a more-or-less continuous sequence across the boundary. Early geological work (e.g. Van Alstine 1948) has been augmented by a mapping programme directed under the auspices of the Department of Mines and Energy in Newfoundland (O'Brien et al. 1977; Strong et al. 1978). A preliminary study of the assemblages of small shelly fossils (Bengtson & Fletcher 1981, 1983) is now being supplemented by a more extensive programme (Landing et al. 1988). The stratigraphical distribution of the trace fossils in the same sequences has been assessed by Crimes & Anderson (1985) and Narbonne & Myrow (1988).

Two areas on the Burin Peninsula are of particular interest with regard to boundary sections: those facing Fortune Bay (see Narbonne et al. 1987) to the south of Grand Bank and those on the east coast of the peninsula near St Lawrence (Fig. 2.2). The boundary sequence may be conveniently taken from the Rencontre Formation, which is represented by a diverse set of clastics. Further north in Fortune Bay and Brunette Island these rocks have been studied in detail by Smith & Hiscott (1984). The Rencontre Formation is succeeded by the sediments of the Chapel Island Formation (Fig. 2.8), which has been divided into five members (e.g. Bengtson & Fletcher 1983; Narbonne et al. 1987). The basal member 1 shows a transition from fluviatile sediments to those apparently deposited in intertidal and shallow-marine conditions. It has yielded a low diversity of trace fossils that, however, includes types such as *Harlaniella*

(Bengtson & Fletcher 1983), which are characteristic of Upper Vendian deposits in the East European Platform (e.g. Palij *et al.* 1979). A more prolific ichnofauna occurs in the overlying member 2, which with member 3 is interpreted as representing a delta-front environment. In member 2 'Palaeozoic-type' trace fossils appear, including scratch-marks e.g. *Monomorphichnus*, that probably represent arthropod activities.

Details of the trace-fossil distribution through the remainder of the Chapel Island Formation and the overlying Random Formation are reviewed by Crimes & Anderson (1985), who also provide sketches of their palaeoenvironmental distributions. One particular point of interest concerns the co-occurrence of ichnotaxa that later in the Phanerozoic are recognized as being characteristic of either shallow or deep water, suggesting that certain patterns of behaviour had their origin in shallow water and only later migrated into deeper environments.

Apart from the occurrence of the organic tubes of *Sabellidites* from the top of member 1 (Narbonne *et al.* 1987; see also Fig. 2.5B) and member 2 and some poorly preserved tubes from members 2 and 3, the earliest shelly fossils are not recorded until member 4 (Fig. 2.8), in which pyritized moulds of hyoliths and the gastropod *Aldanella* (Fig. 2.5 C–E) occur in mudstones shortly above the first laterally persistent limestone. These occurrences mark the lower of the two small shelly fossil assemblages recognized by Bengtson & Fletcher (1983) in south-east Newfoundland. Both are of relatively low diversity and frequently rather poorly preserved. The lower *Aldanella attelborensis* assemblage (Fig. 2.5 C–E) is characterized by the gastropod of that name, *Watsonella*, *Lapworthella ludvigseni* (see Landing 1984), *Fomitchella* cf. *acinaciformis*, *Chancelloria* sp., *Halkieria* sp., and orthothecid hyoliths. This assemblage extends through the Chapel Island Formation, where it was first recorded by Greene & Williams (1974), to the Bonavista Formation, although no shelly fossils have been recorded in the intervening siliciclastics of the Random Formation. More recently, Shergold & Brasier (1986) have modified the concept of this assemblage and erected an *Aldanella attleborensis–Sunnaginia neoimbricata* Zone, but the validity of this zone and related biostratigraphical proposals requires further investigation. The succeeding assemblage of small shelly fossils is characterized by the tubicolous fossil *Coleoloides typicalis*, which gives its name to this assemblage, as well as *Camenella* cf. *baltica*, *Eccentrotheca kanesia*, *Sunnaginia* cf. *imbricata*, various hyoliths, *Watsonella* (Fig. 2.5A), *Halkieria* sp., etc. This assemblage ranges from the upper Bonavista Formation (see Benus & Landing 1984) into the Smith Point Limestone, at the top of which the earliest trilobites (*Callavia* Zone *s.l.*) of this region occur.

At present there is little information on organic-walled microfossils from the Burin sections. However, Downie (1982) reported an acritarch assemblage that he regarded as indicative of either a late Tommotian or more probably Atdabanian age. He did not give a precise locality or stratigraphical details in his paper, but in a letter (pers. comm., 30 June 1986) he informed me that the samples were from the 'Bay View Formation', a unit near St Lawrence that Bengtson & Fletcher (1983, their Fig. 5) equated with the Rencontre Formation and the first three members of the Chapel Island Formation. There seems, therefore, to be a possible discrepancy between the biostratigraphical correlations as based on the palynomorphs (Downie 1982) and small shelly fossils (Bengtson & Fletcher 1983), with the former fossils indicating a younger age.

2.3.2. Other areas

In addition to south-east Newfoundland, sections relevant to a discussion of the boundary are exposed along the length of the Avalon Zone (Fig. 2.1) in Cape Breton Island, Nova Scotia, New Brunswick, Massachusetts, and Rhode Island. In none of these latter areas are the sections either as complete or thick as those of south-east Newfoundland, and in general they are more poorly fossiliferous. Thus, for the most part, detailed intercorrelations are uncertain, but the broad similarities evidently record a related sequence of events along the Avalon Zone that is consistent with an area behaving as a single block from latest Precambrian times onwards (Keppic 1985).

Cape Breton Island, Nova Scotia

The relevant sections, which are somewhat metamorphosed, are exposed along the Mira Valley on the south side of Cape Breton Island (Hutchinson 1952; Weeks 1954; see also Poole *et al.* 1970; North 1971; Rast *et al.* 1976; Rao *et al.* 1986). The basal Cambrian is divided into the Morrison River Formation and overlying MacCodrum Formation (Fig. 2.8), with the former unit resting on the volcanics of the Fourchu Group (Weeks 1954). The upper part of the Morrison River Formation has been widely compared with the Random Formation (Weeks 1954; Hiscott 1982), although the red sandstones and conglomerates of this unit are unfossiliferous. Near the top of the MacCodrum Formation, beds at a single locality (Victoria Bridge) have yielded the trilobite *Strenuella strenua*. This indicates a position within the *Callavia s.l.* Zone, or perhaps *Protolenus* Zone, but the detection of fossils elsewhere in the MacCodrum Formation is rendered difficult because the shales and siltstones have been metamorphosed. Although Hut-

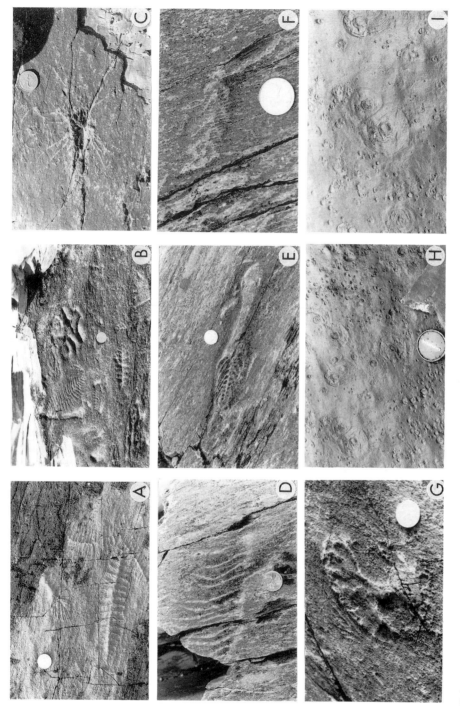

Fig. 2.4. Representative Ediacaran fossils from the Mistaken Point Formation (Conception Group) (A–G) and probable pseudofossils (*Aspidella terranovica*, H and I) from the Fermeuse Formation (St John's Group) of south-eastern Newfoundland. A, bedding plane assemblage of 'spindle' organisms and stalked animals, partially obscured by overlying volcanic ash. B, bedding plane assemblage of 'spindle' organisms and medusoids. C, enigmatic stellate organism. D, enigmatic pectinate organism. E, pennatuloid organism with stalk and holdfast. F, specimen of the pennatuloid *Charnia*. G, enigmatic organism with arborescent body, short stalk, and enlarged holdfast. H, general view of bedding plane that was used in survey of size-frequency of discs, see Fig. 2.3. I, detail of top left-hand corner of H. Diameter of coin is 23 mm (A–G); diameter of lens-cap is 54 mm (H, I).

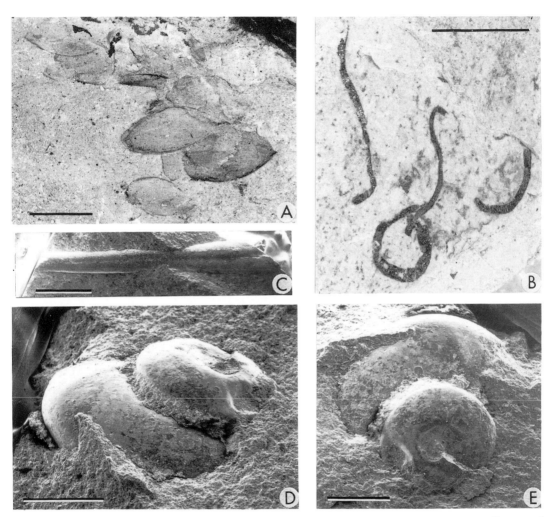

Fig. 2.5. Representative Cambrian fossils from south-eastern Newfoundland. A, group of *Watsonella* sp. aff. *W. crosbyi* (Rostroconchia) from the Bonavista Formation, Smith Sound, GSC 78463. B, *Sabellidites* sp. from member 2 of the Chapel Island Formation, Fortune Head, Burin Peninsula, GSC 78464. C, '*Ladatheca*' *cylindrica*, pyritised specimen showing crushing, from the lower part of member 4 of the Chapel Island Formation, Dantzic Cove, GSC 78465. D, E, *Aldanella attleborensis*, same locality and preservation as C, GSC 78466. A, B photographed under alcohol, C–E are stereoscan electron micrographs. Scale bars equivalent to 5 mm (A, B), 2 mm (C), and 1 mm (D, E).

chinson (1952) included the Morrison River Formation in the *Callavia* Zone *s.l.*, it seems more likely that even some of the overlying MacCodrum Formation lies beneath the *Callavia* Zone *s.l.*, perhaps more-or-less equivalent to the Bonavista Formation and Smith Point Limestone of south-east Newfoundland (Fig. 2.8). With regard to the underlying Fourchu Volcanics, Hutchinson (1952) suggested a possible disconformity (see also Keppie & Halliday 1986), but Weeks (1954) believed them to be conformable with the Morrison River Formation. He equated these volcanics, which also probably occur further to the north in Cape Breton Island (Wiebe 1972), with those within the Musgravetown Group (Bull Arm volcanics), and implied that the succeeding clastics of this group (Big Head Formation to Crown Hill Formation, Fig. 2.8) should be taken as equivalent to the lower part of the Morrison River Formation. However, Anderson (1972) suggested that the correlation of the Morrison River Formation be restricted to the upper part of the Musgravetown Group. As noted above, the Crown Hill Formation may be gradational with the overlying Random Formation in south-east Newfoundland (McCartney 1967), and so it could be argued that the Morrison River Formation shows a similar stratigraphical range, which would also suggest an equivalence to the Chapel Island Formation (Fig. 2.8). However, in the absence of biostratigraphical or reliable radiometric control, this sequence in south Cape Breton could contain undetected unconformities. Some authors (e.g. Strong 1979; O'Brien *et al.* 1983) have indicated an older age for the Fourchu Volcanics, and Keppie & Halliday (1986) obtained some radiometric data that possibly could support this contention (see below). Schenk (1971, p. 1224) referred to a 'unit strongly resembling an indurated varved clay' in this succession, but no other evidence appears to be available to suggest a correlation with glaciogenic sequences elsewhere in the Avalon Zone.

Further to the north-west the Bourinot Group is exposed, and although G. F. Matthew (1903) assigned the two lower formations (Eskasoni and Dugald) to the Lower Cambrian (Etcheminian), the faunas (including vast numbers of inarticulate brachiopods) have been taken as indicating a mid-Cambrian age (Hutchinson 1952). However, it seems conceivable that the Eskasoni Formation may extend into the Lower Cambrian (see also North 1971), and re-study of these faunas may be illuminating. The abundant volcanics within the Bourinot Group (Weeks 1954; Keppie & Dostal 1980; Murphy *et al.* 1985) do not appear to have been dated radiometrically.

Antigonish Highlands, Nova Scotia

On the mainland of Nova Scotia, a section north-west of

Antigonish (Fig. 2.1) has yielded a *Callavia* Zone *s.l.* faunule from near the middle of the Little Hollow Formation (Landing *et al.* 1980). In addition to trilobites the faunule contains a variety of small shelly fossils. These include *Eccentrotheca kanesia, Sunnaginia imbricata, Rhombocorniculum cancellatum, Lapworthella* sp., *Pelagiella*, and possibly *Amphigeisina* and *Turcutheca*. This faunal list not only demonstrates how various elements of the *Coleoloides typicalis* assemblage (see Bengtson & Fletcher 1983) extend into the *Callavia* Zone *s.l.*, but it is also a landmark in the documentation of taxa e.g. *Sunnaginia*, which hitherto had been regarded as diagnostic of the Tommotian and absent from the trilobite-bearing strata.

The Little Hollow Formation overlies the Black John Formation (Fig. 2.8), and together with the Ferrona Formation of probable early Ordovician age, constitutes the Iron Brook Group. The lower quartzite of the Little Hollow Formation and the Black John Formation together have been compared with both the upper Musgravetown Group to Random Formation of southeast Newfoundland, and the Ratcliffe Brook Formation to Glen Falls Formation (Fig. 2.8) of New Brunswick (Landing *et al.* 1980). The possible stratigraphical position of the Glen Falls Formation is returned to below, and in the absence of any faunal control beneath the *Callavia* Zone *s.l.* in the Little Hollow Formation these correlations must be regarded as tentative.

The predominantly volcanic McDonalds Brook Group is regarded as a lateral facies equivalent of the Iron Brook Group (Murphy *et al.* 1985). These authors note the presence of 'Lower Cambrian fossiliferous limestones' in this sequence, but exact palaeontological details do not appear to be available.

New Brunswick

The Cambrian sequences in New Brunswick (Fig. 2.6), especially near the city of St John, have attracted much attention. Although the stratigraphical confusion engendered by the recognition of formational units on the basis of biostratigraphical rather than lithostratigraphical criteria is only now being clarified (Pickerill & Tanoli 1985), it is fortunate that the Lower Cambrian divisions (Hayes & Howell 1937; Alcock 1938) are well founded and have not required extensive revision (see Pickerill & Tanoli 1985).

The stratigraphical sequence established by G. F. Matthew (e.g. 1889, 1895) was reviewed by Hayes & Howell (1937; see also Alcock 1938; Poole *et al.* 1970). Of the available sections, that at Hanford Brook (Fig. 2.6) (G. F. Matthew 1889; see also Walcott 1900) is of special importance in a discussion of boundary sequences. However, other localities including those at

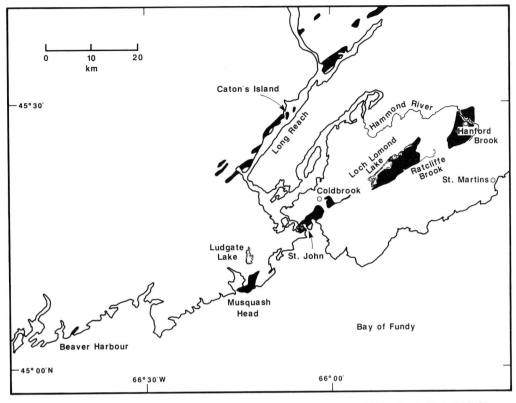

Fig. 2.6. Lower Cambrian outcrops in southern New Brunswick (after Hayes & Howell 1937; Currie 1984, 1986; Greenough *et al.* 1985; Pickerill & Tanoli 1985).

Ratcliffe Brook, the city of St John, along the north-west side of Long Reach and Catons Island, and Musquash Head (see McCutcheon 1981; Currie 1984, 1986; Pickerill & Tanoli 1985) are also relevant (Fig. 2.6). In the Long Reach area the probable Lower Cambrian is associated with volcanics (Greenough *et al.* 1985), but these do not appear to have been dated radiometrically.

The lowest Cambrian unit is the Ratcliffe Brook Formation (Fig. 2.8) (equivalent to the Basal Series as defined by G. F. Matthew 1889), that rests unconformably on the substantially older (see below) Coldbrook Volcanic Group (see Giles & Ruitenberg 1977; Currie 1986). Certain red beds once thought to belong to the Ratcliffe Brook Formation have now been reassigned to the Coldbrook volcanics (Tanoli *et al.* 1985). The basal conglomerate of the Ratcliffe Brook Formation is succeeded by a clastic series of predominantly sandstones and siltstones that Patel (1973, 1975) interprets as representing a terrestrial (including palaeosoils) to shallow-marine transition. The absence of fossils from the lower part of this formation is consistent with fluviatile deposition, but slightly higher in the sequence (especially Beds 1b and 1c) a variety of trace fossils are known (G. F. Matthew 1889; see also Patel 1976). Comparisons between Matthew's figures and the material described from the Chapel Island Formation (Crimes & Anderson 1985) are very tentative, but the possible presence of *Arenicolites*, *Buthotrephis*, *Phycodes* (as *Phycoidella*, but see Häntzschel 1975, p. W187), *Gordia* (as *Palaeochorda*), and *Didymaulichnus* (as *Psammichnites*) are consistent with such a correlation. Hayes & Howell (1937) indicated that body fossils were absent from the Ratcliffe Brook Formation. However, G. F. Matthew (1889) apparently recorded abundant sponge spicules near the base (his Bed 1b), while from the upper part (his Bed 2b) he commented upon occurrences of hyoliths (G. F. Matthew 1893), inarticulate brachiopods, and more significantly *Volborthella* and *Platysolenites*, while Walcott (1900) noted from what is evidently the same beds (his Bed 1c) 'great numbers of a slender species of *Hyolithes* . . . specimens of *Coleoloides* like *C. typicalis* and a large *Iphidea labradorica*?'. If confirmed these occurrences could be stratigraphically significant, although G. F. Matthew later regarded the former record of *Volborthella* as less certain (G. F. Matthew 1895, p. 132).

The overlying and conformable Glen Falls Formation (Fig. 2.8) (equivalent to Band a, Division 1 of G. F. Matthew 1889, see Pickerill & Tanoli 1985, table 52.1) has been equated with the Random Formation (Fig. 2.8) (Hiscott 1982). Possible equivalents of the Glen Falls Formation also occur to the south-west at Beaver Harbour where they are associated with volcanics (Greenough *et al.* 1985). However, if it transpires that the shelly faunas from the Ratcliffe Brook Formation (G. F. Matthew 1889, 1893; Walcott 1900) belong to the *Coleoloides typicalis* assemblage, then comparisons with south-east Newfoundland suggest that the equivalents of the Random Formation might be sought further down within the Ratcliffe Brook Formation.

Much of the Hanford Brook Formation (Fig. 2.8) is regarded as lying within the *Protolenus* Zone *s.l.* G. F. Matthew (1889) divided the formation, which he referred to as Band b of Division 1, into 5 units (or *assises*, b1–b5) which are best seen at Hanford Brook. Of these, b4 and b5 are poorly fossiliferous, while the underlying b2 and b3 yield a *Protolenus* fauna (G. F. Matthew 1889, 1895; see also Walcott 1900, p. 321). The basal division (b1) is characterized by a striking abundance of ostracodes and may also yield trilobites (G. F. Matthew 1889, 1893, 1895). The scheme erected by Hayes & Howell (1937) is broadly similar, although they regarded their *Beyrichona* sandstone as equivalent to beds b1 and b2 of G. F. Matthew (1889), being overlain by the *Protolenus* Shale. Abundant phosphate nodules were noted by W. D. Matthew (1893) in b2 (see also Walcott 1900), and as more scattered occurrences in b3. Elsewhere in the Avalon Zone phosphate deposits are, for the most part, of rather limited extent (see Christie & Sheldon 1986; Notholt & Brasier 1986), especially in comparison with the major Precambrian–Cambrian boundary deposits elsewhere in the world.

If G. F. Matthew's (1889, 1893, 1895) reports of trilobites from the lowest part (his bed b1) of the Hanford Brook Formation are confirmed, this might suggest a position within the *Callavia* Zone *s.l.* If the Glen Falls Formation is correctly equated with the Random Formation, this in turn would indicate a substantial gap in the stratigraphical succession in comparison with south-east Newfoundland (Fig. 2.8), with no obvious equivalent to the *Coleoloides typicalis* assemblage and possibly the upper part of the *Aldanella attleborensis* assemblage as well. However, the contact between the Glen Falls Formation and Hanford Brook Formation appears to be conformable (Pickerill & Tanoli 1985), and thus an alternative possibility is that the Glen Falls Formation is somewhat younger than the Random Formation, perhaps being equivalent to the Smith Point Limestone and/or Bonavista Formation.

In addition to trilobites, brachiopods, molluscs, hyoliths, and ostracodes (G. F. Matthew 1885, 1889, 1891, 1893, 1895; see also Pojeta 1980) *Volborthella* was recorded from the uppermost unit (b5) at Belyea's Landing, south-western end of Long Reach (Fig. 2.6) (G. F. Matthew 1889, 1895). After initial scepticism (Yochelson 1977; see also Miller 1932), the occurrences of *Volborthella* (assumed to refer to this occurrence rather than the more dubious example in the underlying Ratcliffe Brook Formation) appear to be confirmed (Yochelson 1981), although in a letter (7 June 1986) E. L. Yochelson noted that the identification was based on a single specimen, and re-collecting is needed.

Massachusetts and Rhode Island

The succession in the Boston Basin, Massachusetts (Fig. 2.7), long regarded as being of Pennsylvanian age, is now thought probably to range from the late Precambrian to Cambrian (see Billings 1979; Kaye & Zartman 1980; Skehan & Murray 1980; Socci & Smith 1986). The Boston Bay Group (Billings 1976) commences with the Mattapan Volcanics (Fig. 2.8) and continues upwards as a series of volcanics, volcaniclastic sediments and argillites, and includes a tilloid (Squantum Member, Roxbury Conglomerate, Fig. 2.8) that if of glacial origin may be comparable with the Gaskiers Tillite of south-east Newfoundland (Rehmer 1981, but see Smith & Socci 1986). In addition, samples of the Cambridge Argillite (Fig. 2.8) yielded organic-walled microfossils, including *Bavlinella* (Lenk *et al.* 1982), which, as noted above, is particularly characteristic of glaciogenic sediments.

However, in terms of the fossiliferous Lower Cambrian there are only a few restricted localities in Massachusetts (Fig. 2.7) that have sequences relevant to the Boundary in the Avalon Zone; a brief review is given by Theokritoff (1968) and Bell & Alvord (1976). Principal outcrops include those at North Attleboro (Hoppin Hill) and exposures and/or loose beach material adjacent to Boston at Weymouth, Cohasset, Revere, and Nahant. The most celebrated of these localities is at North Attleboro (Shaler & Foerste 1888; Emerson 1917) where the Cambrian rests unconformably (Dowse 1950) on the Hoppin Granite (see below for radiometric dating). The Cambrian consists of a basal quartzite that is succeeded by a sequence (Hoppin Slate, Fig. 2.8) that has two prominent limestone units about 50 and 70 m above the base (Landing & Brett 1982). The palaeontology of these units has received considerable attention (e.g. Shaler 1888; Shaler & Foerste 1888; Foerste *in* Shaler *et al.* 1899; Grabau 1900; Shaw 1950, 1961; Rasetti 1952; Landing & Brett 1982; Landing 1984). The lower limestone (Station 1 of Shaler) is now

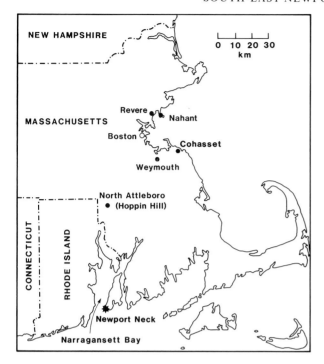

Fig. 2.7. Localities of key Lower Cambrian sections in Massachusetts around Boston and at North Attleboro (dots) and a locality (see Webster *et al.* 1986) at Newport Neck (star), that although unfossiliferous probably represents an equivalent sequence (after Shaw 1961; Theokritoff 1968).

concealed beneath a reservoir, but a temporary draining allowed re-collection (Landing & Brett 1982; Landing 1984). The fauna recovered includes *Aldanella attleborensis, Lapworthella ludvigseni, Sunnaginia imbricata, Eccentrotheca kanesia, Fomitchella* sp. aff. *F. infundibuliformis, Anabarites tripartitus, Watsonella crosbyi, Orthotheca cylindrica, Conotheca mammilata?,* and *Stegotheca* (Landing 1984).

The overlying limestone is a trilobite wackestone. The trilobite assemblage was reviewed by Shaw (1950; see also Rasetti 1952; Shaw 1961), while the small shelly fossils noted by Landing & Brett (1982) include *Rhombocorniculum cancellatum, E. kanesia, Obolella atlantica,* and what were termed '*Milaculum*-like fragments'.

Beds correlative with the sequence at North Attleboro occur at Weymouth and Nahant, where they form the Weymouth Formation (Emerson 1917; Shaw 1961). At Weymouth itself, the lower (Mill Cove) of two fossiliferous horizons is rich in hyoliths and other conical fossils, whereas the overlying unit (Pearl Street, North Weymouth, see Burr 1900) contains trilobites (Grabau 1900; Shimer 1907; Walcott 1910; Shaw 1950; see also Rasetti 1952).

The Cambrian sequences at Nahant (especially East Point, see La Forge 1932) are described by Foerste (1889), Walcott (1890), and Grabau (1900), with the fossiliferous limestones bearing conoidal fossils evi-

dently correlating with the lower limestone horizon of Weymouth and North Attleboro. At least one pebble, however, has yielded the trilobite *Strenuella strenua* (Grabau 1900, p. 611). Comparable occurrences also occur in pebbles from Revere (Clark 1923) and recent drilling in this region has proved *in situ* strata (Kaye & Zartman 1980). Loose material with equivalent conoidal fossils also occurs at Cohasset (Grabau 1900; see also Pojeta 1980).

The similarities of these Massachusetts faunules to those of Newfoundland are relatively clear, in that the lower one is apparently equivalent to the *Aldanella attleborensis* assemblage (see Bengtson & Fletcher 1983), while the upper beds with trilobites fall within *Callavia* Zone *s.l.* However, the stratigraphical distance separating these two fossiliferous units is relatively small (20 m at North Attleboro), and suggests that at least some of the intervening sequence, that bears the *Coleoloides typicalis* assemblage in south-east Newfoundland, is absent (see below).

In passing it is also worth noting that a sequence similar to the sections at Weymouth and North Attleboro is known from Newport Neck, southern Narragansett Bay, Rhode Island (Fig. 2.7). Although this section, which rests unconformably on the underlying Precambrian, has not yielded any fossils (Webster *et al.* 1986), this occurrence is in keeping with the nearby occurrence of Middle Cambrian sediments with Avalonian trilobites (Skehan *et al.* 1978). Skehan & Murray (1980) and Rast & Skehan (1981) review the underlying Precambrian sequence in the southern Narragansett Bay, drawing attention to its similarities with the Massachusetts' sequences containing volcanics and volcaniclastics intruded by granodiorites, as well as the sequence on Anglesey.

2.4. RADIOMETRIC DATING

Debate surrounds the radiometric dating of Precambrian–Cambrian boundary sequences, in that there is a possible discrepancy of 30–40 Ma (or greater) separating some estimates of c.560 (?590) Ma (e.g. Cowie & Cribb 1978), from those values (c.530 Ma) of Odin and co-workers (e.g. Gale 1982; Odin *et al.* 1983; Odin *et al.* 1985; see also Jenkins 1984; Cope & Gibbons 1987). Arguments in favour of each position are given by Cowie & Johnson (1985) and Odin *et al.* (1985), with the former authors stressing the degree of uncertainty. Here discussion, which is taken from Conway Morris (1988), is restricted to radiometric determinations of Avalon Zone rocks, with Table 2.1 summarizing the data that appear to be more-or-less reliable (see below).

Particular attention has been paid to the Holyrood Granite, exposed across about 250 km² at the southern end of Conception Bay on the Avalon Peninsula (Fig. 2.2). Geophysical evidence (Miller & Pittman 1982) suggests that it is a relatively thin body (*c.*2 km). It intrudes volcanics of the Harbour Main Group, and is unconformably overlain by sediments of the Bonavista Formation (McCartney *et al.* 1966). The general consensus (e.g. McCartney *et al.* 1966; Hughes & Brückner 1971; but see Strong & Minatedis 1975) is that the granite is comagmatic beneath the volcanics, and pebbles with a composition similar to the Holyrood Granite occur in conglomerates of the Conception Group. Gale (1982) revised the original determinations of McCartney *et al.* (1966), who used Rb–Sr whole-rock measurements, to give an estimate of 585 ± 15 Ma (cf. Frith & Poole 1972 recalculated value of 594 ± 11 Ma; Dallmeyer *et al.* 1981 recalculated values of 590 ± 11). However, Krogh *et al.* (1988) give a substantially older date for the Holyrood Granite of $620.5^{+2.1}_{-1.8}$ Ma, while King *et al.* (1985) published a date of 610 Ma.

Other radiometric determinations from igneous rocks in south-east Newfoundland include Rb–Sr whole-rock dates of 537 ± 29 and 467 ± 30 Ma for the Harbour Main and Bull Arm Volcanics (Fig. 2.8) (Fairbairn *et al.* 1966; values recalculated by McCartney *et al.* 1966), although these are anomalously young and in the case of the Bull Arm volcanics probably are a reflection of the

extensive metasomatism (Hughes & Malpas 1971). However, O'Driscoll & Gibbons (1980; see also King 1982, p. 103) indicate a minimum U–Pb (zircon) date for these latter volcanics of 584 Ma. More significant, therefore, is the dating of an ignimbrite in the Harbour Main Group (James Cove) at $606^{+3.7}_{-2.9}$ Ma, while an intrusive rhyolite dyke cutting these volcanics gave a date of $585.9^{+3.4}_{-2.4}$ (Krogh *et al.* 1988). This latter date is, of course, closely similar to some estimates of the nearby Holyrood Granite (see above), and considerably younger than some earlier estimates (e.g. Anderson 1972).

For the Love Cove Group (Fig. 2.8) volcanics, which are probably correlative with the Harbour Main Group to the south-east (Fig. 2.8), dates of about 608 ± 25 Ma and 590 ± 30 Ma (U–Pb age on zircons) were obtained by O'Driscoll & Gibbons (1980) and Dallmeyer *et al.* (1981). Furthermore, for the Swift Current granite that intrudes these volcanics, and so could be analogous to the Holyrood Granite, these latter authors obtained a revised date of 580 ± 20 Ma (but see Krogh *et al.* 1988). A U–Pb date of $607.5^{+8.0}_{-3.7}$ Ma (Krogh *et al.* 1988) for the Marystown Group (Fig. 2.8) on the Burin Peninsula also suggests a possible equivalence with the Harbour Main assemblage and Love Cove Group volcanics. A K–Ar date of 565 ± 26 Ma has been given by King (1982) for the Rencontre Formation. Finally, it is worth noting that a granodiorite exposed on Flemish Cap, about 500 km east of the Avalon Peninsula, may be substantially older than the onshore plutonic bodies such as the Holyrood and Swift Current granites. Pelletier (1971) published a K–Ar age from biotite of 615 ± 20 Ma (recalculated by King *et al.* 1985), but these latter authors obtained an age range of 751–833 Ma on the basis of U–Pb analysis of zircons (King *et al.* 1985). Younger ages ranging from 505 ± 48 to 657 ± 29 from K–Ar measurements of hornblende were interpreted as reflecting some resetting by very low metamorphism.

Elsewhere in the Avalon Zone, attention has been focused on the Hoppin Hill Granite and nearby igneous bodies which are believed to be contemporaneous. Dates obtained by Fairbairn *et al.* (1967) on the basis of Rb–Sr whole-rock analyses were recalculated by Gale (1982), who considered only the Northbridge granite to be sufficiently reliable. This computation gives a revised date of 553 ± 10 Ma. However, for the Dedham Granodiorite, Kovach *et al.* (1977) obtained a date of 595 ± 17 Ma on the basis of Rb–Sr whole-rock measurements, while Kaye & Zartman (1980) and Zartman & Naylor (1984) quote a figure of 630 ± 15 Ma on the basis of U–Th–Pb analysis of zircons; these data were omitted from consideration by Gale (1982). Zartman & Naylor (1984) also obtained radiometric analyses for a series of

Table 2.1. Summary of the principal radiometric dates from the late Precambrian of the Avalon Zone, excluding patently anomalous dates or original determinations that subsequently have been revised because of changes in decay constants used.

South-eastern Newfoundland, Avalon Peninsula and Isthmus	South-eastern Newfoundland, northern Burin Peninsula	South-eastern Newfoundland, southern Burin Peninsula	Nova Scotia, Cape Breton Island
	Rencontre Formation 565 ± 26[a]		
Bull Arm Formation 584[a, b]			
Harbour Main Group (volcanics) $585.9^{+3.4}_{-2.4}$[c] $606^{+3.7}_{-2.9}$[c]	Love Cove Group (volcanics)	590 ± 30[e] Marystown Group 608 ± 25[b,e] (volcanics)	$607.5^{+8.0}_{-3.7}$[c] Fourchu Volcanics ?640[h]
Holyrood Granite 585 ± 15[d] 590 ± 11[e] 594 ± 11[f] 610[g] $620.5^{+2.1}_{-1.8}$[c]	Swift Current Granite 580 ± 20[e]		Various granites 548[i]
			Capelin Cove Granite 545 ± 28[h]
			Chéticamp pluton 525 ± 40[j] 550 ± 8[k]
			Loch Lomond Granite 548 ± 18[h] 544 ± 21[l]
			Shunacadie pluton 574 ± 11[m]
			North Branch Baddeck River leucotonalite 614^{+38}_{-4}[k]
Flemish Cap Granodiorite 751–833[g]			

New Brunswick	Massachusetts	Rhode Island
Coldbrook Group (volcanics) 590[n] 640[o] 776 ± 80[p]	Mattapan Volcanics 602 ± 3[r]	
Musquash pluton 615 ± 37[q]	Northbridge Granite 553 ± 10[d]	Narragansett granites 595[u] 603 ± 14[v]
	Dedham Granodiorite 595 ± 17[s] 630 ± 15[r]	Ten Rod Gneiss and Hope Valley Alaskite Gneiss 601 ± 5[w]
	Milford Granite 619 ± 5[t] 624 ± 5[t]	Esmond Granite 621 ± 8[w]

Sources of data: [a]King 1982; [b]O'Driscoll & Gibbons 1980; [c]Krogh et al. 1988; [d]Gale 1982; [e]Dallmeyer et al. 1981; [f]Frith & Poole 1972; [g]King et al. 1985; [h]Keppie & Halliday 1986; [i]Cormier 1972 and Clarke et al. 1980; [j]Barr et al. 1986; [k]Jamieson et al. 1986; [l]Barr et al. 1984; [m]Poole 1980a; [n]Lowdon et al. 1963; [o]Leech et al. 1963; [p]Cormier 1969; [q]Poole 1980b; [r]Kaye & Zartman 1980; [s]Kovach et al. 1977; [t]Zartman & Naylor 1984; [u]Skehan & Murray 1980; [v]Smith & Giletti 1978; [w]Hermes & Zartman 1985. See Fig. 2.8 for position of some of the units listed here, and the text and Conway Morris (1988) for further discussion.

adjacent plutons, including the Milford Granite, and obtained broadly comparable ages of c.620 Ma (Table 2.1). The Dedham Granodiorite is adjacent to the Mattapan Volcanics (Fig. 2.8), but their precise inter-relationships appear to be complex. In places the volcanics lie unconformably on the granodiorites (La Forge 1932; Billings 1979; Skehan & Murray 1980), but Kaye & Zartman (1980) also presented evidence for possible intergradations. In either case, the radiometric dating of the Mattapan Volcanics by Kaye & Zartman (1980) suggested an age broadly comparable with the nearby granodiorites. Specifically, they noted that Rb–Sr methods gave unreliable results (see also Dall-meyer et al. 1981, p. 705), but U–Th–Pb analyses of zircons from a rhyolite provided a date of 602 ± 3 Ma. If the dates of Kaye & Zartman (1980), Zartman & Naylor (1984), and Kovach et al. (1977) are accepted, this suggests this Precambrian volcanic/plutonic sequence may be coeval with the Harbour Main Group and Holyrood Granite of south-east Newfoundland. Finally, it is worth noting that Precambrian granites intruding volcanics in southern Narragansett Bay, Rhode Island (Fig. 2.7) have a similar date of 603 ± 14 Ma (Smith & Giletti 1978, dated as 595 Ma in Skehan & Murray 1980). These outcrops are close to sediments that are the putative equivalents of the Lower Cambrian successions at Weymouth and Hoppin Hill (Webster et al. 1986). Moreover, intrusive bodies to the north (Esmond Granite and related rocks) and west (Ten Rod Granite Gneiss and Hope Valley Alaskite Gneiss) of Narragansett Bay also yielded dates of 621 ± 8 Ma and 601 ± 5 Ma respectively (Hermes & Zartman 1985).

Dating of the Precambrian volcanics forming the Coldbrook Group (Fig. 2.8) in New Brunswick gave anomalously young ages (Fairbairn et al. 1966). How-ever, Cormier (1969) calculated the age of the Cold-brook Group on the basis of Rb–Sr whole-rock analyses to be 750 ± 80 Ma, although Rast et al. (1976, p. 319) regarded this as 'rather doubtful'. On the basis of granite pebbles from a conglomerate within volcanics on Ross Island, adjacent to Grand Manan Island, and located to the south-west of Beaver Harbour (Fig. 2.6), that are tentatively correlated with the Coldbrook Group, K–Ar dates on muscovites gave ages of 590 Ma and 640 Ma (Lowdon et al. 1963, p. 113; Leech et al. 1963, pp. 103–4). Moreover, granites and other igneous rocks that intrude the Coldbrook Group (and Greenhead Group) may well be Precambrian (see O'Brien et al. 1983, p. 205), and possibly contemporaneous with the Coldbrook volcanics (Currie 1986). Poole (1980b) reported unpublished data of W. J. Olszewski and H. E. Gaudette that gave a Rb–Sr date, apparently from the

Musquash pluton near Ludgate Lake (Fig. 2.6), of 615 ± 37 Ma. However, Poole's (1980b) own analysis of this intrusive body gave a date of 439 ± 17 Ma, while for the remaining granites a Cambrian age (526 ± 13 Ma) was obtained by this worker. O'Brien et al. (1983) noted that these, therefore, could be contemporaneous with the nearby volcanics of Cambrian age (see above), but Poole (1980b) concluded that the actual date of intrusion was probably late Precambrian and possibly comagmatic with the Coldbrook volcanics.

Some dates obtained from the Fourchu Volcanics (Fig. 2.8) (Cape Breton Island), which may be more-or-less conformable with the overlying Cambrian (Weeks 1954), are anomalously young (Fairbairn et al. 1966), but more recently Keppie & Halliday (1986) suggested a possible date of c.640 Ma. However, overall their data show considerable scatter, and they conceded that this date 'should be viewed with caution'. If accepted, this would suggest that the Fourchu Volcanics are consider-ably older than many of the other Avalonian extrusives considered here (cf. Fig. 2.8) and could be equivalent to the Coldbrook Volcanic Group (see Keppie 1982). A series of granites on Cape Breton Island, regarded as late Precambrian to Cambrian, gave a series of dates around 548 Ma (Cormier 1972; Clarke et al. 1980) although, using a revised decay constant, two granites (Capelin Cove and Loch Lomond) that cut the Fourchu Volca-nics were dated as 545 ± 28 and 548 ± 18 Ma (see Keppie & Halliday 1986), while Barr et al. (1984) quoted a Rb–Sr date from the Loch Lomond pluton of 544 ± 21 Ma. In addition, Barr et al. (1986) obtained a new date for yet another granite (Chéticamp) of 525 ± 40 Ma, so indicating a possible Cambrian age, although they also questioned whether this granite could be included in the Avalon Zone (see also Barr & Raeside 1986). Furthermore, from U–Pb dating of zircons, Jamieson et al. (1986) obtained a date for the Chéticamp intrusion of 550 ± 8 Ma, while the North Branch Baddeck River leucotonalite was dated as 614^{+38}_{-4} Ma by these workers. Poole (1980a) also presented evidence that some of these Cape Breton Island granites, including the Shunacadie pluton (574 ± 11 Ma), were definitely pre-Middle Cambrian and were probably intruded at a time close to the Precambrian–Cambrian boundary.

Granites from the mainland of Nova Scotia, to the west of the Antigonish Highlands (Fig. 2.1), in the Cobequid Highlands also yield comparable dates using Rb–Sr techniques. Thus, the McCallum Settlement granite is dated at about 575 ± 22 Ma (Gaudette et al. 1983; see also Donohoe et al. 1986), while the nearby Debert Granite gives a date of 596 ± 70 Ma (Donohoe et al. 1986). Although a variety of dates are available from hornblende and micas (K–Ar techniques), the Jeffers Brook Diorite also appears to be of comparable age (Wanless *et al.* 1974; Donohoe *et al.* 1986). In terms of the age of the surrounding rocks intruded by such plutons, Donohoe *et al.* (1986) point out that the sediments adjacent to the Debert granite, previously referred to the Nuttby Formation of early Carboniferous age, could be either late Precambrian or Cambrian, and equivalent to the Warwick Mountain Formation and/or Jeffers Formation. Similarly, Wanless *et al.* (1974) inferred an age of either late Precambrian or earliest Cambrian for the volcanics intruded by the Jeffers Brook Diorite.

In conclusion, radiometric work from across the Avalon Zone points strongly to an igneous episode marked by volcanic activity and acidic intrusions dated at about 590–620 Ma. The complexities of igneous activity over a protracted geological period may account for some of the scatter in dates, but it must be admitted also that a number of analyses may be suspect; special importance is placed here on U–Pb determinations of zircons (e.g. Kaye & Zartman 1980; Krogh *et al.* 1988). Problems such as the inverse dates of older granites intruding apparently younger volcanics in the Avalon Peninsula require further attention (Krogh *et al.* 1988). A key observation is that these igneous rocks are overlain in south-east Newfoundland by about 13 km of sediment, marking a cycle of flyschoid to molassic sedimentation that includes both a tillite and Ediacaran fauna, before the boundary is reached. Thus, although no radiometric dates for Cambrian rocks in the Avalon Zone appear to be available, the dating of this magmatic episode at *c.*600 Ma means that if reasonable rates of sedimentation are accepted, then the boundary is unlikely to be much older that 550 Ma and may be even younger. Such a figure is, therefore, broadly in accord with values proposed by Odin and co-workers (see p. 22).

2.5. CORRELATION OF BOUNDARY SEQUENCES

2.5.1. Avalon Zone

Passing mention of possible correlations between the various Avalon sections was made above, and Fig. 2.8 summarizes these tentative proposals. Despite general lithological similarities, precise correlations in the Late Precambrian and Lower Cambrian are hampered because of the sparse fossil record. It is only with the appearance of trilobites and the recognition of the *Callavia* and *Protolenus* Zones *s.l.*, that intercorrelations begin to have some degree of precision.

The reference sections for the Avalon Zone are, of course, those exposed on the Burin Peninsula, Newfoundland. The two small shelly assemblages identified by Bengtson & Fletcher (1983) beneath the *Callavia* Zone *s.l.* are recognizable on the Avalon Peninsula and Bonavista area, but across most of this area, the lower part of the *Aldanella attelborensis* assemblage is truncated by the regional unconformity that separates the Random Formation from the underlying late Precambrian molassic sequences (Hodgewater Group and Signal Hill Group, Fig. 2.8). It is only in the Bonavista area west of Trinity Bay, that the Musgravetown Group possibly may pass conformably into the Random Formation (McCartney 1967). Thus, if equivalents to the critical sections of the Chapel Island Formation are to be found beyond the Burin Peninsula, it is these latter sections that may repay study.

The biostratigraphical divisions of the Lower Cambrian as recognized in south-east Newfoundland are more-or-less applicable to other regions within the Avalon Zone, but even in Massachusetts where the greatest similarity exists, there are some problems in finding exact equivalence. Here, the trilobite-bearing assemblage belonging to the *Callavia* Zone *s.l.* is separated by a rather thin stratigraphical interval from faunas that appear to belong to the *Aldanella attleborensis* assemblage. As yet no definite equivalent to the *Coleoloides typicalis* assemblage has been identified in Massachusetts. Assuming that trilobites do not extend stratigraphically lower in Massachusetts than in other parts of the Avalon Zone, the reduced thickness of sediment suggests either a very condensed sequence or undetected unconformities.

Minimal faunal control on the clastic sequences in the area around St John, New Brunswick leads to tenuous correlations. The ichnofauna of the Ratcliffe Brook Formation (Fig. 2.8) (G. F. Matthew 1899*a*) may bear comparison with the assemblages recorded in Chapel Island and Random Formations by Crimes & Anderson (1985). The small shelly fossils recorded near the top of the Ratcliffe Brook Formation require re-investigation, but if Walcott's (1900) tentative identification of *Coleoloides* cf. *C. typicalis* is confirmed, this would be important for correlations with south-east Newfoundland. However, as yet reliable correlation would appear to be only possible in the Hanford Brook Formation where trilobites of the *Protolenus* Zone appear. Although Hayes & Howell (1937) appear to have regarded trilobites as probably being absent in the lower part of the Hanford Brook Formation, this is apparently contradicted by Matthew's (1889, 1893, 1895) descrip-

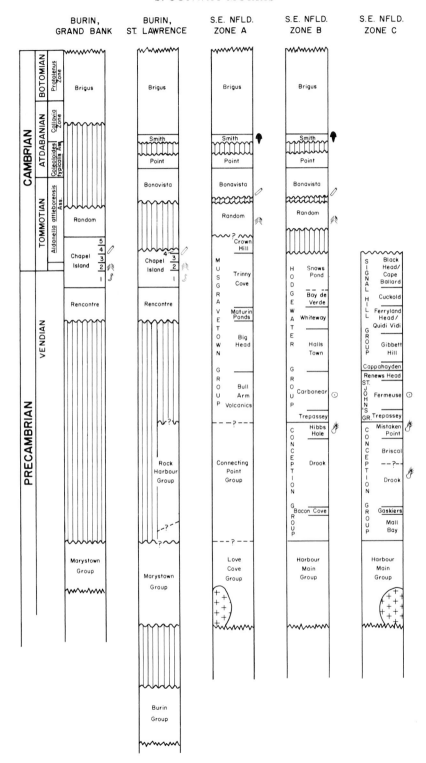

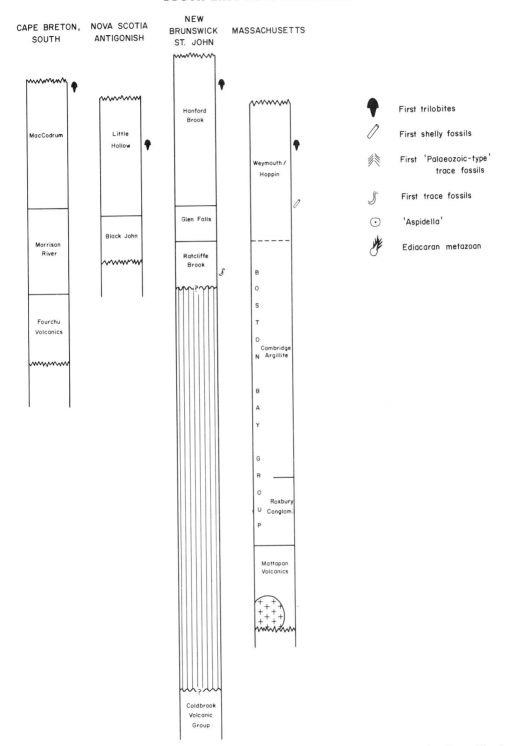

Fig. 2.8. Tentative correlation chart of Precambrian–Cambrian boundary sequences along the Avalon Zone. All units in the geographical columns are of formational status unless indicated otherwise. Based on numerous authors, including Hayes & Howell 1937; Hutchinson 1952; Rast *et al.* 1976; Hoffmann *et al.* 1979; Strong 1979; Strong *et al.* 1979; King 1980; Bengtson & Fletcher 1983; Pickerill & Tanoli 1985.

tion of trilobites. One possibility is that this part of the section is equivalent to the *Callavia* Zone *s.l.* The evidence for the Glen Falls Formation (Fig. 2.8) being correlated with the Random Formation was reviewed above. Much depends on the putative age of the small shelly fossils in the underlying Ratcliffe Brook Formation (Walcott 1900), but if the lithological equivalence of the Glen Falls and Random Formations were to be accepted, this would leave a minimal stratigraphical thickness in which to house any equivalents to the *Coleoloides typicalis* assemblage.

In mainland Nova Scotia a stratigraphical tie is obtained near Antigonish (Fig. 2.1) with the occurrence of an assemblage belonging to the *Callavia* Zone *s.l.* (Landing *et al.* 1980). Although there is no faunal control on the underlying sequence, if the correlation of the Black John Formation and basal part of the Little Hollow Formation with the Random Formation (Fig. 2.8) is accepted, then this too leaves a relatively thin stratigraphical interval (*c.*80 m) in which to accommodate the equivalents of the *Coleoloides typicalis* assemblage. However, as with the Glen Falls Formation this lithological correlation with the Random Formation would require revision.

The correlations between Cape Breton Island and other areas in the Avalon Zone are particularly uncertain. The occurrence of trilobites near the top of MacCodrum Formation may indicate a position within either the *Callavia* Zone *s.l.* or *Protolenus* Zone *s.l.*, but in the absence of any fossils the possible equivalents of the rest of the MacCodrum Formation and underlying Morrison River Formation is based on lithostratigraphical criteria (Fig. 2.8).

A potential difficulty in the biostratigraphical study of the Avalon Lower Cambrian, and indeed many other boundary sequences, is the long-ranging nature of many of the taxa of small shelly fossils. In particular, many of the elements of the *Coleoloides typicalis* assemblage also occur in the overlying *Callavia* Zone *s.l.* (Landing *et al.* 1980; Bengtson & Fletcher 1983). This observation could have some bearing on the apparent difficulties in recognizing the former assemblage in the Avalon Zone outside south-east Newfoundland. As explained above in Massachusetts, and possibly Nova Scotia (Antigonish) and New Brunswick, the interval between the first representatives of the *Callavia* Zone *s.l.* and sediments placed either on biostratigraphical or the admittedly much less secure lithological grounds (i.e. equivalence to the Random Formation) in the *Aldanella attleborensis* association is remarkably thin in comparison with the succession in south-east Newfoundland. The most likely explanation is that rates of rock accumulation in some parts of the Avalon Zone (and possibly the English

Midlands, see Section 2.5.2) were markedly lower than in south-east Newfoundland. However, it is worth considering an alternative possibility that the appearance of trilobites typifying the *Callavia* Zone *s.l.* was diachronous in this region. In any event, caution should be employed in the correlation of sequences on the basis of negative evidence of trilobite absence.

Finally, it is worth noting the potential use of sea-level curves for correlation of boundary sequences, especially where faunal control is poor (Bjørlykke 1982). One such application for local correlations within the Rencontre Formation (Fig. 2.8) has already been attempted by Smith & Hiscott (1984). With a growing number of more detailed stratigraphical and sedimentological analyses (e.g. Patel 1973; Hiscott 1982; Benus & Landing 1984; Landing & Benus 1984, 1988; Crimes & Anderson 1985; Landing *et al.* 1988), a complex pattern of sea-level changes superimposed on the broadly transgressive Lower Cambrian sequences may be expected to emerge, so opening the possibility of testing proposed correlations (Fig. 2.8).

2.5.2. Intercontinental correlations

Correlations of Precambrian–Cambrian boundary sequences around the world continue to generate disagreement and controversy. The place of the south-east Newfoundland sequences (Fig. 2.9), which should be taken as the standard in comparison with the thinner and/or less fossiliferous sections elsewhere in the Avalon Zone, in such a scheme was addressed by Bengtson & Fletcher (1983). These authors suggested that the *Aldanella attleborensis* assemblage was approximately equivalent to the Tommotian Stage as defined in east Siberia, while the appearance of 'Palaeozoic-type' trace fossils in the underlying Member 2 of the Chapel Island Formation would appear to be a reasonable local marker for the Precambrian–Cambrian boundary itself (see also Narbonne & Myrow 1988). The utility of trace fossils for the resolution of boundary sequences is under investigation, and the hope still lies in the small shelly fossils. However, Bengtson & Fletcher (1983) noted that although certain taxa (e.g. *Fomitchella acinaciformis* and *Aldanella attleborensis*) were characteristic of the lower Tommotian in Siberia, they had extended ranges in south-east Newfoundland, so that no meaningful comparisons were possible with the zonal scheme recognized in the Tommotian (Rozanov *et al.* 1969). Bengtson & Fletcher (1983) suggested that other equivalent strata would be the Lontova Series (Estonia) (Fig. 2.9) and Klimontovian (Poland), but no equivalent in terms of a shelly assemblage is identifiable in the English Midlands. In this latter area, however, the lower part of the

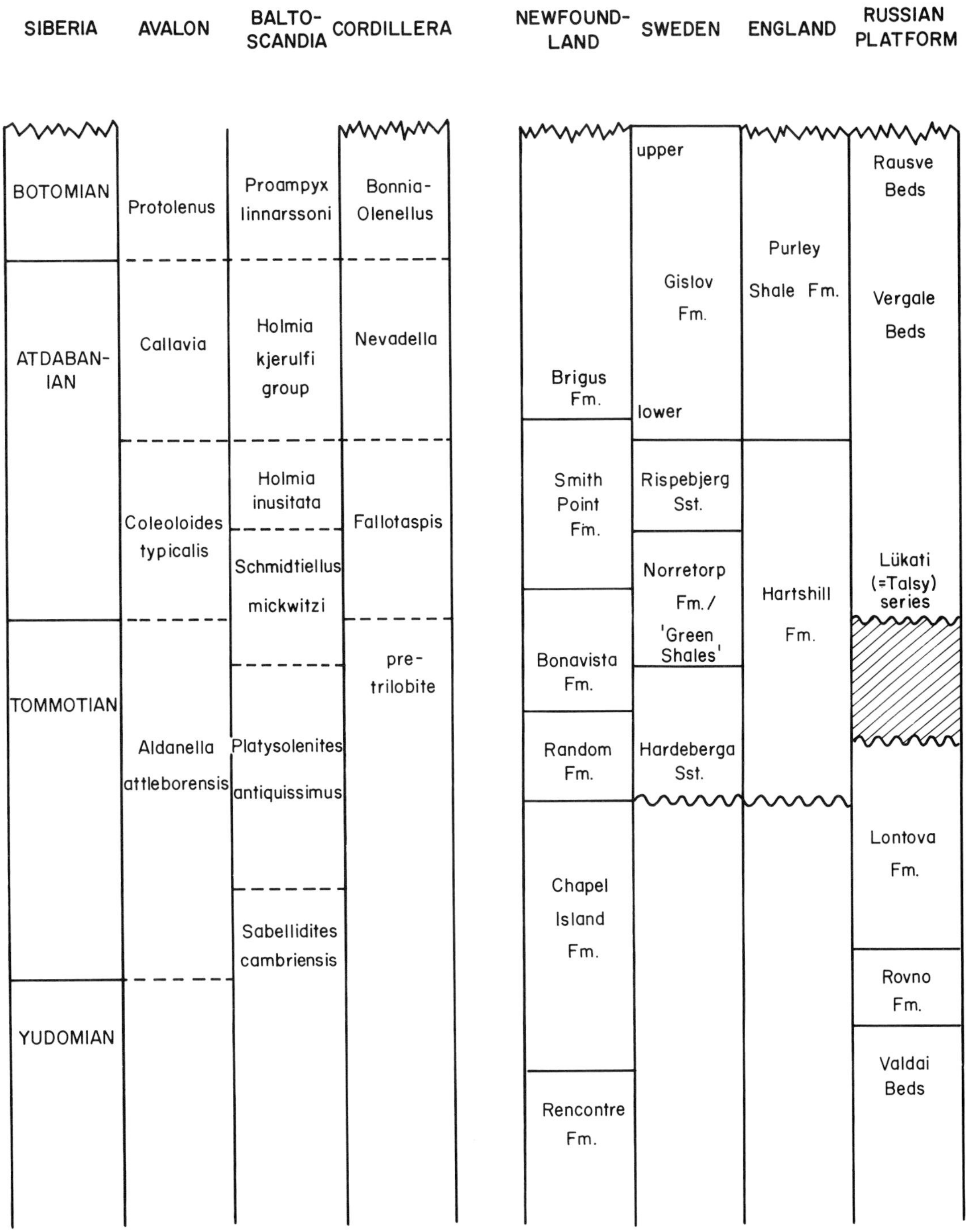

SIBERIA	AVALON	BALTO-SCANDIA	CORDILLERA	NEWFOUND-LAND	SWEDEN	ENGLAND	RUSSIAN PLATFORM
BOTOMIAN	Protolenus	Proampyx linnarssoni	Bonnia-Olenellus	Brigus Fm.	upper		Rausve Beds
ATDABAN-IAN	Callavia	Holmia kjerulfi group	Nevadella		Gislov Fm.	Purley Shale Fm.	Vergale Beds
					lower		
	Coleoloides typicalis	Holmia inusitata	Fallotaspis	Smith Point Fm.	Rispebjerg Sst.	Hartshill Fm.	Lükati (=Talsy) series
		Schmidtiellus mickwitzi			Norretorp Fm./ 'Green Shales'		
TOMMOTIAN			pre-trilobite	Bonavista Fm.			
	Aldanella attleborensis	Platysolenites antiquissimus		Random Fm.	Hardeberga Sst.		Lontova Fm.
		Sabellidites cambriensis		Chapel Island Fm.			Rovno Fm.
YUDOMIAN							Valdai Beds
				Rencontre Fm.			

Fig. 2.9. Tentative correlation chart of the Precambrian–Cambrian boundary sequences in the Avalon Zone with stratigraphical sections in Sweden, England, and the Russian platform, and with the biostratigraphical schemes erected for the Balto-Scandinavian and Cordilleran regions, together with Siberian divisions. Based largely on Bergström 1981; Bergström & Ahlberg 1981; Ahlberg *et al.* 1986.

Hartshill Formation (Brasier *et al.* 1978) could be of equivalent age, as may be the Ringsaker Quartzite and Hardeberga Sandstone (Fig. 2.9) of Scandinavia (Bergström 1981, his Fig. 1). A further guide to correlation might also arise if the occurrence of *Platysolenites* in the Ratcliffe Brook Formation (G. F. Matthew 1889) could be confirmed. This is a long-ranging genus, that appears to span much of the Tommotian and Lower Atdabanian (Bergström 1981, his Fig. 2). On the East European Platform *Platysolenites* is more characteristic of the pre-trilobite sequence (sub-*Holmia*, see Føyn & Glaessner 1979), but does co-occur with *Volborthella* in the lowest trilobite-bearing units belonging to the *Schmidtiellus mickwitzi* (and *Holmia mobergi*) (Bergström 1981, his Fig. 2). Thus if the co-occurrence of *Volborthella* with *Platysolenites* in the Ratcliffe Brook Formation were confirmed (G. F. Matthew 1889), this might narrow further the proposed correlation. However, G. F. Matthew himself cast doubt on the former identification at a later date (G. F. Matthew 1895). Existing evidence suggests at least part of the Ratcliffe Brook Formation (Fig. 2.8) is Tommotian, being equivalent to the *A. attleborensis* assemblage of south-east Newfoundland.

With regard to the *Coleoloides typicalis* assemblage, Bengtson & Fletcher (1983) suggested that equivalent sequences included the upper part of the Hartshill Formation (England), which would include the condensed sequence referred to as the Home Farm Member, the 'Green Shales' (Bornholm), the Talsy Series (Fig. 2.9) and its equivalent, the Lükati Beds, and the Lower *Holmia* Beds of Poland. These units are generally agreed to be Lower Atdabanian and are approximately equivalent to the *Fallotaspis* Zone of the Cordilleran region and the *Schmidtiellus mickwitzi* and *Holmia inusitata* Zones of Scandinavia (see Ahlberg *et al.* 1986). Continuing upwards, the *Callavia* Zone *s.l.* of the Avalonian Zone is probably Upper Atdabanian, and would be approximately equivalent to the *Nevadella* Zone (Cordilleran region), and *Holmia kjerulfi* group (Scandinavia, see Bergström & Ahlberg 1981; Ahlberg *et al.* 1986). This would indicate that the incoming of trilobites, at least in south-east Newfoundland, was relatively late in comparison with many sections beyond the Avalon Zone where the Lower Atdabanian contains fallotaspid trilobites. Finally, the *Protolenus* Zone is regarded as Botomian (Fig. 2.9), and may be equated with the *Bonnia–Olenellus* Zone (Cordilleran region) and *Proampyx linnarssoni* Zone (Scandinavia).

With regard to intercontinental correlation, the occurrence of *Volborthella* in New Brunswick (G. F. Matthew 1889) could also provide some biostratigraphical guidance, as it also occurs widely on the East European platform and Baltic region (Skjeseth 1963;

Føyn & Glaessner 1979; Rozanov 1979), and in California (Lipps & Sylvester 1968; Firby & Durham 1974). Bergström (1981, his Fig. 2) portrayed a stratigraphical range for *Volborthella* of *Schmidtiellus mickwitzi* (*Holmia mobergi*) Zone up to at least the *Holmia kjerulfi* group (see Fig. 2.9), while the Californian occurrences also indicate a lengthy stratigraphical range from the *Fallotaspis* Zone (Andrews Mountain Member, Campito Formation) to the *Bonnia–Olenellus* Zone (base of the Harkless Formation). There is also a proposed Middle Cambrian occurrence from Czechoslovakia (Prantl 1948), but this record appears to be doubtful (Yochelson 1977, p. 441). If the occurrence of *Volborthella* in the upper Hanford Brook Formation (G. F. Matthew 1889) i.e. within the *Protolenus* Zone is confirmed, this would suggest a position near the top of its range.

2.6. CONCLUSIONS

If a stratotype for the Precambrian–Cambrian boundary was to be located in the Avalon Zone, it would almost certainly have to be on the Burin Peninsula. Space prohibits detailed comparison with other candidate stratotypes, and only some of the principal merits and disadvantages require brief discussion.

In favour of this area is a thick sequence of effectively unmetamorphosed sediments, showing a wide variety of lithologies. All the key biological events, i.e. Ediacaran biotas, small shelly fossils, and trilobites, are recorded, and in addition trace fossils are diverse and abundant. The small shelly fossils comprise two distinct assemblages, and moderately secure correlations can be drawn with the Baltic region, East European Platform, and beyond. Rocks suitable for radiometric dating are abundant, and there is convincing evidence for a volcanic/plutonic episode (Harbour Main Group/Holyrood Granite and postulated equivalents, see above) at about 590–620 Ma. The stratigraphical position of these rocks strongly supports the notion that the base of the Cambrian may be placed at about 540–550 Ma.

Arguing against this region as a stratotype candidate, it must be stressed that the Ediacaran biotas are effectively endemic (Anderson & Conway Morris 1982), while the shelly faunas are of a low diversity and archaeocyathans are absent, presumably because of the relatively high-latitude position of the Avalon Zone in the early Cambrian. The small shelly fossils are generally relatively poorly preserved. There is evidence that trilobites appear substantially later in the Avalon Zone region than elsewhere, so that the assumed equivalents of the *Fallotaspis* Zone lack trilobites. Moreover, the interval that corresponds approximately with the *Coleoloides typicalis* assemblage in south-east Newfoundland (Bengtson & Fletcher 1983) has not yet been identified confidently in other areas of the Avalon Zone, although many of the faunal elements occur in the overlying *Callavia* Zone *s.l.* The problem of biostratigraphical utility of many taxa of small shelly fossils, given growing evidence of lengthy ranges in many forms, is only beginning to receive serious attention. The faunas from the Avalon Zone are by no means immune to this difficulty, and various examples are given above (see Landing *et al.* 1980; Bengtson & Fletcher 1983). Lack of space prohibits a detailed discussion of this problem, but in passing it might be mentioned that the recent discovery of anabaritids in Atdabanian strata of South Australia (unpublished observations of the author, S. Bengtson, B. Runnegar, & B. Cooper) is further evidence that at least some of the forms thought to have been restricted to the Tommotian may now be shown to extend into substantially younger rocks.

While biostratigraphical techniques still appear to hold the best hopes of recognizing and correlating the boundary, the rapid development of magnetostratigraphy and isotope stratigraphy offer fresh possibilities. Unfortunately, with regard to a magnetostratigraphy based on the identification of a characteristic pattern of reversed and normal intervals, rocks of the Avalon Zone appear to have been remagnetized systematically during the Acadian Orogeny in the Devonion (Spariosu *et al.* 1983; J. Kirschvink, pers. comm.), and although some apparently original magnetic signatures may be detectable (Seguin 1985), the role of magnetostratigraphy for boundary correlation in this area would seem to be restricted. However, recent work on isotope stratigraphy may offer more hope of testing correlations independent of fossil evidence. Evidence for dramatic shifts in stable carbon isotope ratios near the boundary in carbonates from Morocco (Tucker 1986) and eastern Siberia (Magaritz *et al.* 1986), may or may not reflect contemporary events. An investigation of stable isotope values for the carbonates of the Avalon Zone may help to refine proposed correlations within this area (Fig. 2.8), and with more remote regions.

Acknowledgements

I thank Dr A. W. A. Rushton (British Geological Survey) for reviewing the manuscript and offering helpful comments. M. A. Anderson, T. P. Fletcher, E. Landing, and G. Narbonne kindly introduced me to the boundary sequences in Newfoundland; fieldwork was supported in part by the Royal Society, Cowper–Reed Fund, and the Open University. Phil Perkins computed some pole positions, Sandra Last typed several versions of the paper, and Sheila Ripper drafted the illustrations. Ken Harvey and David Newling assisted with photography and operation of the SEM. This publication is Cambridge Earth Sciences Number 1236.

REFERENCES

Ahlberg, P., Bergström, J., & Johannson, J. (1986). Lower Cambrian olenellid trilobites from the Baltic faunal province. *Geol. Fören. Stockholm Förhandl.*, **108**, 39–56.

Alcock, F. J. (1938). Geology of Saint John region, New Brunswick. *Mem. geol. Surv. Can.*, **216**, 1–65.

Anderson, M. M. (1972). A possible time span for the later Precambrian of the Avalon peninsula, southeastern Newfoundland in the light of worldwide correlation of fossils, tillites, and rock units within the succession. *Can. J. Earth Sci.*, **9**, 1710–26.

Anderson, M. M. (1978). Ediacaran fauna. In *McGraw-Hill Yearbook of Science and Technology*, 1978, (ed. D. N. Lapedes), pp. 146–9. McGraw-Hill, New York.

Anderson, M. M. (1981). The Random Formation of southeastern Newfoundland: a discussion aimed at establishing its age and relationship to bounding formations. *Am. J. Sci.*, **281**, 807–30.

Anderson, M. M. & King, A. F. (1981). Precambrian tillites of the Conception Group on the Avalon peninsula, southeastern Newfoundland. In *Earth's pre-Pleistocene glacial record*, (ed. M. Hambrey & W. B. Harland), pp. 760–7. Cambridge University Press.

Anderson, M. M., Brückner, W. D., King, A. F., & Maher, J. B. (1975). The late Proterozoic 'H. D. Lilly unconformity' at Red Head, northeastern Avalon peninsula, Newfoundland. *Am. J. Sci.*, **275**, 1012–27.

Anderson, M. M., Choubert, G., Faure-Muret, A. & Timofeiev, B. V. (1982). Les couches à *Bavlinella* de part et d'autre de l'Atlantique. *Bull. Soc. géol. Fr.*, (7), **24**, 389–92.

Anderson, M. M. and Conway Morris, S. (1982). A review, with descriptions of four unusual forms, of the soft-bodied fauna of the Conception and St John's Groups (Late-Precambrian), Avalon peninsula, Newfoundland. *Proc. Third N. Am. Paleont. Conv.*, **1**, 1–8.

Barr, S. M., Sangster, D. F., & Cormier, R. F. (1984). Petrology of early Cambrian and Devono-Carboniferous intrusions in the Loch Lomond complex, southeastern Cape Breton Island, Nova Scotia. *Geol. Surv. Pap. Can.*, **84–1A**, 203–11.

Barr, S. M., MacDonald, A. S., & Blenkinsop, J. (1986). The Chéticamp pluton: a Cambrian granodioritic intrusion in the western Cape Breton Highlands, Nova Scotia. *Can. J. Earth Sci.*, **23**, 1686–99.

Barr, S. M. & Raeside, R. P. (1986). Pre-Carboniferous tectonostratigraphic subdivisions of Cape Breton, Nova Scotia. *Mar. Sedim. & Atlan. Geol.*, **22**, 252–63.

Bell, K. G. & Alvord, D. C. (1976). Pre-Silurian stratigraphy of northeastern Massachusetts. *Geol. Soc. Am. Mem.*, **148**, 179–216.

Bengtson, S. & Fletcher, T. P. (1981). The succession of skeletal fossils in the basal Lower Cambrian of southeastern Newfoundland. *US geol. Surv. Open File Rep.*, **81–743**, p. 18.

Bengtson, S. & Fletcher, T. P. (1983). The oldest sequence of skeletal fossils in the Lower Cambrian of southeastern Newfoundland. *Can. J. Earth Sci.*, **20**, 525–36.

Benus, A. P. & Landing, E. (1984). Depositional environment and biofacies of the Bonavista Formation (early Cambrian [Tommotian–Lower Atdabanian]), eastern Newfoundland. *Geol. Soc. Am. Abstr. Prog.*, **16**, p. 3.

Bergström, J. (1981). Lower Cambrian shelly faunas and biostratigraphy in Scandinavia. *US geol. Surv. Open File Rep.*, **81–743**, 22–5.

Bergström, J. & Ahlberg, P. (1981). Uppermost Lower Cambrian biostratigraphy in Scania, Sweden. *Geol. Fören. Stockh. Förhandl.*, **103**, 193–214.

Billings, M. P. (1976). Geology of the Boston Basin. *Mem. Geol. Soc. Am.*, **146**, 5–30.

Billings, M. P. (1979). Boston Basin, Massachusetts. *US geol. Surv. Prof. Pap.*, **1110A**, A15–120.

Bjørlykke, K. (1982). Correlation of late Precambrian and early Palaeozoic sequences by eustatic sea-level changes and the selection of the Precambrian-Cambrian boundary. *Precambr. Res.*, **17**, 99–104.

Bland, B. H. (1984). *Arumberia* Glaessner & Walter, a review of its potential for correlation in the region of the Precambrian–Cambrian boundary. *Geol. Mag.*, **121**, 625–33.

Brasier, M. D., Hewitt, R. A. & Brasier, C. J. (1978). On the Late Precambrian–Early Cambrian Hartshill Formation of Warwickshire. *Geol. Mag.*, **115**, 21–36.

Brückner, W. D. (1969). Geology of the eastern part of Avalon peninsula, Newfoundland; a summary. *Mem. Am. Assoc. Pet. Geol.*, **12**, 130–8.

Brückner, W. D. & Anderson, M. M. (1971). Late Precambrian glacial deposits in southeastern Newfoundland—a preliminary note. *Proc. geol. Assoc., Can.*, **24**, 95–102.

Burr, H. T. (1900). A new Lower Cambrian fauna from eastern Massachusetts. *Am Geol.*, **25**, 41–50.

Choubert, B. (1935). Recherches sur la genèse des chaînes paleozoïques et antécambriennes. *Revue Géogr. phys. Géol. dyn.*, **8**, 5–50.

Christie, R. L. & Sheldon, R. P. (1986). Proterozoic and Cambrian phosphorites—regional review: North America. In *Phosphate deposits of the world* (eds P. J. Cook & J. H. Shergold), Vol. 1. pp. 101–17. Cambridge University Press.

Clark, T. H. (1923). New fossils from the vicinity of Boston. *Proc. Boston Soc. natur. Hist.*, **36**, 473–85.

Clarke, D. B., Barr, S. M., & Donohoe, H. V. (1980). Granitoid and other plutonic rocks of Nova Scotia. In *Proceedings 'The Caledonides in the USA'*, (ed. D. R. Wones). Dep. Geol. Sci., Virginia Polytech. Inst., State Univ. Mem., 2, pp. 107–11.

Conway Morris, S. (1987). The search for the Precambrian–Cambrian boundary. *Am. Scient.*, **75**, 157–67.

Conway Morris, S. (1988). Radiometric dating of the Precambrian–Cambrian Boundary in the Avalon Zone. *Bull. NY State Mus.*, **463**, 53–8.

Conway Morris, S. & Rushton, A. W. A. (in press). Precambrian to Tremadoc biotas in the Caledonides. *Spec. Publ. geol. Soc. Lond.*

Cope, J. C. W. & Gibbons, W. (1987). New evidence for the relative age of the Ercall Granophyre and its bearing on the Precambrian–Cambrian Boundary in southern Britain. *Geol. J.*, **22**, 53–60.

Cormier, R. F. (1969). Radiometric dating of the Coldbrook Group of southern New Brunswick, Canada. *Can. J. Earth Sci.*, **6**, 393–8.

Cormier, R. F. (1972). Radiometric ages of granitic rocks, Cape Breton Island, Nova Scotia. *Can. J. Earth Sci.*, **9**, 1074–86.

Cowie, J. W. & Cribb, S. J. (1978). The Cambrian System. *Am. Assoc. Pet. Geol. Stud. Geol.*, **6**, 355–62.

Cowie, J. W. & Johnson, M. R. W. (1985). Late Precambrian and Cambrian geological time-scale. In *The chronology of the geological record*, (ed. N. J. Snelling). Geol. Soc. Lond. Mem., 10, pp. 47–64. Blackwell Scientific, Oxford.

Crimes, T. P. & Anderson, M. M. (1985). Trace fossils from late Precambrian–early Cambrian strata of southeastern Newfoundland (Canada): temporal and environmental implications. *J. Paleont.*, **59**, 310–43.

Currie, K. L. (1984). A reconsideration of some geological relations near Saint John, New Brunswick. *Geol. Surv. Pap. Can.*, **84–1A**, 193–201.

Currie, K. L. (1986). The boundaries of the Avalon tectonostratigraphic zone, Musquash Harbour–Loch Alva region, southern New Brunswick. *Geol. Surv. Can. Pap.*, **86–1A**, 333–4.

Dallmeyer, R. D., Odom, A. L., O'Driscoll, C. F., & Hussey, E. M. (1981). Geochronology of the Swift Current granite and host volcanic rocks of the Love Cove Group, southwestern Avalon Zone, Newfoundland: evidence of a late Proterozoic volcanic–subvolcanic association. *Can. J. Earth Sci.*, **18**, 699–707.

Donohoe, H. V., Halliday, A. N., & Keppie, J. D. (1986). Two Rb–Sr whole rock isochrons from plutons in the Cobequid Highlands, Nova Scotia, Canada. *Mar. Sedim. & Atlant. Geol.*, **22**, 148–54.

Downie, C. (1982). Lower Cambrian acritarchs from Scotland, Norway, Greenland and Canada. *Trans. R. Soc. Edinb.: Earth Sci.*, **72**, 257–85.

Dowse, A. M. (1950). New evidence on the Cambrian contact at Hoppin Hill, North Attleboro, Massachusetts. *Am. J. Sci.*, **248**, 95–9.

Edhorn, A. S. (1977). Early Cambrian algae croppers. *Can. J. Earth Sci.*, **14**, 1014–20.

Edhorn, A. S. & Anderson, M. M. (1977). Algal remains in the Lower Cambrian Bonavista Formation, Conception Bay, southeastern Newfoundland. In *Fossil algae*, (ed. E. Flugel), pp. 113–23. Springer-Verlag, Berlin.

Emerson, B. K. (1917). Geology of Massachusetts and Rhode Island. *US geol. Surv. Bull.*, **597**, 1–289.

Fairbairn, H. W., Bottino, M. L., Pinson, W. H., & Hurley, P. M. (1966). Whole-rock age and initial $^{87}Sr/^{86}Sr$ of volcanics underlying fossiliferous Lower Cambrian in the Atlantic provinces of Canada. *Can. J. Earth Sci.*, **3**, 509–21.

Fairbairn, H. W., Moorbath, S., Ramo, A. O., Pinson, W. H., & Hurley, P. M. (1967). Rb–Sr age of granitic rocks of southeastern Massachusetts and the age of the Lower Cambrian at Hoppin Hill. *Earth Planet. Sci. Lett.*, **2**, 321–8.

Firby, J. B. & Durham, J. W. (1974). Molluscan radula from earliest Cambrian. *J. Paleont.*, **48**, 1109–19.

Foerste, A. F. (1889). The paleontological horizon of the limestone at Nahant, Mass. *Proc. Boston Soc. natur. Hist.*, **24**, 261–3.

Føyn, S. & Glaessner, M. F. (1979). *Platysolenites*, other animal fossils, and the Precambrian–Cambrian transition in Norway. *Norsk geol. Tidsskr.*, **59**, 25–46.

Frith, R. A. & Poole, W. H. (1972). Late Precambrian rocks of eastern Avalon peninsula, Newfoundland—a volcanic island complex: Discussion. *Can. J. Earth Sci.*, **9**, 1058–9.

Gale, N. H. (1982). Numerical dating of Caledonian times (Cambrian to Silurian). In *Numerical dating in stratigraphy*, (ed. G. S. Odin), pp. 467–86. John Wiley, Chichester.

Gardiner, S. & Hiscott, R. N. (1984). Sedimentology and basin analysis of the Precambrian Conception Group, Avalon Zone, Newfoundland. *Geol. Surv. Can. Pap.*, **84–1A**, 213–6.

Gaudette, H. E., Olszewski, W. J., & Donohoe, H. V. (1983). Age and origin of the basement rocks, Cobequid Highlands, Nova Scotia. *Geol. Soc. Am. Abstr. Prog.*, **15**, p. 136.

Giles, P. S. & Ruitenberg, A. A. (1977). Stratigraphy, paleogeography, and tectonic setting of the Coldbrook Group in the Caledonia Highlands of southern New Brunswick. *Can. J. Earth Sci.*, **14**, 1263–75.

Glaessner, M. F. (1984). *The dawn of animal life, a biohistorical study*. Cambridge University Press.

Glaessner, M. F. & Walter, M. R. (1975). New Precambrian fossils from the Arumbera sandstone, Northern Territory, Australia. *Alcheringa*, **1**, 59–69.

Grabau, A. W. (1900). Palaeontology of the Cambrian terranes of the Boston basin. *Occ. Pap. Boston Soc. natur. Hist.*, **4**, 601–94.

Greene, B. & Williams, H. (1974). New fossil localities and the base of the Cambrian in southeastern Newfoundland. *Can. J. Earth Sci.*, **11**, 319–23.

Greenough, J. D., McCutcheon, S. R., & Papezik, V. S. (1985). Petrology and geochemistry of Cambrian volcanic rocks from the Avalon Zone in New Brunswick. *Can. J. Earth Sci.*, **22**, 881–92.

Greenough, J. D. & Papezik, V. S. (1985). Petrology and geochemistry of Cambrian volcanic rocks from the Avalon peninsula, Newfoundland. *Can. J. Earth Sci.*, **22**, 1594–1601.

Greenough, J. D. & Papezik, V. S. (1986). Acado-Baltic volcanism in eastern North America and western Europe: implications for Cambrian tectonism. *Mar. Sedim. & Atlant. Geol.*, **22**, 240–51.

Hagstrum, J. T., Van der Voo, R., Auvray, B., & Bonhommet, N. (1980). Eocambrian–Cambrian palaeomagnetism of the Armorican Massif, France. *Geophys. J. R. astron. Soc.*, **61**, 489–517.

Häntzschel, W. (1975). Miscellanea. Supplement 1. Trace fossils and problematica. In *Treatise on invertebrate paleontology*, (ed. C. Teichert). Part W. Geological Society of America and University of Kansas Press, Lawrence, Kansas.

Haworth, R. T. & Lefort, J. P. (1979). Geophysical evidence for the extent of the Avalon Zone in Atlantic Canada. *Can. J. Earth Sci.*, **16**, 552–67.

Hayes, A. O. & Howell, B. F. (1937). Geology of Saint John, New Brunswick. *Spec. Pap. geol. Soc. Am.*, **5**, ix, 1–146.

Hermes, O. D. & Zartman, R. E. (1985). Late Proterozoic and Devonian plutonic terrane within the Avalon Zone of Rhode Island. *Bull. geol. Soc. Am.*, **96**, 272–82.

Hiscott, R. N. (1981). Stratigraphy and sedimentology of the late Proterozoic Rock Harbour Group, Flat Islands, Placentia Bay, Newfoundland Avalon Zone. *Can. J. Earth Sci.*, **18**, 495–508.

Hiscott, R. N. (1982). Tidal deposits of the Lower Cambrian Random Formation, eastern Newfoundland: facies and paleoenvironments. *Can. J. Earth Sci.*, **19**, 2028–46.

Hofmann, H. J. (1971). Precambrian fossils, pseudofossils, and problematica in Canada. *Bull. geol. Surv. Can.*, **189**, 1–146.

Hofmann, H. J., Hill, J., & King, A. F. (1979). Late Precambrian microfossils, southeastern Newfoundland. *Geol. Surv. Can. Pap.*, **79–1B**, 83–98.

Hughes, C. J. (1970). The late Precambrian Avalonian Orogeny in Avalon, southeastern Newfoundland. *Am. J. Sci.*, **269**, 183–90.

Hughes, C. J. (1972). Geology of the Avalon peninsula, Newfoundland, and its possible correspondence with Morocco. *Notes et Mém. Serv. Mines Carte géol. Maroc.*, **236**, 265–75.

Hughes, C. J. & Brückner, W. D. (1971). Late Precambrian rocks of the eastern Avalon peninsula, Newfoundland—a volcanic island complex. *Can. J. Earth Sci.*, **8**, 899–915.

Hughes, C. J. & Malpas, J. G. (1971). Metasomatism in the late Precambrian Bull Arm Formation in southeastern Newfoundland: recognition and implications. *Proc. geol. Assoc. Can.*, **24**, 85–93.

Hutchinson, R. D. (1952). The stratigraphy and trilobite faunas of the Cambrian rocks of Cape Breton Island, Nova Scotia. *Mem. geol. Surv. Can.*, **263**, v, 1–124.

Hutchinson, R. D. (1953). Geology of the Harbour Grace map-area, Newfoundland. *Mem. geol. Surv. Can.*, **275**, v, 1–43.

Hutchinson, R. C. (1962). Cambrian stratigraphy and trilobite faunas of southeastern Newfoundland. *Bull. geol. Surv. Can.*, **88**, 1–156.

Irving, E. & Strong, D. F. (1985). Paleomagnetism of rocks from the Burin Peninsula, Newfoundland: hypothesis of the late Paleozoic displacement of Acadia criticized. *J. geophys. Res.*, **90**, 1949–63.

Jamieson, R. A., van Breeman, O., Sullivan, R. W., & Currie, K. L. (1986). The age of igneous and metamorphic events in the western Cape Breton Highlands, Nova Scotia. *Can. J. Earth Sci.*, **23**, 1891–1901.

Jenkins, R. J. F. (1984). Ediacaran events: boundary relationships and correlation of key sections, espcially in 'Armorica'. *Geol. Mag.*, **121**, 635–43.

Jenness, S. E. (1963). Terra Nova and Bonavista map-areas, Newfoundland. *Mem. geol. Surv. Can.*, **327**, 1–184.

Johnson, R. J. E. & Van der Voo, R. (1985). Middle Cambrian paleomagnetism of the Avalon terrane in Cape Breton Island, Nova Scotia. *Tectonics*, **4**, 629–51.

Johnson, R. J. E. & Van der Voo, R. (1986). Paleomagnetism of the Late Precambrian Fourchu Group, Cape Breton Island, Nova Scotia. *Can. J. Earth Sci.*, **23**, 1673–85.

Kaye, C. A. & Zartman, R. F. (1980). A late Proterozoic Z to Cambrian age for the stratified rocks of the Boston Basin, Massachusetts, U.S.A. In *Proceedings 'The Caledonides in the USA'*, (ed. D. R. Wones), Dept. Geol. Sci. Virginia Polytech. Inst., State Univ. Mem., 2 pp. 257–61.

Keppie, J. D. (1982). The Minas geofracture. *Spec. Pap. geol. Assoc. Can.*, **24**, 263–80.

Keppie, J. D. (1985). The Appalachian collage. In *The Caledonide Orogen—Scandinavia and related areas*, (eds D. G. Gee & B. A. Sturt), pp. 1217–26. John Wiley, Chichester.

Keppie, J. D. & Dostal, J. (1980). Paleozoic volcanic rocks of Nova Scotia. In *Proceedings 'The Caledonides in the USA'*, (ed. D. R. Wones), Dept. Geol. Sci. Virginia Polytech. Inst., State Univ. Mem., 2, pp. 249–56.

Keppie, J. D. & Halliday, A. N. (1986). Rb–Sr isotopic data from three suites of igneous rocks, Cape Breton Island, Nova Scotia. *Mar. Sedim. & Atlant. Geol.*, **22**, 162–71.

King, A. F. (1980). The birth of the Caledonides: late Precambrian rocks of the Avalon peninsula, Newfoundland, and their correlatives in the Appalachian orogen. In *Proceedings 'The Caledonides in the USA'*, (ed. D. R. Wones), Dept. Geol. Sci. Virginia Polytech. Inst., State Univ., Mem., 2, pp. 3–8.

King, A. F. (ed.) (1982). *Field Guide for Avalon and Meguma Zones.* The Caledonide Orogen IGCP Project 27, NATO Advanced Study Institute, Atlantic Canada, August 1982. Memorial University, St Johns, Newfoundland.

King, L. H., Fader, G. B. J., Poole, W. H., & Wanless, R. K. (1985). Geological setting and age of the Flemish Cap granodiorite, east of the Grand Banks of Newfoundland. *Can. J. Earth Sci.*, **22**, 1286–98.

King, L. H., Fader, G. B. J., Jenkins, W. A. M., & King, E. L. (1986). Occurrence and regional geological setting of Paleozoic rocks on the Grand Banks of Newfoundland. *Can. J. Earth Sci.*, **23**, 504–26.

Knoll, A. H. & Swett, K. (1985). Micropalaeontology of the late Proterozoic Veteranen Group, Spitsbergen. *Palaeontology*, **28**, 451–73.

Kovach, A., Hurley, P. M., & Fairbairn, H. W. (1977). Rb–Sr whole rock age determinations of the Dedham granodiorite, eastern Massachusetts. *Am. J. Sci.*, **277**, 905–12.

Krogh, T. E., Strong, D. F., O'Brien, S. J., & Papezik, V. (1988). Precise U–Pb zircon dates from the Avalon Terrane in Newfoundland. *Can. J. Earth Sci.*, **25**, 442–53.

La Forge, L. (1932). Geology of the Boston area, Massachusetts. *Bull. US geol. Surv.*, **839**, v, 1–105.

Landing, E. (1984). Skeleton of lapworthellids and the supragenic classification of tommotiids (early and middle Cambrian phosphatic problematica). *J. Paleont.*, **58**, 1380–98.

Landing, E. & Benus, A. P. (1984). Lithofacies belts of the Smith Point Limestone (Lower Cambrian, eastern Newfoundland) and the lowest occurrence of trilobites. *Geol. Soc. Am. Abstr. Prog.*, **16**, p. 45.

Landing, E. & Benus, A. P. (1988). Stratigraphy of the Bonavista Group, southeastern Newfoundland: growth faults and the distribution of the sub-trilobitic Lower Cambrian. *Bull. NY State Mus.*, **463**, 59–71.

Landing, E. & Brett, C. E. (1982). Lower Cambrian of eastern Massachusetts: microfaunal sequence and the oldest known borings. *Geol. Soc. Am. Abstr. Prog.*, **14**, p. 33.

Landing, E., Nowlan, G. S., & Fletcher, T. P. (1980). A microfauna associated with Early Cambrian trilobites of the *Callavia* Zone, northern Antigonish Highlands, Nova Scotia. *Can. J. Earth Sci.*, **17**, 400–18.

Landing, E., Narbonne, G. M., Myrow, P., Benus, A. P., & Anderson, M. M. (1988). Faunas and depositional environments of the Upper Precambrian through Lower Cambrian, southeastern Newfoundland. *Bull. NY State Mus.*, **463**, 18–52.

Leech, G. B., Lowdon, J. A., Stockwell, C. H., & Wanless, R. K. (1963). Age determinations and geological studies (including isotopic ages—Report 4). *Geol. Surv. Can. Pap.*, **63–17**, 1–140.

Lenk, C., Strother, P. K., Kaye, C. A., & Barghoorn, E. S. (1982). Precambrian age of the Boston Basin: New evidence from microfossils. *Science*, **216**, 619–20.

Lilly, H. D. (1966). Late Precambrian and Appalachian tectonics in the light of submarine exploration on the Great Bank of Newfoundland and in the Gulf of St Lawrence. Preliminary views. *Am. J. Sci.*, **264**, 569–74.

Lipps, J. H. & Sylvester, A. G. (1968). The enigmatic Cambrian fossil *Volborthella* and its occurrence in California. *J. Paleont.*, **42**, 329–36.

Lowdon, J. A., Stockwell, C. H., Tipper, H. W., & Wanless, R. K. (1963). Age determinations and geological studies (including isotopic ages—Report 3). *Geol. Surv. Can. Pap.*, **62–17**, 1–140.

McCartney, W. D. (1967). Whitbourne map-area, Newfoundland. *Mem. geol. Soc. Can.*, **341**, 1–135.

McCartney, W. D. (1969). Geology of the Avalon peninsula, southeast Newfoundland. *Mem. Am. Assoc. Pet. Geol.*, **12**, 115–29.

McCartney, W. D., Poole, W. H., Wanless, R. K., Williams, H., & Loveridge, W. D. (1966). Rb/Sr age and geological setting of the Holyrood Granite, southeastern Newfoundland. *Can. J. Earth Sci.*, **3**, 947–57.

McCutcheon, S. R. (1981). Revised stratigraphy of the Long Reach area, southern New Brunswick: evidence for major northwestward-directed Acadian thrusting. *Can. J. Earth Sci.*, **18**, 646–56.

Magaritz, M., Holser, W. T., & Kirschvink, J. L. (1986). Carbon-isotope events across the Precambrian/Cambrian boundary on the Siberian platform. *Nature*, **320**, 258–9.

Marcou, J. (1889). Canadian geological classification for the province of Canada [*sic*]. *Proc. Boston Soc. Natur. Hist.*, **24**, 54–83.

Marcou, J. (1890). The Lower and Middle Taconic of Europe and North America. I–III. *Am. Geol.*, **5**, 357–75, 79–102, 221–33.

Matthew, G. F. (1885). Illustrations of the fauna of the St. John Group continued. No. III—Descriptions of new genera and species (including a description of a new species of *Solenopleura* by J. F. Whiteaves). *Proc. Trans. R. Soc. Can., Sect. IV*, **3**, 29–84.

Matthew, G. F. (1888a). On *Psammichnites* and the early trilobites of the Cambrian rocks in eastern Canada. *Am. Geol.*, **2**, 1–9.

Matthew, G. F. (1888b). On the classification of the Cambrian rocks in Acadia. *Can. Rec. Sci.*, **3**, 71–81.

Matthew, G. F. (1889). On Cambrian organisms in Acadia. *Proc. Trans. R. Soc. Can. Sect. IV*, **7** (2nd Ser.), 135–62.

Matthew, G. F. (1891). Illustrations of the fauna of the St. John's Group, no. V. *Proc. Trans. R. Soc. Can.*, **IV**, (8), 123–66.

Matthew, G. F. (1892). On the diffusion and sequence of the Cambrian faunas—Presidential address for the year. *Proc. Trans. R. Soc. Can. Sect.*, *IV*, **10**, 3–16.

Matthew, G. F. (1893). Illustrations of the fauna of the St John Group, VIII. *Proc. Trans. R. Soc. Can. Sect. IV*, **11**, 85–129.

Matthew, G. F. (1895). The Protolenus fauna. *Trans. NY Acad. Sci.*, **14**, 101–53.

Matthew, G. F. (1899a). A Palaeozoic terrane beneath the Cambrian. *Ann. NY Acad. Sci.*, **12**, 41–56.

Matthew, G. F. (1899b). The Etcheminian fauna of Smith Sound, Newfoundland. *Trans. R. Soc. Can. Sect. IV* (2nd Ser.), **5**, 97–123.

Matthew, G. F. (1903). *Report on the Cambrian rocks of Cape Breton*. Geological Survey of Canada, Ottawa.

Matthew, G. F. (1908). Geological cycles in the maritime provinces of Canada. *Proc. Trans. R. Soc. Can.*, **2** (Ser. 3), 121–43.

Matthew, W. D. (1893). On phosphate nodules from the Cambrian of southern New Brunswick. *Trans. NY Acad. Sci.*, **12**, 108–20.

Miller, A. K. (1932). The mixochoanitic cephalopods. *Iowa Univ. Stud. natur. Hist.*, **14**, 1–62.

Miller, H. G. (1987). A geophysical interpretation of the onshore and offshore geology of the southern Avalon Terrane, Newfoundland. *Can. J. Earth Sci.*, **24**, 60–9.

Miller, H. G. & Pittman, D. A. (1982). Geophysical constraints on the thickness of the Holyrood Pluton, Avalon Peninsula, Newfoundland. *Mar. Sedim. & Atlant. Geol.*, **18**, 75–82.

Murphy, J. B., Cameron, K., Dostal, J., Keppie, J. D., & Hynes, A. J. (1985). Cambrian volcanism in Nova Scotia, Canada. *Can. J. Earth Sci.*, **22**, 599–606.

Narbonne, G. M. & Myrow, P. (1988). Trace fossil biostratigraphy in the Precambrian–Cambrian boundary interval. *Bull. NY State Mus.*, **463**, 72–6.

Narbonne, G. M., Myrow, P. M., Landing, E., & Anderson, M. M. (1987). A candidate stratotype for the Precambrian–Cambrian boundary, Fortune Head, Burin Peninsula, southeastern Newfoundland. *Can. J. Earth. Sci.*, **24**, 1277–93.

Nautiyal, A. C. (1976). First record of filamentous algal remains from the late Precambrian rocks of Random Island (Trinity Bay), eastern Newfoundland, Canada. *Current Sci. [India]*, **45**, 609–11.

Nixon, G. T. & Papezik, V. S. (1979). Late Precambrian ash-flow tuffs and associated rocks of the Harbour Main Group near Colliers, eastern Newfoundland: chemistry and magmatic affinites. *Can. J. Earth Sci.*, **16**, 167–81.

North, F. K. (1971). The Cambrian of Canada and Alaska. In *Cambrian of the New World*, (ed. C. H. Holland), pp. 219–324. Wiley-Interscience, London.

Notholt, A. J. G. & Brasier, M. D. (1986). Proterozoic and Cambrian phosphorites—regional review: Europe. In *Phosphate deposits of the world*, (eds P. J. Cook & J. H. Shergold), Vol. 1. pp. 91–100. Cambridge University Press.

O'Brien, S. J., Strong, P. G., & Evans, J. L. (1977). The geology of the Grand Bank (IM/4) and Lamaline (IL/13) map areas, Burin peninsula, Newfoundland. *Miner. Dev. Div. Dept. Mines & Energy, Govt Newfl & Labrador, Rep.*, **77-7**, 1–16.

O'Brien, S. J. Wardle, R. J., & King, A. F. (1983). The Avalon Zone: a pan-African terrane in the Appalachian orogen of Canada. *Geol. J.*, **18**, 195–222.

Odin, G. S. *et al.* (1983). Numerical dating of Precambrian–Cambrian boundary. *Nature*, **301**, 21–3.

Odin, G. S., Gale, N. H., & Doré, F. (1985). Radiometric dating of the late Precambrian times. In *The chronology of the geological record*, (ed. N. J. Snelling), Geol. Soc. Lond. Mem., 10, pp. 65–72. Blackwell Scientific, Oxford.

O'Driscoll, C. F. & Gibbons, R. V. (eds) (1980). Geochronology report—Newfoundland and Labrador. In *Current Res. Miner. Dev. Div. Dep. Mines & Energy, St Johns, Rep.*, **80–1**, 143–6.

Palij, V. M., Posti, E., & Fedonkin, M. A. (1979). Soft-bodied Metazoa and trace fossils of Vendian and Lower Cambrian. In *Upper Precambrian and Cambrian paleontology of the East European platform*, (eds B. M. Keller & A. Yu. Rozanov), pp. 49–82. Akad. Nauk, Moscow. [pp. 56–94 in English edition 1983, eds A. Urbanek & A. Yu. Rozanov, Wydawnictwa Geologiczne, Warszawa.]

Papezik, V. S. (1970). Petrochemistry of volcanic rocks of the Harbour Main Group, Avalon peninsula, Newfoundland. *Can. J. Earth Sci.*, **7**, 1485–98.

Papezik, V. S. (1972). Late Precambrian ignimbrites in eastern Newfoundland and their tectonic significance. *24th Int. geol. Congr.*, **1**, 147–52.

Papezik, V. S. (1973). Detrital garnet and muscovite in Late Precambrian sandstone near St John's, Newfoundland, and their significance. *Can. J. Earth Sci.*, **10**, 430–2.

Papezik, V. S. (1980). Volcanic rocks of Newfoundland: a review. In *Proceedings 'The Caledonides in the USA'*, (ed. D. R. Wones), Dep. Geol. Sci. Virginia Polytech. Inst., State Univ. Mem., 2, pp. 245–8.

Patel, I. M. (1973). Sedimentology of the Ratcliffe Brook Formation (Lower Cambrian?) in southeastern New Brunswick. *Geol. Soc. Am. Abstr. Prog.*, **5**, 206–7.

Patel, I. M. (1975). The Precambrian–Cambrian boundary in southern New Brunswick. *Geol. Soc. Am. Abstr. Prog.*, **7**, p. 104.

Patel, I. M. (1976). Lower Cambrian of southern New Brunswick and its correlation with successions in northeastern Appalachians and parts of Europe. *Geol. Soc. Am. Abstr. Prog.*, **8**, p. 243.

Pelletier, B. R. (1971). A granodiorite drill-core from the Flemish Cap, eastern Canadian continental margin. *Can. J. Earth Sci.*, **8**, 1499–1503.

Perigo, R., Van der Voo, R., Auvray, B., & Bonhommet, N. (1983). Palaeomagnetism of late Precambrian–Cambrian volcanics and intrusives of the Armorican Massif, France. *Geophys. J. R. astron. Soc.*, **75**, 235–60.

Perroud, H. (1985). Synthèse des résultats paléomagnétiques sur le massif Armoricain. *Hercynica*, **1**, 65–71.

Pickerill, R. K. & Tanoli, S. K. (1985). Revised lithostratigraphy of the Cambro-Ordovician Saint John Group, southern New Brunsiwck—a preliminary report. *Geol. Surv. Can. Pap.*, **85–1B**, 441–9.

Piqué, A. (1981). Northwestern Africa and the Avalonian Plate: Relations during late Precambrian and late Paleozoic time. *Geology*, **9**, 319–22.

Pojeta, J. (1980). Molluscan phylogeny. *Tulane Stud. Geol. Paleont.*, **16**, 55–80.

Poole, W. H. (1980*a*). Rb–Sr age of Shunacadie pluton, central Cape Breton Island, Nova Scotia. *Geol. Surv. Can., Pap.* **80–1C**, pp. 165–9.

Poole, W. H. (1980*b*). Rb–Sr age of some granitic rocks between Ludgate Lake and Negro Narbour, southwestern New Brunswick. *Geol. Surv. Can. Pap.*, **80–1C**, pp. 170–3.

Poole, W. H., Sanford, B. V., Williams, H., & Kelley, D. G. (1970). Geology of southeastern Canada. In *Geology and economic minerals of Canada*, (ed. R. J. W. Douglas), pp. 228–304, Geol. Surv. Can. Econ. Geol. Rep., 1.

Prantl, F. (1948). On the occurrence of the genus *Volborthella* Schmidt in Bohemia (Nautiloidea). *Sb. nár. Mus. Praze*, **4(B)**, 3–13.

Rao, K. V., Seguin, M. K., & Deutsch, E. R. (1981). Paleomagnetism of Siluro-Devonian and Cambrian granitic rocks from the Avalon Zone in Cape Breton Island, Nova Scotia. *Can. J. Earth Sci.*, **18**, 1187–1210.

Rao, K. V., Seguin, M. K., & Deutsch, E. R. (1986). Paleomagnetism of Early Cambrian redbeds on Cape Breton Island, Nova Scotia. *Can. J. Earth Sci.*, **23**, 1233–42.

Rasetti, F. (1952). Revision of the North American trilobites of the family Eodiscidae. *J. Paleont.*, **26**, 434–51.

Rast, N. (1980). The Avalonian plate in the northern Appalachians and Caledonides. In *Proceedings 'The Caledonides in the USA'*, (ed. D. R. Wones), Dept. Geol. Sci. Virginia Polytech. Inst. State Univ. Mem., 2, pp. 63–6.

Rast, N., O'Brien, B. H., & Wardle, R. J. (1976). Relationships between Precambrian and lower Palaeozoic rocks of the 'Avalon platform' in New Brunswick, the northeast Appalachians and the British Isles. *Tectonophysics*, **30**, 315–38.

Rast, N. & Skehan, J. W. (1981). Possible correlation of Precambrian rocks of Newport, Rhode Island, with those of Anglesey, Wales. *Geology*, **9**, 596–601.

Rast, N. & Skehan, J. W. (1983). The evolution of the Avalonian plate. *Tectonophysics*, **100**, 257–86.

Rehmer, J. (1981). The Squantum tilloid member of the Roxbury Conglomerate of Boston, Massachusetts. In

Earth's pre-Pleistocene glacial record, (eds M. J. Hambrey & W. B. Harland), pp. 756–9. Cambridge University Press.

Rodgers, J. (1967). Chronology of tectonic movements in the Appalachian region of eastern North America. *Am. J. Sci.*, **265**, 408–27.

Rose, E. R. (1952). Torbay map-area, Newfoundland. *Mem. geol. Surv. Can.*, **265**, v, 1–64.

Rozanov, A. Yu. (1979). *Volborthella*. In *Upper Precambrian paleontology of the East European platform*, (eds B. M. Keller & A. Yu. Rozanov), pp. 93–4. Akad. Nauk, Moscow. [pp. 111–13 in English edition (1983), eds A. Urbanek & A. Yu. Rozanov, Wydawnictwa Geologiczne, Warszawa.]

Rozanov, A. Yu. *et al.* (1969). The Tommotian Stage and the Cambrian lower boundary problem. *Trudy Geol. Inst. Akad. Nauk SSSR*, **206**, 1–380. [English edition (1981) Amerind Publishing Co., New Delhi.]

Schenk, P. E. (1971). Southeastern Atlantic Canada, northwestern Africa, and continental drift. *Can. J. Earth Sci.*, **8**, 1218–51.

Seguin, M. K. (1985). Paleomagnetism of the Avalonian Finn Hill sequence of eastern Newfoundland, Canada. *Mar. Sedim. & Atlant. Geol.*, **21**, 55–68.

Seguin, M. K. & Lortie, J. (1986). Palaeomagnetism of porphyritic stocks and rhyolite sills, Avalon Zone of eastern Newfoundland. *Phys. Earth & Planet. Inter.*, **43**, 148–59.

Shaler, N. S. (1888). On the geology of the Cambrian district of Bristol County, Massachusetts. *Bull. Mus. comp. Zool. Harv.*, **16**, 13–26.

Shaler, N. S. & Foerste, A. F. (1888). Preliminary description of North Attleborough fossils. *Bull. Mus. Comp. Zool. Harv.*, **16**, 27–41.

Shaler, N. S., Woodworth, J. B., & Foerste, A. F. (1899). Geology of the Narragansett Basin. *Monogr. US geol. Surv.*, **33**, 1–402.

Shaw, A. B. (1950). A revision of several Early Cambrian trilobites from eastern Massachusetts. *J. Paleont.*, **24**, 577–90.

Shaw, A. B. (1961). Cambrian of south-eastern and northwestern New England. *20th Int. geol. Congr.*, **3**, 433–71.

Shergold, J. H. & Brasier, M. D. (1986). Proterozoic and Cambrian phosphorites—specialist studies: biochronology of Proterozoic and Cambrian phosphorites. In *Phosphate deposits of the world*, (eds P. J. Cook & J. H. Shergold), Vol. 1. pp. 295–326. Cambridge University Press.

Shimer, H. W. (1907). An almost complete specimen of *Strenuella strenua* (Billings). *Am. J. Sci.*, **23**, (4th Ser.), 199–201, 319.

Skehan, J. W. & Murray, D. P. (1980). A model for the evolution of the eastern margin (EM) of the northern Appalachians. In *Proceedings 'The Caledonides in the USA'*, (ed. D. R. Wones), Dept. Geol. Sci. Virginia Polytech. Inst. State Univ. Mem., 2, pp. 67–72.

Skehan, J. W., Murray, D. P., Palmer, A. R., Smith, A. T., & Belt, E. S. (1978). Significance of fossiliferous Middle Cambrian rocks of Rhode Island to the history of the Avalonian microcontinent. *Geology*, **6**, 694–8.

Skjeseth, S. (1963). Contributions to the geology of the Mjøsa districts and the classical Sparagmite area in southern Norway. *Norsk geol. Tidsskr.*, **220**, 1–126.

Smith, B. M. & Giletti, B. J. (1978). Rb–Sr whole rock study of the deformed porphyritic granitic rocks of Aquideck and Conanicut Islands, Rhode Island. *Geol. Soc. Am. Abstr. Prog.*, **10**, p. 86.

Smith, G. W. & Socci, A. P. (1986). Late Precambrian diamictites of the Boston Basin, Boston, Massachusetts. *Geol. Soc. Am. Abstr. Prog.*, **18**, p. 66.

Smith, S. A. & Hiscott, R. N. (1984). Latest Precambrian to early Cambrian basin evolution, Fortune Bay, Newfoundland: fault-bounded basin to platform. *Can. J. Earth Sci.*, **21**, 1379–92.

Socci, A. D. & Smith, G. W. (1986). Depositional history of late Precambrian sedimentary rocks in the Boston Basin, Boston, Massachusetts: a preliminary model. *Geol. Soc. Am. Abstr. Prog.* **18**, p. 67.

Spariosu, D. J., Kent, D. V., Opdyke, N. D., & Patel, I. M. (1983). A fold test on the Cambrian Ratcliffe Brook Formation, southern New Brunswick. *EOS, Trans. Am. Geophys. Union*, **64**, p. 218.

Strong, D. F. (1979). Proterozoic tectonics of northwestern Gondwanaland: new evidence from eastern Newfoundland. *Tectonophysics*, **54**, 81–101.

Strong, D. F. & Minatedis, D. G. (1975). Geochemistry and tectonic setting of the late Precambrian Holyrood plutonic series of eastern Newfoundland. *Lithos*, **8**, 283–95.

Strong, D. F., O'Brien, S. J., Taylor, S. W., Strong, P. G., & Wilton, D. H. (1978). Geology of the Marystown (1M/13) and St Lawrence map sheets, Newfoundland. *Miner. Dev. Div., Dept. Mines & Energy, Govt Newfl & Labrador, Rep.* **77–8**, 1–81.

Strong, D. F. O'Brien, S. J., Taylor, S. W., Strong, P. G., & Wilton, D. H. (1978). Aborted Proterozoic rifting in eastern Newfoundland. *Can. J. Earth Sci.*, **15**, 117–31.

Tanoli, S. K., Pickerill, R. K., & Currie, K. L. (1985). Distinction of Eocambrian and Lower Cambrian redbeds, Saint John area, southern New Brunswick. *Geol. Surv. Can. Pap.*, **85–1A**, 699–702.

Theokritoff, G. (1968). Cambrian biogeography and biostratigraphy in New England. In *Studies of Appalachian geology, northern and maritime* (eds E-an Zen, W. S. White, J. B. Hadley, & J. B. Thompson), pp. 9–22, Wiley-Interscience, New York.

Thorpe, R. S. *et al.* (1984). Crustal growth and late Precambrian–early Palaeozoic plate tectonic evolution of England and Wales. *J. geol. Soc. Lond.*, **141**, 521–36.

Timofeyev, B. V., Choubert, G., & Faure-Muret, A. (1980). Acritarchs of the Precambrian in mobile zones. *Earth-Sci. Rev.*, **16**, 249–55.

Tucker, M. E. (1986). Carbon isotope excursions in Precambrian/Cambrian boundary beds, Morocco. *Nature*, **319**, 48–50.

Van Alstine, R. E. (1948). Geology and mineral deposits of the St Lawrence area, Burin peninsula, Newfoundland. *Bull. geol. Surv. Newfld.*, **23**, 1–64.

Walcott, C. D. (1890). The fauna of the Lower Cambrian or *Olenellus* Zone. *Rep. US geol. Surv.*, **10**, 511–774.

Walcott, C. D. (1900). Lower Cambrian terrane in the Atlantic province. *Proc. Washington Acad. Sci.*, **1**, 301–39.

Walcott, C. D. (1910). *Olenellus* and other genera of the Mesonacidae. *Smithson. Misc. Collns*, **53**, 231–422.

Wanless, R. K., Stevens, R. D., Lachance, G. R., & Delabio, R. N. (1974). Age determinations and geological studies. K–Ar isotopic ages, Report 11. *Geol. Surv. Can. Pap.*, **73–2**, 1–139.

Webster, M. J., Skehan, J. W., & Landing, E. (1986). Newly discovered Lower Cambrian rocks of the Newport Basin, southeastern Rhode Island. *Geol. Soc. Am. Abstr. Prog.*, **18**, p. 75.

Weeks, L. J. (1954). Southeast Cape Breton Island, Nova Scotia. *Mem. geol. Surv. Can.*, **277**, 1–112.

Wiebe, R. A. (1972). Igneous and tectonic events in northeastern Cape Breton Island, Nova Scotia. *Can. J. Earth Sci.*, **9**, 1262–77.

Williams, H. & Hatcher, R. D. (1982). Suspects terranes and accretionary history of the Appalachian orogen. *Geology*, **10**, 530–6.

Williams, H. & Hatcher, R. D. (1983). Appalachian suspect terranes. *Mem. geol. Soc. Am.*, **158**, 33–53.

Williams, H. & King, A. F. (1979). Trepassey map area, Newfoundland. *Mem. geol. Surv. Can.*, **389**, 1–24.

Wilson, J. T. (1966). Did the Atlantic close and then re-open? *Nature*, **211**, 676–81.

Yochelson, E. L. (1977). Agmata, a proposed extinct phylum of early Cambrian age. *J. Paleont.*, **51**, 437–54.

Yochelson, E. L. (1981). A survey of *Salterella* (Phylum Agmata). *US geol. Surv. Open File Rep.*, **81–743**, 244–8.

Zartman, R. E. & Naylor, R. S. (1984). Structural implications of some radiometric ages of igneous rocks in southeastern New England. *Bull. geol. Soc. Am.*, **95**, 522–39.

3

China and the Palaeotethyan Belt (India, Pakistan, Iran, Kazakhstan, and Mongolia)

M. D. Brasier

3.1. INTRODUCTION

Precambrian–Cambrian boundary sequences that often bear commercial-grade phosphorites occur across a wide belt from southern China to Iran (Fig. 3.1). At present, the main outcrops are found either to the west and south of the Greater Himalayas (i.e. in Iran, Pakistan, and India) or to the north and east of this mountain chain (i.e. Kazakhstan, Mongolia, and China). Current research tends to indicate a fairly distinctive sequence of faunal and lithological events across the platform sequences of this belt, herein termed the Palaeotethyan Belt, for which the Meishucun section of East Yunnan in China provides a standard reference section. The biostratigraphy of phosphorites across Asia has been reviewed in some detail by Shergold & Brasier (1986) and only the main characteristics of important sections will be outlined below, followed by a summary for the Palaeotethyan Belt.

The palaeogeographical and tectonic setting of these fossiliferous deposits needs to be studied further. Deposition to the north and east of the Himalayas seems to have taken place on the margins of separate platforms (i.e. the Siberian, Kazakhstanian, North China–Tarim, and Yangtze Platforms). The Indian, Pakistanian, and Iranian strata seem, however, to have accumulated on the tropical, northern margin of Gondwana (e.g. Parrish et al. 1986). The palaeogeography of Gondwana at this time is uncertain and the Yangtze Platform of southern China may also have formed part of this supercontinent in the Cambrian (op cit. p. 290).

3.2. YANGTZE PLATFORM, CHINA

Central and southern China contain many important Precambrian–Cambrian boundary successions deposited around the margins of the old Yangtze Platform. These sequences and their palaeontology have been reviewed in Chinese (e.g. Xiang et al. 1981; Xing et al. 1983; Qian 1983; Luo et al. 1984) and more briefly in English summaries (Xing & Luo 1984; Xing et al. 1984). An English account of the Cambrian System in China, updated from Xiang et al. (1981), is in preparation (Brasier in press). A great deal of information has been generated during the last few years, resulting in stratigraphical and palaeontological nomenclatures that are in a state of flux. The present account therefore represents a necessary simplification that may soon become outdated.

The Meishucun section of Eastern Yunnan is, at the time of writing, still the stratotype candidate for the Precambrian–Cambrian boundary, though the decision has been deferred (Cowie 1985). Other sections provide evidence of nearly homotaxial sequences over a fairly wide area. Each of the successions has subtle distinctions of fauna and lithology, resulting in differences of zonal and lithostratigraphical nomenclature. In recent years it has been Chinese practice to place the Sinian–Cambrian boundary at the base of the *Anabarites–Protohertzina* (= *Anabarites–Circotheca–Protohertzina*) Zone, referred to here as Zone I. This has also been called the *Anabarites–Protohertzina–Barbitositheca* Zone by Qian (1984). Because the earliest appearance of this first skeletal fauna (China Marker 'A') may be subject to downward revision, recently Qian (1984) and others have preferred to take the base at the first appearance of the Zone II assemblage, locally known as the *Paragloborilus–Siphogonuchites* (= *Paragloborilus–Siphogonuchites–Lapworthella*) Zone, distinguished by a diverse and widespread assemblage of gastropods and monoplacophorans (China Marker 'B'; Luo et al. 1984). This has been called the *Siphogonuchites–Paragloborilus–Sachites* Zone by Qian (1984). The third zone, of *Sinosachites* (= *Eonovitatus–Sinosachites–Ebianotheca* Zone) generally overlies a widespread disconformity that may contain a hiatus. This zonal assemblage is poorly fossiliferous and the nominated species do not seem to be useful for wide international correlation. Better indices may be the chancelloriids of the *Allonnia tripodophora* Doré & Reid group and *Archiasterella pentactina* Sdzuy, plus the tommotiid *Tannuolina multi-*

Fig. 3.1. Map of the present distribution of Precambrian–Cambrian boundary sections in Asia, the Middle East, and Europe referred to in the text. Shaded area represents the Himalayan and Alpine fold belts. Ringed dots indicate areas with *Rhombocorniculum.*

forata Fonin & Smirnova. This has been called the *Sinosachites–Tannuolina* Zone by Qian (1984).

Zones I, II, and III have been combined into a Meishucunian Stage (e.g. Xing *et al.* 1984), although this is inconsistent with taking the system boundary as the base of Zone II. To avoid confusion, the base of Zone I at Marker 'A' is here retained as the base of the Meishucunian Stage and of the Cambrian System, as in Brasier (in press). It should also be noted that the Qiongzhusi Formation begins at the base of Zone III, contemporaneous with the start of the Qiongzhusian

Stage of Xiang *et al.* (1981), but the Qiongzhusian Stage *sensu* Luo *et al.* (1982, 1984) begins higher, with the first trilobites. It should further be noted that sub-trilobite faunas from the lowest Qiongzhusian Stage *sensu* Luo *et al.* (1982) are retained by some in Zone III. Trilobites appear in Zone IV (the *Parabadiella* Zone) and Zone V (the *Eoredlichia* Zone). These and other Chinese trilobite zones can now be correlated across the country, but their strongly provincial nature has not yet allowed firm correlations far beyond the borders. A brief summary of the succession in each area is given below.

3.2.1. Eastern Yunnan

This region lies on the south-western margin of the Yangtze Platform, for which the Meishucun section at the Kunyang Phosphorite Mine, Jinning County, provides the candidate Precambrian–Cambrian section.

A comparable boundary section at Wangjiawan also exposes a clear sequence of underlying Sinian strata. The nature of the Meishucun succession is shown in Fig. 3.2, with data on the faunas taken from the above sources, with some minor taxonomic revisions after Qian *et al.* (1985). Other important works on these faunas include Qian (1977), Jiang (1980), Luo *et al.*

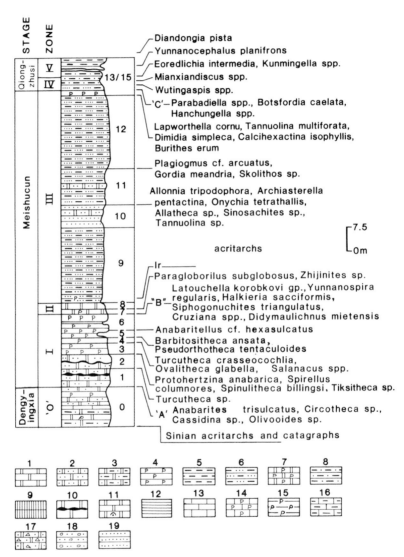

Fig. 3.2. The Meishucun section, Jinning, Yunnan in China, showing the first-appearance datum of selected fossil taxa, Markers A, B, C, and the iridium anomaly (Ir). Key for Figs 3.2 to 3.7 is as follows: 1, dolomite; 2, sandy dolomite; 3, sandy argillaceous dolomite; 4, phosphorite; 5, argillite; 6, argillaceous siltstone; 7, phosphatic dolomite; 8, sandy argillite; 9, massive chert; 10, dolomite with chert bands; 11, fenestral (bird's-eye) dolomite; 12, shale; 13, limestone; 14, phosphatic limestone; 15, phosphatic siltstone and mudstone; 16, argillaceous limestone; 17, arenaceous–rudaceous dolomite; 18, sandstone with pebbles; 19, sandstone. Based on Xing *et al.* (1984) and sources in the text.

(1980, 1982, 1984), Luo (1981), Xing & Luo (1984), and Xing *et al.* (1983, 1984).

Sinian strata

These are widely developed in this region, with continental Nantuo tillites and shales followed by mixed lithologies of the Lunasi or Wangjiawan Formations, by algal dolomites of the Donglongtan Formation, and shales and carbonates of the Dengying (= Yuhucun) Formation. While the upper part of the Dengying Formation contains small shelly fossils, the lower part contains undisputed Sinian units of the Jiucheng and Baiyanshao Members.

The Jiucheng Member (*c*.40 m) comprises argillaceous dolomites to shales at Wangjiawan, and black carbonaceous siltstones at Meishucun. These rocks locally contain abundant macroscopic, carbonaceous algal remains, ascribed to the *Chuaria* and *Vendotaenia* groups or to other problematical taxa. The succeeding Baiyanshao Member (*c*.176 m) is represented by grey crystalline dolomites, intercalated with algal dolomites and cherts. The Xiaowaitoushan Member (8.2 m to 11.2 m) which follows, consists of grey arenaceous dolomites with chert bands, characterized by the presence of rare skeletal fossils (see below).

The several cycles of shale to dolomite compare with the Grand Cycles of North America (e.g. Aitken 1981), and cycle tops may therefore prove useful for correlation. Those at the top of the Donglangtan Member (and also in the Dahai Member: see below) show fenestral features consistent with intertidal–supratidal conditions.

Zone I

The first small shelly fossils appear 0.8 m above the base of the Xiaowaitoushan Member of Meishucun, marking the base of Zone I and the China 'A' datum (Fig. 3.2). The lithology above and below the marker remains uniformly dolomitic, but there is a subtle change from flaggy-bedded to large-scale cross-bedded units at the marker. Thus the rare small shelly fossils appearing in Marker 'A' are from a transported shell-lag. At Wangjiawan, the fauna first appears in a thin phosphatic lens above finely burrowed dolomites and below medium- to thick-bedded phosphorite and dolomite units. The fauna from Beds 1 and 2 notably contains tubes of *Anabarites trisulcatus* Miss., *Hyolithellus* sp.; hyolith '*Conotheca*' (or '*Circotheca*') *longiconica* Qian; globomorph *Olivooides* sp.; ?monoplacophoran *Cassidina* sp., and ?brachiopod *Artimycta pusilla* Liu & Jiang.

Bed 2 is dolomitic with several units of large-scale cross-bedding near the top. The Zhongyicun Member

(11.6 m) marks a change to phosphorite and/or siliciclastic deposition, though no significant change takes place in the fauna at first. At Meishucun, the junction is an irregular, phosphorite-coated dolomitic hardground with up to 20 cm relief (top of Bed 2), overlain by a unit of dolomitic and phosphatic pebbles in a dolomitic phosphorite matrix; i.e. a discontinuity surface (base of Bed 3) is present. The following then appear through Beds 3 to 5: hyoliths described as '*Conotheca*' (or '*Circotheca*') *multisulcata* (Qian) and related species, plus *Turcutheca crasseocochlia* (Syss.), *Ovalitheca glabella* Syss.; tubes of *Anabaritellus* cf. *hexasulcatus* Miss., *Pseudorthotheca tentaculoides* Qian & Jiang; protoconodont *Protohertzina anabarica* Miss.; conodontomorph *Salanacus* spp.; conulariid *Barbitositheca ansata* Qian & Jiang; and ?skeletal alga *Spirellus columnores* Jiang. Qian (1983) also records *Tiksitheca huangshandongensis* Qian from *c*. Bed 3.

Bed 3 is typically a thick-bedded dolomitic phosphorite to phosphatic dolomite, with the branching bilobed trace of *Sellaulichnus meishucunensis* Jiang. Bed 4 marks a change to thin-bedded alternations of dolomite and phosphorite lenses. Bed 5 is sandy and fossiliferous at the base and passes up into brown shale. Bed 6 marks a return to dark, thin-bedded phosphorite.

Lithological changes through Zone I suggest that peritidal conditions (Beds 1 and 2) changed to deepening subtidal conditions (Beds 3 to 5), followed by slight shallowing (Bed 6). The relative uniformity of the fauna through these changes is noteworthy, although waterworn, phosphatized preservation suggests transport of the remains.

Zone II

A marked faunal change occurs within the phosphorites of the Zhongyicun Member, *c*.2.55 m from their top, at the base of Bed 7, marking also the base of Zone II and the China 'B' datum. Bed 7 shows an abrupt lithological change, from thin-bedded phosphorite (top of Bed 6) to massive, large-scale cross-bedded dolomitic phosphorite (lower Bed 7) with a richly fossiliferous shell-lag in the bottomsets at the base. Here appear diverse monoplacophorans, e.g. *Latouchella korobkovi* Vostokova and related forms including *Maidipingoconus maidipingensis* (Yu), *Igorella hamata* Yu; gastropods such as *Yunnanospira regularis* Jiang; conodontiforms, e.g. *Yunnanonodus doleres* Wang & Jiang; siphogonuchitids, e.g. *Siphogonuchites triangulatus* Qian; tommotiids, e.g. *Yunnanotheca kunyangensis* Qian & Jiang; and brachiopods, e.g. '*Aldanotreta*' sp. The systematics of this assemblage are under revision. In addition, the bilobed track of *Didymaulichnus mietensis* Young, and arthropod traces of *Cruziana* spp. and *Rusophycus cardiopetalus* Crimes & Jiang are found.

In Bed 8, the phosphorite lithology passes up into massive dolomites of the Dahai Member (1–2 m). Fossiliferous phosphatic hardgrounds occur at several levels within fenestral dolomites that bear chert nodules near the top. The somewhat similar fauna includes *Paragloborilus subglobosus* (He). It seems possible that this part of the succession is condensed. The top is weathered, and overlain by a phosphatic skin and a limonitic to manganiferous crust, interpreted by some as evidence for shoaling and emergence (Xiang *et al.* 1981; Luo *et al.* 1984).

The allochthonous molluscan fauna brought in by Marker 'B' clearly coincided with a shallowing phase, ultimately leading to the peritidal Dahai Member and emergence. Sedimentological conditions at this time seem to be similar to those noted in Beds 1 and 2.

Zone III

This zone spans the lower part of the Qiongzhusi Formation, placed in the Badaowan Member (54 m). The lower part consists of black siltstones and shales, while the middle and upper parts comprise grey siltstones intercalated with silty dolomite.

The basal part of Bed 9 is characterized by a 0.4 m-thick layer of glauconitic and siliceous phosphorite and clay. This clay layer has yielded iridium and carbon isotope anomalies that can be traced to the Yangtze Gorges (Hsü *et al.* 1985), but is unfossiliferous apart from undiagnostic acritarchs. Higher in Bed 9 are found large septarian siderite nodules. Bed 11 in the middle part of the member (*c.*28 m above the base) yields small shelly fossils, including chancelloriids *Allonnia tripodophora* Doré & Reid (= *A. erromenosa* Jiang), *Chancelloria* sp., *Archiasterella* cf. *pentactina* Sdzuy; wiwaxiid *Sinosachites* sp.; hyolith *Allatheca* sp.; and tommotiid *Tannuolina* sp. The trace fossils *Plagiogmus* cf. *arcuatus* Roedel, *Gordia meandria* Jiang, and *Skolithos* sp. appear abundantly within the upper 16 m of the member in Bed 12.

A thin (0.2 m) layer of grey-black sandy, rudaceous, brecciated bioclastic phosphorite occurs at the base of Bed 13, low in the upper or Yu'anshan Member of the Qiongzhusi Formation. Trilobites are lacking, but the following are notably present: tommotiids of the *Lapworthella cornu* Wiman group, and *Tannuolina multiforata* Fonin & Smirnova; tubes of *Coleoloides typicalis* Walcott (or *C. qiongzhusiensis* (Qian)); chancelloriids *Allonnia tripodophora* (= *A. erromenosa*), *Archiasterella pentactina*, *Dimidia simpleca* Jiang; sponge *Calcihexactina isophyllis* Jiang; and hyolith *Burithes* cf. *erum* Miss.

Zones IV and V

Within the 16.6 m-thick carbonaceous and argillaceous siltstones of Bed 13 can be found three successive assemblages. The lowest (2.4 m from the base of the bed) yields the *Parabadiella* Zone (Zone IV) fauna: trilobites *Parabadiella conica* Luo, *P. yunnanensis* Luo; bradoriid crustaceans *Meishucunella processa* Jiang, *Bajiella dalongtanensis* Jiang, *Hanchungella tenuis* Huo, and *H. shanglingshanensis* Huo; hyoliths *Aimitus circupluteus* Qian and *Ambrolinevitus ventricosus* Qian; brachiopod *Botsfordia caelata* (Hall); chancelloriid *Allonnia tripodophora*; and acritarchs including *Baltisphaeridium* sp. The middle assemblage (10.6 m from the base) yields *Wutingaspis kunyangesnsis* Luo, while the upper assemblage has *Tsunyidiscus badaowensis* (Luo).

The Yu'anshan Member (72 m) passes upwards from these dark shales into yellow-green shales of the *Eoredlichia* Zone (Zone V): trilobites *Eoredlichia intermedia* (Lu), *E. walcotti* (Mansuy), *Yunnanocephalus planifrons* Luo, and *Kuanyangia* sp.; bradoriid crustaceans *Kunmingella maxima* Huo, *K. douvillei* (Mansuy), *K. parva* Huo, *Monotella viviosa* Lee; homopod *Isoxys aurita* Jiang; hyolith *Ambrolinevitus ventricosus*, and brachiopod *Diandongia pista* Rong.

These strata are unconformably overlain by Middle Devonian rocks at Meishucun. Elsewhere in the region, however, the sequence passes up into Canglangpuian clastics (with the *Yiliangella–Yunnanaspis–Drepanuroides* Zones in the lower part and the *Palaeolenus* Zone in the upper part) and Longwangmiaoan carbonates and clastics (with the *Hoffetella–Redlichia murakamii* Zone).

3.2.2. South-western Sichuan

Another important section is found at Maidiping, Mount Emei, in south-western Sichuan (Fig. 3.3). Important works on boundary faunas from this province include Yin *et al.* (1980*a*,*b*), Xing *et al.* (1983, 1984), Yang *et al.* (1983), and Yang & He (1984).

Late Sinian algae ('catagraphs') and acritarchs are here found in the Mao'ergang Member in the upper part of the dolomitic Dengying (= Hongchunping) Formation. Stromatolitic dolomites with chert bands appear below the Zone 1 assemblage in the upper part of this member.

The Zone I assemblage, with ?hyolith *Spinulitheca billingsi* (Syss.), protoconodont *Protohertzina anabarica* Miss., globomorph *Archaeooides* sp., and tube *Hyolithellus tenuis* Miss. appears in the upper part of the Mao'ergang dolomites (Bed 31,5), here joined by the

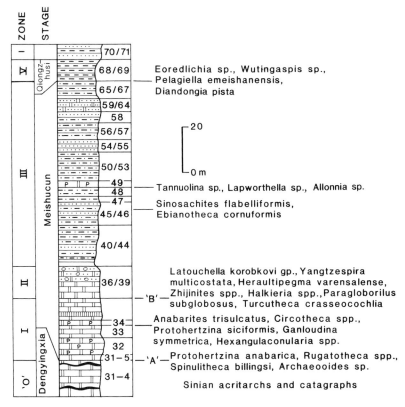

Fig. 3.3. The Maidiping section, Emei, Sichuan in China, showing the first-appearance datum for selected fossil taxa and Markers A and B. Key as for Figure 3.2. Based on Xing *et al.* (1984) and sources in the text.

tube *Rugatotheca*. Marker 'A' is here a thin, irregular pocket of granular phosphatic dolomite overlying an irregular phosphatized hardground surface. This hardground developed over a dolomite with laminar fenestrae and passes upwards into fenestral dolomite with wispy algal laminations of Bed 31,5, and crinkly laminations of Bed 32,1.

Phosphorites appear near the base of the Maidiping Member. As in Meishucun, this change is marked by an irregular phosphatized disconformity surface, here passing up into festoon-bedded and rippled phosphorite–dolomite and chert of Bed 32,2, bearing a similar fauna to beds below. The phosphatic content increases abruptly in the lower part of Bed 32,3, passing up into dolomites of Bed 33; and this is repeated in Beds 34 to 35,6. Bed 34 contains elements of the underlying fauna, plus tubes *Anabarites trisulcatus*, *Lobiochrea* cf. *natella* Valkov & Syssoiev; hyolith '*Circotheca*' spp.; protoconodonts *Protohertzina siciformis* Miss., *Ganloudina symmetrica* He; and conulariids *Quadrisiphogonuchites* (= *Eoconularia* of some authors) and *Hexangulaconularia* spp. In

other sections, *A. trisculatus* appears with the first skeletal fauna.

A significant faunal change from Zone I to Zone II assemblages takes place from Beds 36 to 39 in the upper part of the Mao'ergang dolomites, which become arenaceous to rudaceous at the top. The faunal change accompanies a change from massive fenestral dolomite to thin-bedded argillaceous phosphatic dolomite. Here are reported monoplacophorans of the *Latouchella korobkovi* group; gastropods such as *Yangtzespira multicostata* He and *Archaeospira* sp.; rostroconch *Heraultipegma varensalense* (Cobbold); and numerous other molluscs, plus wiwaxiid *Halkieria* spp., cambroclavid *Zhijinites* spp., hyolith *Paragloborilus subglobosus* (He), and other elements of the second zonal assemblage.

As at Meishucun, the change from Zone II to Zone III assemblages takes place over a major disconformity between brown ?karstic-weathered dolomites (of the Maidiping Member) and pebbly sandstones overlain by black carbonaceous shales, at the base of the lower member of the Jiulaodong Formation. This formation mainly comprises black, laminated argillaceous siltstones but fossils occur at several less siliceous levels. Chancelloriids, wiwaxiid *Sinosachites*, and hyolith *Ebianotheca* sp. occur in the glauconitic and phosphatic dolomites and clays of Beds 45 to 49. *Lapworthella rete* Yue and *Tannuolina* sp. also occur in the latter (Yue 1987) along with *Allonnia tripodophora* (Brasier & Ashouri, unpublished data). Meandering traces of *Scolicia* type are common at about the level of Beds 50 to 56. A Qiongzhusian assemblage, with gastropod *Pelagiella emeishanensis* He, brachiopod *Diandongia pista*, hyoliths *Linevitus* and *Sulcavitus*, and trilobites *Mianxiandiscus*, *Eoredlichia* and *Wutingaspis*, occurs in the upper member (Beds 68–69). Higher strata in western Sichuan contain Upper Canglangpuian strata with *Palaeolenus* and Longwangmiaoan beds with *Redlichia chinensis* Walcott.

Small shelly faunas with conulariids have also been described from Leibo County, south of Emeishan (He 1984) and the Nanjiang area of northern Sichuan (Yang et al. 1983, Yang & He 1984). The distinction between Zones I and II is not yet clear in the latter area.

3.2.3. Western Guizhou

The best sections in this region are found at Gezhongwu and Wuzhishan in Zhijin County. Stratigraphy and palaeontology have been reviewed by Xing et al. (1983, 1984), Wang et al. (1984), and Qian & Yin (1984a). At Gezhongwu (Fig. 3.4), Zone I is relatively thin, occurring above chert-banded dolomites, in the upper dolomite beds of the Maolongjing Member of the Dengying Formation. These also contain *Zhijinites* sp. (usually indicative of Zone II). Phosphorites appear close above, in the Gezhongwu Member, with little faunal change at the base. A faunal change, interpreted as the turnover from Zones I to II (Xing et al. 1984) takes place from Bed 15, and the new assemblage ranges throughout phosphorites up to Bed 20. This Zone II assemblage does not contain diverse molluscs, as elsewhere, but bears *Lapworthella orientalis* He or *L. gezhongwuensis* Qian & Yin, *Halkieria* spp., *Zhijinites* spp. and, at Wuzhishan, *Paragloborilus*. A more profound change takes place from Bed 21, where dark carbonaceous siltstones and cherty phosphorites of the

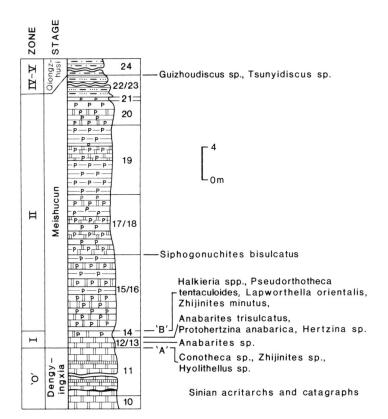

Fig. 3.4. The Gezhongwu section, Zhijin County, Guizhou in China, showing the first-appearance datum of selected fossil taxa and Markers A and B. Key as for Figure 3.2. Based on Xing et al. (1984) and sources in the text.

Qiongzhusi (= Nuititang) Formation appear. Bed 24 yields the trilobite *Guizhoudiscus* sp. associated with *Tsunyidiscus* sp. (i.e. Zone IV or V). Further east, these trilobites may be underlain by thicker shales or sandstones bearing *Palaeoacmaea* sp. (Xing *et al.* 1983). These are herein taken to be local developments of the lower Qiongzhusi Formation (*contra* Chinese authors). The fossil may actually be *Tannuolina* sp. (cf. Qian *et al.* 1985) and correlation with Zone III is indicated. Thus Zones III–V seem to be condensed and incomplete in the west of the area. In northern Guizhou is found a very full succession of higher trilobite zones of Canglangpuian and Longwangmiaoan age.

3.2.4. South-western Shaanxi

Two sections across the Sinian–Cambrian boundary from this north-western part of the Yangtze Platform should be mentioned: those of Shizhonggou (Fig. 3.5)

and of Yuanjiaping (Fig. 3.6) in the Kuanchuanpu area of Ningqiang County (e.g. Xing *et al.* 1983, 1984).

At Shizhonggou, globomorph *Olivooides multisulcatus* is found in chert-banded dolomites below the main occurrences of Zone I faunas, which appear with phosphorites and cherts in the upper part of the Dengying Formation. The typical elements of *Anabarites trisulcatus*, *Protohertzina anabarica*, 'Circotheca' spp., and *Cambrotubulus* sp. are here joined by *Siphogonuchites triangulatus* and *Trapezochites* sp. (usually found in Zone II). The assemblage becomes more impoverished upwards, as phosphorites give way to cherts and argillaceous limestones. The appearance of *Turcutheca* sp. in Bed 27 cannot be diagnostic of Zone II (*contra* Xing *et al.* 1984) since this is reported also in Zone I at Meishucun. The presence of Zone II is therefore equivocal. The disconformable change from carbonate to shale of the Qiongzhusi Formation is present from Bed 32, while at Mofangyan, this junction is marked by a thin limonitic clay bed.

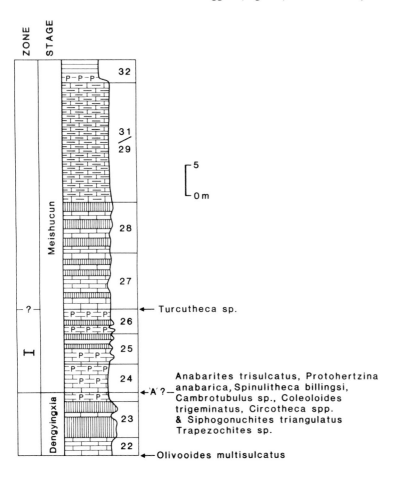

Anabarites trisulcatus, Protohertzina anabarica, Spinulitheca billingsi, Cambrotubulus sp., Coleoloides trigeminatus, Circotheca spp. & Siphogonuchites triangulatus Trapezochites sp.

Olivooides multisulcatus

Fig. 3.5. The Shizhonggou section, Ningqiang, Shaanxi in China, showing the first-appearance datum of selected fossil taxa and the possible level of Marker A. Key as for Fig. 3.2. Based on Xing *et al.* (1984) and sources in the text.

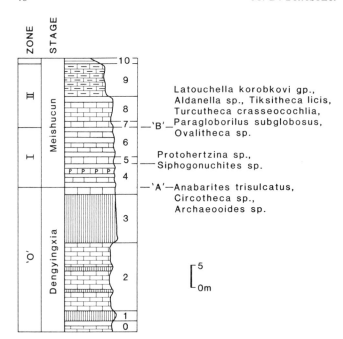

Fig. 3.6. The Yuanjiaping section, Ningqiang, Shaanxi in China, showing the first-appearance datum of selected fossil taxa and Markers A and B. Key as for Fig. 3.2. Based on Luo *et al.* (1984) and sources in the text.

The Yuanjiaping section has only a few fossiliferous horizons but these are more clearly differentiated into two zones (Fig. 3.6). Cherty dolomites here pass up into dolomites bearing a simple Zone I assemblage with *Anabarites trisulcatus*, '*Circotheca*' sp., and *Archaeooides* sp., which range through a thin phosphatic layer in Bed 4. A richer assemblage in Bed 5 has siphogonuchitids and *Protohertzina* sp., while Bed 6 is barren. The Zone II assemblage appears abundantly in Bed 7, with mono-placophorans of the *Latouchella korobkovi* group, gastropod *Aldanella* sp., plus anabaritid tubes of *Tiksitheca licis* Miss. and *T. korobkovi*. The presence of the latter may not be of diagnostic significance since it also occurs with Zone I assemblages in Sichuan (Xing *et al.* 1983). An abrupt change from argillaceous limestone to shale of the Qiongzhusi Formation in Bed 10 suggests correlation with Zone III. The bradoriid crustaceans *Hanchungella* and *Liangshanella* occur in the middle to upper part of the lower member of the Qiongzhusi Formation (i.e. Zone IV), while the upper member contains the trilobites *Eoredlichia*, *Wutingaspis*, *Yunnanocephalus*, *Parabadiella*, and *Mianxiandiscus* spp., plus the bradoriid *Kunmingella* and brachiopod *Lingulella* (i.e. Zone V). The overlying Xiannudong Formation is of Canglangpuian age with archaeocyathans of Lower Botomian affinity in the opinion of Rozanov & Sokolov (1984).

3.2.5. Western Hubei

Sections showing a similar sequence of events are seen in the Yanjiahe and Jijiapo areas of the Yichang district in the eastern Yangtze Gorges of western Hubei (e.g. Fig. 3.7). Here, strata around the boundary are mainly composed of dark-grey to black limestones, cherts, and carbonaceous shales, placed in the Tianzhushan Member of the Dengying Formation, and represent a rather condensed succession. Aspects of stratigraphy and palaeontology have been reviewed by Qian *et al.* (1979), Xing *et al.* (1983, 1984), and Chen (1984). The Dengying Formation of Yichang yields a petalonamean referred to *Charnia*, and algal species of *Vendotaenia* and *Tyrasotaenia* (Xing & Ding 1985), and the 'Xilingxia Fauna' of tubes *Shipaitubulus* and *Sinotubulites* (Chen *et al.* 1981); spicules of sponges are also reported from the underlying Duoshantuo Formation (Xing & Ding 1985). In the Huangjiatang section, dolomites with siliceous nodules (Bed 4) pass up into dolomites of Bed 5 with a sparse fauna of hyoliths *Turcutheca crasseocochlia* and *Ovalitheca multicosta* Qian, of probable Zone I age. At other localities at this level are found *Anabarites trisulcatus*, *Protohertzina anabarica*, '*Circotheca*' sp., *Turcutheca crasseocochlia*, *Spinulitheca billingsi*, *Hyolithellus tenuis*, ?*Archaeooides* sp., and *Chancelloria* sp. An abundant fauna in Bed 6 occurs in a sandy rudaceous

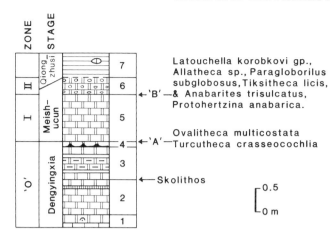

Latouchella korobkovi gp.,
Allatheca sp., Paragloborilus
subglobosus, Tiksitheca licis,
←—'B'—& Anabarites trisulcatus,
Protohertzina anabarica.

←'A'—Ovalitheca multicostata
Turcutheca crasseocochlia

←——Skolithos

Fig. 3.7. The Huangjiatang section, Yichang, Hubei in China, showing the first-appearance datum of selected fossil taxa and Markers A and B. Key as for Fig. 3.2. Based on Luo *et al.* (1984) and sources in the text.

dolomite and appears to represent an admixutre of elements typical of Zone I, e.g. *Anabarites trisulcatus* and *Protohertzina anabarica*, with those typical of Zone II, e.g. monoplacophorans of the *Latouchella korobkovi* group, *Obtusoconus multicostatus* (Qian), hyoliths *Allatheca* sp., and *Paragloborilus subglobosus* (He), plus gastropod *Aldanella yanjiaheensis* Chen Ping at other localities. As elsewhere, this facies terminates abruptly and is overlain by black carbonaceous shales of the Shuijingtuo Formation. The contact is locally phosphatic and probably marks a major hiatus, above which Hsü *et al.* (1985) found ferruginous clays with iridium and carbon isotope anomalies. The trilobites *Tsunyidiscus* and ?*Zhenbaspis* occur in black shales shortly above this, indicating a late Qiongzhusian age, while *Hupeidiscus orientalis* (also known from the Botomian of Kazakhstan) occurs near the top of the stage. Thus Zone III, if present, may be highly condensed. Higher strata contain a succession of Canglangpuian and Longwangmiaoan trilobites.

3.2.6. Northern Hubei

Further north in Fangxian County of Hubei, thin phosphatic beds of the Xihaoping Member, Dengying Formation, also contain small shelly fossils (Qian & Zhang 1983). Here, however, they have affinities with faunas in west Xinjiang (Tibet) and southern Kazakhstan. Lower beds contain sclerites *Cambroclavus fangxianensis* Qian & Zhang, *Isoclavus bilobus* Qian & Zhang, and *Paraformichella orientalis* Qian & Zhang; chancelloriids ?*Allonnia* sp., *Chancelloria floriformis* Qian & Zhang; wiwaxiid *Sachites* sp.; tommotiid *Lapworthella hubeiensis* Qian & Zhang; tubes *Cambrotubulus decurvatus* Miss. and *Pseudorthotheca* sp.;

hyoliths *Turcutheca crasseocochlia* (Syss.), *Paragloborilus circulatus* Qian & Zhang, *P. mirus* He, and Circotothecidae. Higher horizons contain similar elements, plus conoidal *Rhombocorniculum cancellatum* (Cobbold) and hyoliths *Adyshevitheca* cf. *adyshevi* Mambetov and *Microcornus parvulus* Mambetov. The lower assemblage compares with the *Cambroclavus* Subzone and the higher one with the *Adyshevitheca* Subzone of the *Rhombocorniculum cancellatum* Zone in Kazakhstan (e.g. Missarzhevsky & Mambetov 1981).

3.3. ANHUI AND HENAN

A considerable hiatus separates Sinian from Lower Cambrian deposits on the North China Platform, so that the earliest skeletal fossils of Anhui and Henan are younger than those of the Yangtze Platform.

A succession from Huainan and Huoqiu Counties of Anhui Province contains small shelly fossils below late Canglangpuian trilobites *Palaeolenus* (*Megapalaeolenus*) *fengyangensis* (Chu) and *Redlichia* cf. *nobilis* Walcott. According to Xiao & Zhou (1984) and Zhou & Xiao (1984) this lower assemblage includes trilobites *Hsuaspis* spp., brachiopods *Obolus*, *Kutorgina*, and *Lingulella* spp., hyoliths, monoplacophorans, and gastropods. Notable forms include ceratoconid latouchellids and lax-spired pelagiellid *Auriculaspira* (or *Auriculatespira*) spp. An early to mid-Canglangpuian age is indicated.

Another 'first' assemblage of small shelly fossils of late aspect occurs in the Xinji Formation of Fangcheng County, Henan Province. Here, *Auriculaspira* is joined by blade-like ?protoconodont elements referred to *Henaniodus* and *Bioistodina* spp. (He *et al.* 1984).

3.4. XINJIANG

Precambrian–Cambrian boundary sections of Xinjiang, north-west China, are relatively poorly known. Outcrops here may be placed either within the Tianshan–Hinggan stratigraphical region or the Tarim stratigraphical region of Xiang *et al.* (1981). The former is characterized by shelf-margin to geosynclinal facies with cosmopolitan faunas, while the latter is of platform type, with Yangtze affinities.

A notable section in the Tianshan–Hinggan region occurs in the western part of northern Tianshan (Xiang *et al.* 1985). Here, the Linkuanggou Formation (30–40 m thickness) is composed of dark-grey to black, thin-bedded siltstones intercalated with limestone nodules, thin-bedded limestones, and limestone lenses, and bears multiple layers of phosphorite with poorly preserved and unidentified sponge spicules, hyoliths, brachiopods, gastropods, and trilobites. Redlichiids locally appear about 1 m above the lowest phosphorite bed, while *Calodiscus* appears in the overlying Huocheng Formation (ibid.).

Sequences from the Tarim Platform contain small shelly fossil faunas with abundant cambroclavi: elements, such as those described from the Yurtus Formation of the Aksu–Wushi region of north-west Xinjiang (Qian & Xiao 1984). Here, lower phosphatic beds contain tubular '*Circotheca*' sp., *Hyolithellus* sp., and *Polycladium* sp. The second unit has gastropods *Aldanella* sp. and *Poraconus* sp., plus various cambroclavid elements, while hyoliths *Ovalitheca* and *Paragloborilis* spp. appear in the third unit. Shortly above these appears the trilobite *Schizhudiscus*, indicating a late Qiongzhusian age. A similar lithological sequence, with *Schizhudiscus* above phosphatic and carbonaceous rocks, seems to be present in the correlated Xorbulak Formation of Kensayi Pass (e.g. Xiang *et al.* 1981). The same formation is noted to contain two small shelly faunas by Xing *et al.* (1984): firstly a *Circotheca–Aldanella–Ovalitheca* assemblage; secondly a *Cambroclavus–Aetheosachites–Poraconus* assemblage.

These distinctive assemblages cannot yet be correlated firmly with those of the Yangtze Platform.

3.5. LESSER HIMALAYA, INDIA

The 350-km-long 'Krol Belt' has for long been a highly controversial portion of Indian Himalayan stratigraphy. It extends from Solan in the north-west to Nainital in the south-east, with exposures of the Krol and Tal Formations in the Mussoorie and Garwhal Synclines of Lesser (Garwhal) Himalaya, in the central sector of the belt. These formations have variously been given Precambrian to Cretaceous ages (see Bhatt *et al.* 1983; Ahluwalia & Gupta 1984; Azmi 1985). Much of the problem has been caused by a parallel unconformable contact with Cretaceous rocks and by misinterpretation of the microfossil *Spirellus* and associated conoidal microfossils from the Chert–Phosphorite Member at the base of the Lower Tal Formation. Brasier & P. Singh (1987) have redescribed the material of P. Singh & Shukla (1981) from this level in the Maldeota Mine, Mussoorie Syncline, and found that none of the elements have an Ordovician or younger character. They compare most closely with assemblages found in phosphorites of the Zhongyicun Member at Meishucun and in rocks of similar age in south-west Sichuan, China, as well as in phosphorites of the Chulaktau Formation in Southern Kazakhstan.

3.5.1. The Krol-Tal succession

Fig. 3.8 summarizes the succession, noting fossil records herein thought to be relatively reliable. This Precambrian–Cambrian boundary succession lies unconformably below the Upper Cretaceous Manikot Shell Limestone Formation (formerly in the Upper Tal Formation). Older formations and their ages are discussed below.

The Krol Formation

Dolomites of the Krol Formation contain several reported algal floras and microfaunas. The Krol B Member contains the calcareous algae *Renalcis* and *Oleckmia* according to Gansser (1974), which if correct implies a Nemakit–Daldynian or younger age (e.g. Riding & Voronova 1984); the records may be doubtful, however. Banerjee (1986) notes thin phosphorite bands within the Upper Krol algal dolomites of Nainital, and I. B. Singh & Rai (1983) mention rare, unfigured, finds of small phosphatic tubes and rounded forms in Krol D carbonates of the Nainital area. An assemblage from within a dolomite–shale–phosphorite–shale intercalation in the Krol D dolomites at Durmala, reported by Azmi & Pancholi (1983), contains ?skeletal alga *Spirellus shankari* (P. Singh & Shukla), tubes of *Coleoloides typicalis* Walcott and doubtful *Anabarites* sp., protoconodonts of the *Protohertzina anabarica* Missarzhevsky group ('unguliform' and 'siciform' elements), and several conical fragments (?hyoliths). Oncolites comparable with *Osagia*, *Volvatella*, and *Vesicularites* from this member compare with assemblage IV of the Yudomian in Siberia (Tewari 1985).

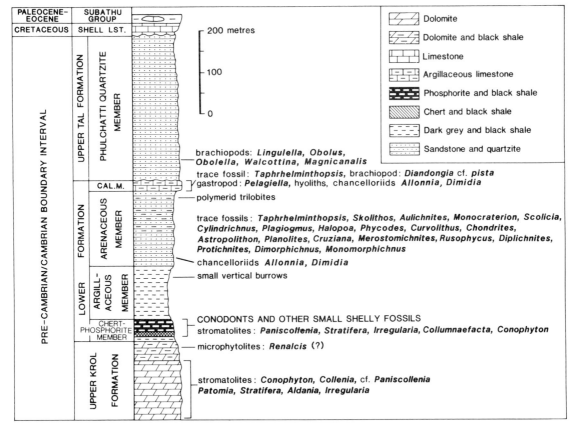

Fig. 3.8. Composite section for the Mussoorie and Garwhal synclines, Uttar Pradesh, Lesser Himalaya of India, showing the main fossil horizons. After Brasier & P. Singh (1987) and sources in the text.

Massive grey dolomites in the lower part of the Krol E Member are reported to contain stromatolites *Collenia*, cf. *Paniscollenia, Conophyton, Patomia, Stratifera, Aldania, Irregularia*, and various Tungussida (I. B. Singh 1983; I. B. Singh & Rai 1983). These may be compared with stromatolitic assemblages from the upper Yudoma Formation on the Siberian Platform (e.g. Krylov *et al.* in Rozanov *et al.* 1969). An uppermost Precambrian? age is also thought possible for stromatolites of the upper Krol Formation by Dr M. E. Raaben (*in* Banerjee 1986).

Argillaceous dolomites and algal lenses occur at the top of the Krol E Member. They contain problematic structures referred to archaeocyathans, *Renalcis*, and/or *Epiphyton* (I. B. Singh & Rai 1983) although these are not accepted by Brasier & Singh (1987)

The Lower Tal Formation: Chert–Phosphorite Member

Overlying the Krol E dolomites is the Lower Tal Formation, divided into four members (Fig. 3.8). A

conformable and continuous sequence is indicated, with littoral algal mats of Krol times gradually submerged by the stagnant, organic- and phosphate-rich lagoons of Lower Tal times. The lowest or Chert–Phosphorite Member shows a variable lithology, which at Maldeota changes upwards from carbonaceous dolomites, to carbonaceous siliceous dolomites and 'cherts', to carbonaceous dolomitic pelletal phosphorites. These locally contain small domal stromatolites and algal mats that are diagenetically silicified or phosphatized. Highly negative $\delta^{13}C$ values (−15 to −44‰) are found in these beds (Banerjee 1986). The stromatolitic forms *Columnaefacta vulgaris* Sidorov, *Aldania, Paniscollenia, Stratifera*, and *Irregularia* are tentatively identified by I. B. Singh & Rai (1983), and suggest a Tommotian age in the opinion of Tewari (1984). At Durmala, forms resembling *Conophyton* occur (Banerjee 1986).

Small shelly fossils appear in the Chert–Phosphorite Member at many localities. Brasier & Singh (1987) recognized the presence of the following: conoidal phosphatic microfossils *Maldeotaia bandalica* Singh & Shukla, *Protohertzina anabarica* Missarzhevsky group (including anabariform, unguliform, 'broad', hertziniform, and siciform elements), trumpet-shaped elements, acicular elements A & B; hyoliths ?*Conotheca* sp., *Ovalitheca* cf. *multicostata* Qian, allathecid sp. A; ?conulariids *Barbitositheca ansata* Qian & Jiang, *Hexangulaconularia* cf. *formosa* He; tubular problematica *Coleoloides* aff. *typicalis* Walcott, *Hyolithellus* aff. *insolitus* Grigorieva, *Hyolithellus* cf. *isiticus* Missarzhevsky, *Hyolithellus vladimirovae* Missarzhevsky; ?skeletal algae *Spirellus shankeri* (Singh & Shukla), *Olivooides multisulcatus* Qian emend. Xing & Yue. Further forms that are suspected from the illustrated material of other authors include: ?*Salanacus* sp. (= *Hirsutodontus* cf. *simplex* (Druce & Jones) of Azmi 1983, pp. 385–6, pl. 2, figs 1,6); ?*Maikhanella* sp. (*Maikhanella* sp. of Bhatt *et al.* 1985, p. 97, pl. II, figs 16,16a). Dr. P. Kalia (pers. comm. 1986) has also found sponge spicules in 'spongiolites' from the base of the Chert–Phosphorite Member and *Protospongia*-like spicules in the middle of the member. Kumar *et al.* (1987) have recorded a similar assemblage from the Chert–Phosphorite Member of the Ganga Valley, Lesser Himalaya. Of the forms that they report, *Spirellus shankari* and *Olivooides* sp. are clearly present. These microfossils from the Chert–Phosphorite Member suggest correlation with Zone I of China, especially those beds above its base bearing *Barbitositheca* and *Hexangulaconularia* spp.

The top metre of the Chert–Phosphorite Member marks the appearance of more ferroan sediments, with phosphatic–pyritic columnar stromatolites and fenestral dolomites (P. Kalia pers. comm. 1986). If this shoaling event is correlated with the top of the Dahai Member in Zone II of Meishucun (e.g. Brasier & P. Singh 1987), then both Marker B and Zone II faunas appear to be poorly characterized or barren here.

The Lower Tal Formation: Argillaceous and Arenaceous Members

The overlying Argillaceous Member consists of black shales with bands, lenticles, or large concretions of carbonate. These yield small vertical burrows (I. B. Singh & Rai 1983).

Abundant and diverse trace fossils in the Arenaceous Member include many forms typical of undoubtedly Lower Cambrian deposits (Fig. 3.8, I. B. Singh & Rai 1983; I. B. Singh *et al.* 1984). Polymerid trilobite impressions appear about 2 m below the top of this member (Rai & I. B. Singh 1983). A laminated siltstone in the lower part of the Arenaceous Member of the Ganga Valley has yielded chancelloriids *Allonnia erro-*

menosa Jiang, *Dimidia* sp. cf. *D. simpleca* Jiang, and hyolithelminthes *Hyolithellus tenuis* Miss. Grey micaceous siltstones in the upper part of the member contain *A. erromenosa* and *Dimidia* sp. cf. *D. simpleca* in association with trace fossils *Skolithos* sp. and *Taphrhelminthopsis circularis* Crimes (Kumar *et al.* 1987). Both lithofacies and biofacies compare with the Qiongzhusi Formation and Zones III to IV in China (Brasier & P. Singh 1987).

The Lower Tal Formation: Calcareous Member

The Calcareous Member is a grey calcareous siltstone, weathering to a brown colour. Locally, it contains lenses and burrow fillings of angular quartz grains and fragments of phosphatic brachiopods. Small shelly fossils have been recorded by Kumar *et al.* (1983, 1987) from this member at several localities: hyoliths, poriferids, acrotretaceans attributed to *Diandongia* cf. *pista* Rong, and gastropods of the *Pelagiella lorenzi* Kobayashi group. This fauna invites comparison with Zone V faunas in Sichuan and Yunnan (Brasier & P. Singh 1987).

The Upper Tal Formation

The Upper Tal Formation comprises the large-scale cross-bedded Phulchatti or Tal Quartzite, laid down during regressive conditions (Tewari 1985). A tentative Silurian–Devonian age was suggested by Azmi *et al.* (1981), Azmi & Joshi (1983), and Azmi (1983), although Banerjee (1986) suggested an Ordovician age. An early Cambrian age now seems more likely from reports of *Lingulella* sp., *Obolus* sp., *Obolella* sp., *Walcottina* sp., and *Magnicanalis* sp. and *Redlichia noetlingi* (Redlich) in shales about 52 m above the Calcareous Member, on the southern limb of the Mussoorie Syncline (Tripathi *et al.* 1984; Kumar pers. comm. 1987). Lithological and faunal comparison with the Canglangpuian of China is suggested (Brasier & P. Singh 1987).

3.6. NORTH WEST FRONTIER, PAKISTAN

Dolomite–phosphate–clastic sequences reappear again in the Hazara division of the North West Frontier Province of Pakistan. Here they occur on the south-western side of the main Himalayan boundary fault. Outcrops in the region of Abbottabad show a succession from dolomites and clastics of the Abbottabad Formation, to sandstones and siltstones of the Hazira Formation, with *c.*8 m of phosphatic–glauconitic sandstones at the base of the latter (Hasan 1986).

The Abbottabad Formation in the type area comprises a lower unit of sandstone, variegated shale, quartzite, and dolomitic limestone; a middle unit of quartzose sandstone, with subordinate shale and limestone, bearing phosphate at the top; and an upper unit of cherty dolomite, with subordinate limestone and shale. Phosphate is developed close to the top of this unit. Dutro (*in* Shah 1977) has identified the phosphatic tube *Hyolithellus*, associated with a hyolith in the upper part of this formation. Lithostratigraphical comparison with comparable sequences from China to Iran (herein) suggests a possible Zone I age for the upper part of the Abbottabad Formation.

The Hazira Formation has yielded *Allonnia tripodophora*, *Archiasterella pentactina*, 'stauractinid' sponge spicules, *Lapworthella* sp., *Rushtonia* sp., and hyoliths (Fuchs & Mostler 1972). The latter were compared with *Circotheca*? and *Linevitus*? by Rushton (1973). The fauna therefore suggests correlation with Zones III to V of China. The presence of Zones I and II has yet to be demonstrated.

3.7. GREATER HIMALAYA

Precambrian–Cambrian boundary sections can be traced further north into the Greater Himalayan basins of Kumaun, Zanskar-Spiti, and Kashmir, as summarized by Shah (1982) and Kumar *et al.* (1984). Here, the Cambrian is popularly referred to as of 'Tethyan' facies, as distinct from the southern 'Lesser Himalayan' facies described in Section 3.5. The region exposes the best Cambrian trilobite sequence in the Himalaya.

The late Precambrian of Spiti and Kashmir is characterized by laminated dark-grey to grey argillites with subordinate quartzites, deposited in shallow-water environments. The succeeding Lolab Formation of Kashmir (and its equivalents elsewhere) consists of quartzites with subordinate shales, with abundant trace fossils in the lower part. At the base of the Lolab Valley section of Kashmir are found *Planolites* spp. and *Bergaueria* sp., while 1400 m up occur *Skolithos* sp., *Monomorphichnus* sp., *Phycodes* sp., and *Gordia* sp. (Kumar *et al.* 1984). Beds with trilobites *Redlichia* cf. *noetlingi* (Redlich) and brachiopods *Neobolus* sp. and *Botsfordia granulata* appear about 2800 m above the base (ibid.) indicating a probable Canglangpuian age. The Pohru Valley of Kashmir has many more Cambrian trilobites with Chinese and cosmopolitan affinities, but forms older then *Redlichia* are also lacking; below this are found *Rusophycus* and *Cruziana* traces (Shah 1982). The latter suggests that the Precambrian–Cambrian boundary should be placed below the zone bearing *Rusophycus*. Trace fossils of presumed early Cambrian age also occur in the Spiti Valley region, but as yet these

cannot be calibrated against a skeletal fossil record (Kumar *et al.* 1984). This thick clastic belt of so-called 'Tethyan' facies continues across the borders into China, in the 'Himalayan region' of Xiang *et al.* (1981). Here again, the first trilobite is *Redlichia noetlingi*.

3.8. ELBURZ MOUNTAINS, IRAN

An important but little-known succession of Precambrian–Cambrian boundary rocks outcrops in the Elburz (or Alborz) Mts of northern Iran. The history of research has been reviewed recently by Wolfart (1981). Dr Bahaeddin Hamdi of the Geological Survey of Iran has kindly provided material and information from Hamdi *et al.* (in prep.) towards the following outline (Fig. 3.9).

The fossiliferous Precambrian–Cambrian boundary section belongs mainly to the Soltanieh Formation which outcrops over a 400 km distance, through the Elburz Mts and into parts of east and central Iran, where they form prominent escarpments. In the Soltanieh Mts, the Bayandor Formation is found beneath the Soltanieh Formation, but, in the central Elburz area, the latter disconformably overlies the older, Kahar Formation, which contains the Precambrian acritarchs: *Protoleiosphaeridium* cf. *leguminiforme*, *P. angulatum*, *Protoleiosphaeridium* sp., and *Lophosphaeridium* sp. The finds of an Ediacarian fauna in the Rizu and Esfordi Formations of central Iran, including *Persimedusites*, *Charnia*, and other unidentifiable forms doubtfully resembling *Dickinsonia* and *Spriggina* (see Glaessner 1984, pp. 99–100) also probably come from a horizon below the Soltanieh Formation (Dr Hamdi, pers. comm. 1986).

In the Zanjan–Soltanieh area, the Doran Granite intrudes the Kahar Formation and is overlain by the Bayandor Formation, which comprises purple sandstones and shales with thin dolomite intercalations; archaeocyathans reported from these beds are questionable (Wolfart 1981).

The Soltanieh Formation is typically a thick dolomite sequence, forming conspicuous cliffs. This facies passes laterally into evaporitic rocks in the Kerman area to the south, which may, in turn, partly correlate with the Hormuz Salt Formation of south Iran and similar sequences in eastern Arabia and in Pakistan. At the type area, south-east of the town of Soltanieh, the Soltanieh Formation can be subdivided into five main members, in ascending order according to Hamdi and Golshani (1983): the Lower Dolomite Member; the Lower (or Chopoghlu, or Chapoghlu) Shale Member; the Middle Dolomite Member; the Upper Shale Member; and the Upper Dolomite Member. The composite description

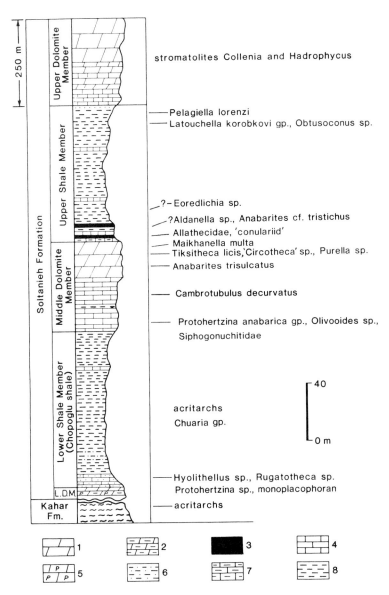

Fig. 3.9. Composite section for the Dalir and Valiabad sections of the Elburz Mts, northern Iran, showing the first-appearance datum of selected fossil taxa. Key is as follows: 1, dolomite; 2, argillaceous dolomite; 3, phosphorite; 4, limestone; 5, phosphatic dolomite; 6, sandy argillite; 7, argillaceous limestone; 8, argillite. Based on unpublished data of Hamdi *et al.* (in prep.), and sources in the text.

given below is based largely on sections at Dalir and Valiabad.

3.8.1. Lower Dolomite Member

Where present, this unit comprises from a few metres to 120 m of yellow, well-bedded, recrystallized dolomites, cherty dolomites, and limestones. In the Valiabad section, thin phosphatic beds yield phosphatic tubes, herein identified as *Hyolithellus* sp. aff. *tenuis* Miss., forms resembling *Rugatotheca* sp., *Protohertzina* sp., ?siphogonuchitids and poorly preserved ?monoplacophorans. The composition of this early assemblage is being studied. In some localities, this unit is missing so that the Lower Shale comes to rest directly on the Kahar Formation.

3.8.2. Lower Shale (or Chopghlu) Member

This unit (c.130–247 m) may intergrade with the underlying dolomite, and consists of dark green-grey to dark-grey shales, with interbeds of limestone. The following acritarchs have been reported by Seger (1977) from the 'Chopoghlu Shale Member' (= Lower Shale Member of Hamdi) in the western Elburz Mountains, north of Alamut: *Protoleiosphaeridium* cf. *duricorum*, *P. nervatum*, *Orygmatosphaeridium* cf. *tubiginosum*, *Leioligotriletum compactum*, *Botryoligotriletum* sp., *Laminarites* sp., and *Caryopsphaeroides* sp. These were considered to be of Vendian age. Specimens of the large discoidal algal vesicle *Chuaria* Walcott are common (e.g. *Fermoria* of Stocklin *et al.* 1964, *Chuaria* of Eisenack *in* Assereto 1966). Even larger ellipsoidal vesicles occur, comparable with those described from the Pusa Shale of Spain (*Beltanelloides* of Brasier *et al.* 1979). The latter are probably examples of *Shouhsenia* sp., a form known from below the Sinian system to near its top in China (Xing *et al.* 1985). The Lower Shale and the Pusa Shale assemblages both have a 'transitional' Precambrian–Cambrian character, with *Vendotaenia–Chuaria* type floras associated either with minute trace fossils of Phanerozoic type (in Spain, Brasier *et al.* 1979) or above the first small shelly fossils (in Iran, Hamdi *et al.* in prep.).

3.8.3. Middle Dolomite Member

This comprises about 78 m of limestones, dolomites and phosphatic layers, with abundant small shelly fossils in some sections. Currently available data suggest that their order of first appearance is as shown in Fig. 3.9. Abundant and well-preserved small shelly fossils include

tubes of *Hyolithellus* spp., and protoconodonts of the *Protohertzina anabarica* group. Less common elements include tubes of *Anabarites trisulcatus* Miss., *Cambrotubulus decurvatus* Miss.; protoconodont *?Hertzina* sp.; wiwaxiids *Palaeosulcachites irregularis* Qian, *P. biformis* Qian, *Lopochites latazonalis* Qian, *Trapezochites* sp.; and globomorphs *Archaeooides* sp., *Olivooides* sp. This assemblage compares with that from the lower part of the Manykayan Stage, or Nemakit–Daldyn Formation of Anabar in Siberia (cf. Missarzhevsky 1983).

Beds near the top of the member contain more abundant *A. trisulcatus*, *C. decurvatus* and siphogonuchitid wiwaxiids, plus problematical ?molluscs *Maikhanella multa* Zhegallo, *Purella* sp., tube *Tiksitheca licis* Miss. The hyoliths *Conotheca longiconica* (Qian) and *C. subcurvata* (Yu) are also abundant. This assemblage compares with the first shelly fossil zone in Mongolia, the upper Manykayan of Anabar, and Zone I of China.

3.8.4. Upper Shale Member

This unit (40–212 m) tends to be coarser grained and thicker bedded than the Lower Shale Member. The base comprises a few metres of calcareous shale to black phosphatic limestone containing abundant and diverse tubular fossils and molluscs, whose order of appearance is indicated in Fig. 3.9. The fauna is provisionally identified as follows: tubes *Hyolithellus* sp., *Anabarites trisulcatus*, *A.* cf. *tristichus* Miss., *Cambrotubulus decurvatus*, *Tiksitheca licis* Miss.; hyoliths *Turcutheca* sp., *Conotheca obesa* Qian and Allathecidae; siphogonuchitids *Palaeosulcachites* sp., *Lopochites* sp., *Siphogonuchites triangulatus* Qian; protoconodonts of the *Protohertzina anabarica* group; molluscs resembling the 'uncoiled' forms of *Aldanella* sp. reported from Massachusetts by Landing & Brett (1982). This assemblage appears to carry over from that found in the upper part of the dolomite.

Verbal reports of trilobites *Eoredlichia* and *Wutingaspis* sp. from 0.2 to 15 m above the base of the phosphatic beds have not yet been confirmed by further collecting (Dr Hamdi, pers. comm. 1986). The details of this fauna have yet to be worked out, but elements of Zones I and II of China are clearly present here.

The upper part of this member contains abundant and diverse phosphatized molluscs. Specimens of the *Latouchella korobkovi* group, including 'close-coiled' *Yangtzespira* sp., 'lax-coiled' *Bemella* sp., and 'uncoiled' *Cyrtoconus* sp. and *Obtusoconus* sp. appear abundantly about 20 m from the top. The gastropod *Pelagiella lorenzi* (Kobayashi) appears in the top 10 m, suggesting correlation with Zone V of China.

3.8.5. Upper Dolomite Member

The thickness of this unit ranges from 710 m in the type area to 250–300 m in the Dalir–Valiabad region. Stromatolites *Collenia spissa* Fenton & Fenton and *Hadrophycus immanis* Fenton & Fenton are moderately common in the upper part of the member.

The Soltanieh Formation is succeeded by the Barut, Lalun–Zagun, and Mila formations, of which the latter yields late Middle Cambrian to Upper Cambrian trilobites (see Wolfart 1981).

3.9. MALY KARATAU, KAZAKHSTAN

Phosphatic sequences of Precambrian–Cambrian boundary age are found throughout the Ulutau–Tienshan region of central and southern Kazakhstan, of which the platform sequence of Maly Karatau in southern Kazakhstan is the most important. The lithostratigraphy of this region has been described by Bushinsky (1966) and Eganov & Sovetov (1979), while the palaeontology and biostratigraphy are outlined by Missarzhevsky (1973), Ergaliev & Pokrovskaya (1977), and Missarzhevsky & Mambetov (1981). The sequence is of moderate thickness, with condensed facies and hiatuses (Fig. 3.10), but the fauna provides an important link between Palaeotethyan, Siberian, and Acado-Baltic successions.

3.9.1. Kyrshabakty Formation

The Kyrshabakty Formation (or Suite; thickness of 0–300 m) comprises a basal conglomerate plus lower clastic sediments lacking faunal remains. These are overlain, however, by dolomite and glauconitic limestones associated with fine-grained phosphorites, and bearing sparse remains of *Protohertzina anabarica* Miss. and *P. unguliformis* Miss. This 'Lower Dolomite' unit *c*.2–25 m) has been referred to the *P. anabarica* Zone by Missarzhevsky & Mambetov (1981), which also contains the Upper Vendian microphytolites *Vesicularites concretus* Zhur., *V. rectus* Zhur., and *V. nubecularites morulus* Zhur., plus the small shelly fossil *Chancelloria* sp. (Ergaliev & Pokrovskaya 1977). This may be either regarded as an impoverished assemblage of Chinese Zone I type, or as a sparse fauna of early Manykayan type.

3.9.2. Chuluktau Formation

The Lower Dolomite is overlain with erosional contact by the Chuluktau Formation (0–100 m). At the base is the Siliceous Member (*c*.3–30 m) which is, in turn, overlain by the Productive Member (*c*.0–40 m), bearing two grainstone phosphorite layers of commercial grade, separated by a shaly member.

The Siliceous Member has yielded protoconodont *Protohertzina siciformis* Miss.; tubes of *Anabarites signatus* Mamb., *Tiksitheca licis* Miss., *Hyolithellus rectus* Mamb., *Pseudorthotheca costata* Mamb.; and hyolith *Allatheca* sp. This assemblage is probably of Zone I type.

Anabarites trisulcatus and *Tiksitheca licis* range into the lower part of the Productive Member, and the following are found through this unit: protoconodonts and conodontiforms *Protohertzina siciformis*, *Formitchella* aff. *infundibuliformis* Miss., and *Kijacus kijanicus* Miss.; tubes of *Pseudorthotheca costata*, *Hyolithellus vladimirovae* Miss.; and the monoplacophoran *Shabaktiella shabaktiensis* Miss. The latter belongs to the *Latouchella korobkovi* group, suggesting correlation with Zone II of China.

The following appear in the upper part of the phosphorite, and range into the overlying limestone: tubes of *Torellella biconvexa* Miss.; hyoliths *Uniformitheca ovaliformis* Mamb., *U. aladzharica* Mamb., *Conotheca mammilata* Miss.; and chancelloriid *Chancelloria* sp. The whole assemblage from the base of the chert to the top of the phosphorite has been placed in the *Pseudorthotheca costata* Zone, figured as being from *c*.3 to 52 m thick (Missarzhevsky & Mambetov 1981).

This phosphatic sequence is capped by a stromatolitic Iron–Manganese Carbonate Member (*c*.1–5 m), containing tubes of *Hyolithellus vladimirovae*, *Torellella biconvexa;* hyoliths *Uniformitheca* spp. and *Conotheca mammilata;* wiwaxiids *Halkieria sacciformis* (Meshkova)., *H. denlanatiformis* Mamb. and *H. amorphe* (Meshkova); tommotiids *Camenella korolevi* Miss. and *Bercutia cristata* Miss.; globomorph *Geresia rugosa* Miss.; and *Cambroclavus clavus* Mambetov. This assemblage is placed in the *Bercutia cristata* Zone, figured as being from *c*.0.5–5 m thick (Missarzhevsky & Mambetov 1981). *Cambroclavus* allows comparison with Zone III–V faunas of Fangxian in China, while *Torellella biconvexa* is typical of strata higher than middle Tommotian in Siberia.

3.9.3. Shabakty Formation

This unit disconformably overlies the Chulaktau Formation, and consists of dolomites and limestones with some pelletal phosphorites near the base. The top of the Iron–Manganese Carbonate Member is locally eroded and the succeeding hiatus may span part of the Lower

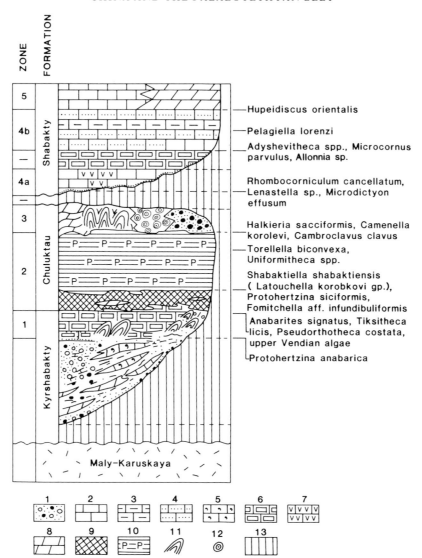

Fig. 3.10. Composite section for Maly Karatau, southern Kazakhstan, USSR, showing the first-appearance datum of selected fossil taxa. Key is as follows: 1, sandstone and conglomerate; 2, limestone; 3, argillaceous limestone; 4, sandy limestone; 5, glauconitic limestone; 6, dolomite with patches; 7, siliceous limestone; 8, dolomite; 9, chert; 10, phosphorite; 11, stromatolites; 12, oncolites; 13, hiatuses. Based on Missarzhevsky & Mambetov (1982).

Atdabanian (e.g. Missarzhevsky 1982; Rozanov & Sokolov 1984) with the Shabakty Formation transgressive over Precambrian Rocks in parts of the Karatau Basin (Missarzhevsky & Mambetov 1981). Several successive assemblages have been recognized within this suite, with trilobites appearing in the upper part. The *Rhombocorniculum cancellatum* Zone spans the lower, sub-trilobitic, Subzones of *Cambroclavus* and *Adyshe-*

vitheca respectively, figured as being *c.*6–48 m thick (Missarzhevsky & Mambetov 1981). The *Hebediscus orientalis* and *Ushbaspis limbata* Zones span the upper part of the suite.

The *Cambroclavus* Subzone contains the following: hyoliths *Uniformitheca aladzharica*, *Uniformitheca* sp. 1, *Asijatheca probata* Mamb., and *Asijatheca* sp. 1; tubes of *Hyolithellus vladimirovae*, *Torellella biconvexa*, and

Pseudorthotheca filosa Cobbold; monoplacophorans *Yochelcionella recta* Miss., *Igorella talassica* Miss.; protoconodonts *Amphigeisina renae* Miss., *Glauderia multifida* Miss., *?Oneotodus triangulus* Mamb. & Miss., *Rhombocorniculum cancellatum* (Cobbold), and *Protohertzina cultrata* Miss.; tommotiid *Ninella serebrjannikovae* Miss.; wiwaxiids *Halkieria amorphe*, *H. curvativa* Mamb., *H. trianguliformis* Mamb., *H. sacciformis*; the cambroclavids *Cambroclavus clavus*, *C. antis* Mamb., *C. undulatus* Mamb., and *Pseudoclavus singularis* Mamb.; *Karatubulus nodosus* Miss.; sponges *Lenastella mucronata* Miss., *L. umbonata* Miss., and *L. araniformis* Miss.; button-like *Mobergella scutata* Miss.; and net-like *Microdictyon effusum* Bengtson, Matthews & Missarzhevsky.

The *Adyshevitheca* Subzone has the following small shelly fossils: hyoliths *Conotheca corniformis* Mamb., *Adyshevitheca adyshevi* Mamb., *A. utchbasica* Mamb., *Asijatheca probata*, *Microcornus parvulus* Mamb., *M. talassicus* Mamb., *Burithes elongatus* Miss., and *Laticornus curtus* Mamb.; tubes of *Hyolithellus vladimirovae* Miss., *H. vitricus* Mamb., *Torellella explicata* Mamb. & Miss., and *Koksuja costulifera* Miss.; molluscs *Purella insueta* Miss., *Beshtashella tortilis* Miss., *Protowenella plena* Miss., *Pelagiella lorenzi* (Kobayashi), *Igorella talassica*, and *Stenothecoides minutus* Mamb.; protoconodonts *Amphigeisina renae*, *Glauderia multifidus*; tommotiids *Ninella serebrjannikovae* and Kellanellidae gen. et sp. indet.; *Cambroclavus clavus* and *Mongolitubulus squamifer* Miss.; sponges *Lenastella mucronata*, *L. umbonata*, *L. araniformis*; chancelloriid *Allonnia* sp.; net-like *Microdictyon effusum*; globomorph *Gaparella porosa*; problematicum *Resegia glandiformis* Miss.; and button-like *Mobergella scutata*. This assemblage may appear with trilobites *Ushbaspis* spp. I and II, Redlichiidae genus and species indeterminate, plus brachiopods *Linnarssonia constans* Koneva and *Lingulella* sp. (Missarzhevsky & Mambetov 1981).

Succeeding beds (*c*.3–5 m thick) bear the trilobite *Hupeidiscus orientalis* (Chang), which indicates a late Qiongzhusian age (Zone V). A higher assemblage spans 25–30 m of strata, with *H. orientalis* plus *Calodiscus korolevi* Erg., *Ushbaspis limbata* Pokr., and *Redlichia* sp., which suggest a Canglangpuian age. A younger assemblage (40–45 m) with *Redlichia chinensis* Walcott and *Kootenia gimmelfarbi* Erg., indicates a Lonwangmiaoan age. The brachiopods *Linnarssonia constans* and *Botsfordia caelata* (Hall) are common in this assemblage.

3.10. SALANYGOL, MONGOLIA

Although the Khubsugul Basin in the northern part of the People's Republic of Mongolia contains a comparable but thicker succession of phosphatic and dolomitic rocks, more richly fossiliferous sections are found further west at Salanygol, north of the town of Altai, as outlined by Voronin *et al.* (1982). Here is found a thick, folded succession of late Precambrian to Lower Cambrian rocks with palaeontological characteristics that allow comparisons between China, Iran (and other Palaeotethyan sections) and the Siberian Platform (Fig. 3.11). Some confusion has arisen, however, from the presence of a significant unconformity within the sequence, with the largely Botomian Chairchanskaya Formation overstepping the eroded Tommotian Bayangol Formation and the Atdabanian Salanygol Formation, and thereby incorporating reworked fossils. Thus the upper ranges of small shelly fossils in Voronin *et al.* (1982) need revision and doubtful primary occurrences are excluded here.

3.10.1. Tsaganolom Formation

This comprises a transgressive clastic–carbonate cycle, some 1137 m thick. The base is 180 m of tuffaceous grey sandstones with conglomeratic layers. The middle part consists of *c*.560 m of conglomerates and finer clastics, passing up into dolomites and limestones with oncolites. In the upper part is some 397 m of dark-grey and black limestones with siliceous patches and dolomite beds near the top. Upper Vendian microphytolites include: *Vesicularites concretus* Zhur., *V. irregularis* (Reitl.), *Ambigulomellatus horridus* Zhur., *Volvatella zonalis* Nar., and *V. vadosa* Zhur.

3.10.2. Bayangol Formation

This is a unit of alternating clastics and carbonates, some 1300 m thick. The order of appearance of fossils through this very thick succession broadly repeats the Zone I to Zone II succession of China, and suggests that the latter can be correlated with part of the Lower to Middle Tommotian of the Siberian Platform. Assemblages of the Bayangol Formation will be described under zonal headings.

Tiksitheca licis–Maikhanella multa *Zone*
The basal 40 m of the Bayangol Formation comprises barren siltstone, and the first skeletal fossils appear near the base of overlying grey limestones, some 120 m thick. At levels G to I are found: tubes *Anabarites trisulcatus* Miss., *Cambrotubulus decurvatus* Miss., *Tiksitheca licis* Miss., and *Hyolithellus* cf. *vladimirovae* Miss.; conodontiforms *Salanacus vorononi* Grig., *S. cristata* Grig., and *S. cornuta* Grig.; siphogonuchitids *Lopochites curtus*

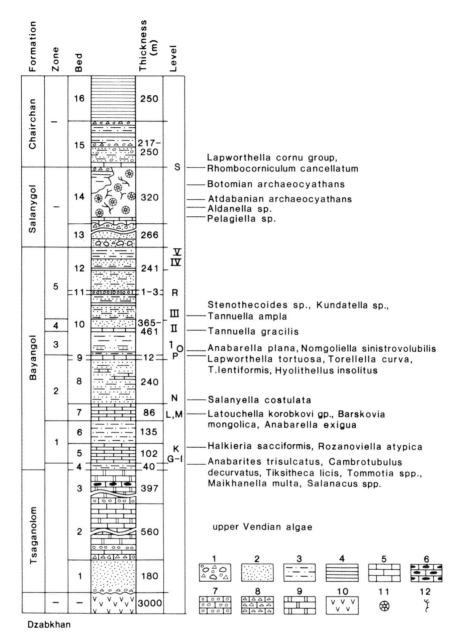

Fig. 3.11. The Salanygol section, Mongolian People's Republic, showing the first-appearance datum of selected fossil taxa. Key is as follows: 1, conglomerate and breccia; 2, sandstone; 3, siltstone and argillaceous sandstone; 4, argillite; 5, limestone; 6, limestone with chert; 7, oncolitic limestone; 8, limestone breccia; 9, dolomite; 10, effusive igneous rocks; 11, archaeocyathans; 12, biohermal algae. Based on Voronin *et al.* (1982).

Grig. and *Palaeosulcachites biformis* Qian; tommotiids *Tommotia applanata* Grig. and *T.* cf. *baltica* Bengtson; ?mollusc *Maikhanella multa* Zheg.; and algae *Korilophyton inopinatum* (Voron.), *Renalcis polymorphus* (Masl.), *Angulocellaria anisotoma* Vologd., and *Gemma maculosa* Voron. & Drosd. The skeletal assemblage has affinities with Zone I of China.

About 100 m higher, at level K, are found *Anabarites trisulcatus*, *Cambrotubulus decurvatus*, and *Lopochites curtus*, plus new elements: monoplacophoran *Rozanoviella atypica* Miss.; wiwaxiid *Halkieria sacciformis* (Mesh.); and tubes of Coleolidae. The algal fauna is similar to the above except that *Renalcis polymorphus* is replaced by *R. rotundus* (Drosd.). Included in the upper part of this zone are 135 m of siltstones and sandstones.

Missarzhevsky (1982) also mentions the presence of ?monoplacophoran *Purella atypica* and protoconodont *Protohertzina unguliformis* in this zone.

Isanella compressa *Zone*

This zone is marked by the appearance of diverse molluscs. An 86 m-thick unit of grey argillaceous limestone contains the following at levels L and M, near the middle: monoplacophorans *Ilsanella compressa* Zheg., *Latouchella korobkovi* (Vost.), *L. sibirica* (Vost.), *L. minuta* Zheg., *Anabarella exigua* Zheg., and *Rozanoviella atypica* Miss.; gastropods *Barskovia mongolica* Zheg., *Nomgoliella sinistrovolubilis* Miss.; plus wiwaxiids *Halkieria sacciformis* and *H. costulatus* (Mesh.); globomorph *Archaeooides granulatus* Qian; and alga *Tubomorphophyton* sp. The appearances of the *Latouchella korobkovi* group and aldanellid *Barskovia* suggest comparison with Zone II of China and the Tommotian of Siberia. Above this unit lies 240 m of fine-grained sandstones and siltstones with microfauna in limestone lenses at level N: monoplacophorans *Salanyella costulata* Miss., *Latouchella korobkovi* and *Rozanoviella atypica*; wiwaxiid *Halkieria sacciformis*; tube *Anabarites trisulcatus*; and globomorph *Archaeooides* sp. Algae in the upper part of the bed include: *Renalcis rotundus*, *R. granulatus* (Korde), *R. novus* Voron., *R. ex gr. polymorphus*, and *Epiphyton* sp. 1.

About 326 m above the base of the zone there occurs a thin (12 m) oncolitic limestone bed at level P, with algae *Botominella lineata* Reitl., *Gordonophyton* sp., and *Renalcis* sp.

Just above level P, the assemblage notably contains forms typical of the late *sunnaginicus* to *regularis* Zones of the Siberian Platform: tommotiid *Lapworthella tortuosa* Miss.; tubes *Hyolithellus tenuis* Miss., *H. insolitus* Grig., *Torellella lentiformis* (Syss.), and *T. curva* Miss.; wiwaxiid *Halkieria proboscidea* (Mesh.); hyolith *Exilitheca multa* Syss. These are accompanied by

monoplacophoran *Ilsanella compressa*; gastropods *Nomgoliella sinistrovolubilis* Miss., *Kairkania rotata* Miss., and *K. evoluta* Zheg. The remains of *Torellella lentiformis* plus *Lapworthella tortuosa* suggest correlation with the *regularis* Zone of the Tommotian in Siberia.

Anabarella plana *Zone*

This is located entirely within a thick unit (<461 m) of alternating siltstones and sandstones. At level O, near the base of the zone, are found: monoplacophorans *Ilsanella compressa* Zheg., *Latouchella korobkovi*, *L. minuta*, *Anabarella plana* Vost., *A. exigua* Zheg., and *Rozanoviella atypica* Miss.; gastropod *Nomgoliella sinistrovolubilis* Miss.; hyoliths *Ovalitheca aplicata* Syss. and *O. mongolica* Syss.; siphogonuchitid *Lopochites curtus;* and tubes of Coleolidae. A similar assemblage occurs near the top of the zone at level 1, although *Ilsanella compressa* is missing and *Lopochites curtus* is replaced by *L. latazonalis* Qian. The molluscs *Salanyella costulata* and *Barskovia mongolica* reappear here.

Tannuella gracilis *Zone*

Underlying elements range up into this zone, which at marker level II contains the cap-shaped monoplacophoran *Tannuella gracilis* Zheg. plus monoplacophorans *Bemella jacutica* (Miss.), *Rozanoviella atypica*, *Latouchella sibirica*, *Anabarella plana*, and *Salanyella costulata;* gastropod *Nomgoliella sinistrovolubilis*; siphogonuchitid *Lopochites latazonalis*, *Siphogonuchites pusilliformis*, and wiwaxiid *Halkieria sacciformis*; globomorph *Archaeooides granulatus* Qian; hyolith *Ovalitheca aplicata;* and tubes of Coleolidae.

Stenothecoides *Zone*

The problematical ?mollusc *Stenothecoides* sp. and ?orthid *Kundatella* sp. appear as distinctive elements in this zone, spanning the upper part of the siltstone–sandstone alternations in levels II, R, IV, and V. Other elements include the molluscs *Rozanoviella atypica*, *Latouchella minuta*, *L. gobiica* Zheg., *Salanyella costulate*, *Tannuella ampla* Zheg., *T. gracilis*, and *Kairkhania rotata*; wiwaxiids *Halkieria costulatus*, *H. sacciformis*; siphogonuchitid *Lopochites latazonalis;* and hyoliths *Ovalitheca aplicata*, *O. mongolica*, and *O. glabella* Syssoiev.

3.10.3. Salanygol Formation

A new depositional cycle is initiated with this formation, which comprises a lower unit (266 m) of conglomerates and pebbly sandstones without fauna and an upper unit (<320 m) of algal–archaeocyathan limestones. This bears a rich assemblage of Middle to Upper Atdabanian archaeocyathans such as *Alataucyathus jaroschevitschi* Zhur., *Tabulacyathellus bidzhaensis* Miss., and *Pretiosocyathus subtilis* Roz., with gastropods *Pelagiella* sp. and *Aldanella* sp. The upper part of the limestones contains an early Botomian archaeocyathan assemblage, with *Soanocyathus admirandus* Roz., *Dokidocyathus tuvaensis* Roz., and *Flindersicyathus latus* (Vologd.). This assemblage ranges up into the overlying Chairchanskaya Formation.

3.10.4. Chairchanskaya Formation

The basal contact of this unit is an erosional unconformity over the underlying units, with conglomeratic sandstones and reworked fossils near the base. The formation comprises a lower unit (217–250 m) of sandstones and conglomerates and an upper unit (< 250 m) of shales. The presence of the tommotiids of the *Lapworthella cornu* group and conoidal *Rhombocorniculum cancellatum* (Cobb.) is noteworthy for correlation with the Upper Atdabanian–Lower Botomian in Siberia and the Acado-Baltic, although they may have been reworked from underlying beds.

3.11. CORRELATION

Before discussing the methodology and scheme of correlation adopted here, it is useful to comment upon the different views on correlation of candidate marker horizons for the Precambrian–Cambrian boundary at Meishucun, especially with Mongolian and Siberian sections. Some of the recent criticisms are also examined here.

3.11.1. Correlating Marker A (basal Zone I)

The base of Zone I at Marker A has generally been taken as the base of the Meishucunian Stage and the Cambrian System in China in recent years (e.g. Xiang *et al.* 1981; Luo *et al.* 1982, 1984; Xing *et al.* 1983, 1984; Xing & Luo 1984; Xing 1985; Yang Zunyi *et al.* 1986). This datum will be retained for the base in the forthcoming 'Cambrian System of China' (Brasier in press).

The correlations suggested by Rozanov (1984) and Rozanov & Sokolov (1984) place both Marker A (base of Chinese Zone I) and the first assemblage of Mongolia at the level of the basal Tommotian *Aldanocyathus sunnaginicus* Zone of the Aldan–Lena region in Siberia. This was also the correlation first loosely implied between Siberia and China by Qian (1977). In favour of this stratigraphically 'high' view is the occurrence of

supposed *Turcutheca crasseocochlia* and supposed inarticulate brachiopods near or at the base of Zone I in Meishucun; and of *Tiksitheca licis* and *Tommotia* cf. *baltica* at the base in Mongolia. Another form that may support this correlation is '*Spinulitheca billingsi*' (e.g. Rozanov & Sokolov 1982), known from Marker A in Maidiping of Sichuan, Ningqiang of Shaanxi, Bed 3 at Meishucun (all in China), and Bed 8 at Ulakhan–Sulugur of Siberia. The systematics of these annulated tubes require revision, however.

A different view was expressed by Missarzhevsky (1982, 1983). He suggests a correlation of Marker A and the base of the first zone in Mongolia with about the base of the middle Manykayan Stage (*Purella cristata* Zone) of the Anabar region in Siberia, i.e. somewhat below the basal Tommotian. In favour of this stratigraphically 'low view' is the first appearance of *Cassidina* sp. and related *Maikhanella* sp. below *Latouchella korobkovi* and aldanellid faunas in each area.

A sub-Tommotian correlation for Marker A is also implied by Luo *et al.* (1980), Xing & Luo (1984), and Jiang (1985), though Xing *et al.* (1984) suggest it may lie above most of the *Anabarites trisulcatus* Zone of Anabar. They raise the correlation of the first Mongolian zone, however, to the level of basal Tommotian. This is presumably done on the presence of *Tiksitheca licis*—a fossil not present at Meishucun.

Qian (1984), however, notes the presence of *T. licis* within Zone I of China, and *Tiksitheca* sp. is reported from the base of the Zhongyicun Member (Qian 1983) and at similar levels elsewhere in Yunnan (e.g. Luo *et al.* 1982, pp. 149, 151, & 153). Even so, Qian (1984) places Marker A below the Tommotian (i.e. at an unspecified level within the Manykayan) because of simple orthothecid hyoliths and the supposed lack of the *Anabarites trisulcatus*–*Protohertzina anabarica* fauna in the basal Tommotian of the Aldan–Lena region.

3.11.2. Correlating Marker B (basal Zone II)

The suggestion that Marker B may provide a workable datum for the boundary, arose from the Bristol meeting of 1983 and a subsequent favourable ballot of voting members of the Working Group (Cowie 1984). Concern has been expressed over correlation, however, resulting in a deferred decision (Cowie 1985).

Luo *et al.* (1980) and Jiang (1985) have correlated Marker B with the base of the Middle Tommotian *regularis* Zone in Siberia. Xing & Luo (1984) also implied a correlation of Marker B with a point above the base of the Tommotian and Mongolian first faunas. Luo has since correlated Marker B with the base of the Tommotian, Bed 8 at Ulukhan–Sulugur (in the unpublished Circular–Newsletter of IGCP Project 29 for 1984) and a similar correlation is implied by Xing *et al.* (1984). Less certain, Qian (1984) has suggested that the base of the Tommotian may correlate with either Zone II or III of the Meishucunian.

Missarzhevsky (1983) initially correlated Marker B with the datum bearing the first appearance of *Latouchella korobkovi* and *Barskovia* in the second zone of Mongolia; this, in turn, is correlated with the *Anabarella plana* Zone of the Anabar Shield, regarded by him as being of earliest Tommotian age.

Rozanov (1984, and *in* Kirschvink & Rozanov 1984) and Rozanov & Sokolov (1984) dissent from this view, however. They correlate the second, *Latouchella korobkovi* fauna of Mongolia with the second, *Dokidocyathus regularis* Zone of the Aldan–Lena region of Siberia; the *Latouchella korobkovi* fauna of Anabar is correlated with the upper part of the first, *sunnaginicus* Zone. This may follow partly from the assertion that *T. licis* begins in the basal Tommotian and partly from the lack of *L. korobkovi* until *regularis* times in the Aldan–Lena area. It does occur in the *sunnaginicus* Zone of the Maya and Yudoma Rivers, however (e.g. Rozanov *et al.* 1969, pl. IV. fig. 17). The *regularis* Zone correlation for the second fauna in Mongolia may also be supported by the presence of algae *Tubomorphophyton* and *Epiphyton* at this level (Voronin *et al.* 1982; see also Riding & Voronova 1984), though there may be some circular reasoning here. Rozanov has effectively declined to correlate Marker B, merely showing a generalized correlation for the Meishucunian.

This dissent was expressed more fully by Bengtson *et al.* in an unpublished article of limited circulation (IUGS/IGCP PЄ–Є Boundary Working Group Circular–Newsletter of June 1984). The points they raised at that time were as follows.

1. 'it has not been shown that the 2nd Faunal Assemblage at Meishucun is, or is time-equivalent to, the earliest known such assemblage'. This is an appeal to the decision at the Cambridge meeting (Cowie 1978) that the boundary stratotype point 'should be placed as close as is practicable to the base of the oldest stratigraphical unit to yield Tommotian (*sensu lato*) fossil assemblages.' Tommotian (*sensu lato*) may therefore be taken to include Manykayan (= Nemakit–Daldynian) assemblages.

Bengtson, Missarzhevsky, and Rozanov then argued that 'If the Cambridge decision is to be adhered to we should be looking for the oldest unit *anywhere* to yield such an assemblage'. Although this is the working hypothesis of Missarzhevsky (e.g. 1983) it is not the practical view of Rozanov (1984), Rozanov & Sokolov

(1984), or Sokolov & Fedonkin (1984). These authors exclude Manykayan assemblages from the Cambrian. By the time of the 1983 Bristol meeting it was clear, too, that voting indicated a preference for a datum with correlatable faunas below and above the boundary (e.g. Cowie 1984).

The writer views these earlier decisions as historical benchmarks, vulnerable to the rising tide of new data.

2. 'In addition to a few forms known from the Tommotian of Siberia', the second faunal assemblage 'contains taxa (e.g. *Zhijinites*, *Quadrotheca*, and even fragments described as *Buccanotheca*, which, judging from the published illustrations, could represent genal spines of trilobites) elsewhere known only from Atdabanian and later deposits, as well as representatives of a large number (possibly exaggerated due to the effects of preservation) of nominal taxa most of which are extremely difficult to identify elsewhere'.

This comment can now be answered in part. It may be that *Zhijinites* intergrades with *Cambroclavus* in Atdabanian assemblages bearing *Rhombocorniculum cancellatum* in southern Kazakhstan and Fangxian of Hubei (see above), but only *Zhijinites* is present in Zone II (Qian & Yin 1984) and the imputation of an Atdabanian age is unproven. The quadrate tubes of *Quadrotheca* sp. occur with those of *Tetratheca* sp. (e.g. Luo *et al.* 1982, p. 151); the latter is an Atdabanian hyolith in Siberia, but the veracity of this comparison is unproven, however, and *Quadrotheca* does not seem to be cited in recent compilations from Meishucun (e.g. Luo *et al.* 1984). Hyoliths do not have a proven record for international correlation and the trend from simple, circular to lipped, geometrical tubes seen in several regions may be ecological and diachronous. The comparison of *Buccanotheca* with a trilobite spine cannot be confirmed by microscopical observations (Cowie, pers. comm. 1986) and this is, in any case, a palaeobiological and not a stratigraphical point. The final comment, on poor preservation of nominal taxa, is fair but it is no longer true to say that they remain difficult to identify elsewhere.

3. 'Even the 1st Faunal Assemblage contains forms (diverse brachiopods, *Barbitositheca*) that have a suspiciously 'young' aspect when compared to what is known from the oldest skeletalized fossil assemblage in Siberia.' The comment about diverse brachiopods seems to be an error, since only two taxa are usually cited ('*Artimycta*' and '*Ramenta*') and their brachiopodal nature is doubtful. The 'young' aspect of *Barbitositheca* is also misleading. This taxonomic group consistently appears in the first Zone from Yunnan, southern Sichuan, northern

Sichuan, and the Lesser Himalayas, and Qian (1984) elevates it to an index for the zone. Its primitive morphology in comparison with *Hexagulaconularia*, *Hexaconularia*, and other conulariids is clear enough. Stratigraphy must be practical; it cannot be based on evolutionary ideologies applied to unfamiliar groups.

4. 'It is quite conceivable that the presence at Meishucun of possible Tommotian forms together with possible Atdabanian forms results from stratigraphic admixture or condensation, in view of the fact that this assemblage occurs mainly as phosphatic moulds in an interval of only 1.4 m of phosphorites, cherts, and phosphorite- and manganese-bearing dolomite'. The presence of possible Atdabanian forms has been questioned above as unproven. Even so, the problem of condensation is real, compounded by the practice of combining all faunal data from 'Bed 7' together, and likewise for 'Bed 8' (e.g. Luo *et al.* 1982, 1984; Xing *et al.* 1983, 1984). Bed 7 seems to have been deposited fairly rapidly in subtidal to peritidal conditions, with shell-lags marking the abrupt base of a cross-bedded dune at Meishucun, of a ?storm unit at Wangjiawan, and an influx of siliciclastic dolomite at Maidiping. Bed 8 contains fenestral dolomites with phosphatic hard-grounds and evidence for final prolonged emergence. The span covered by this bed is unclear, but could be considerable. The problem here, then, is that there is no good fossil datum, above Marker B and below the break in sedimentation, to provide supporting control on correlation. There are, however, the geochemical markers of negative $\delta^{13}C$ and the iridium spike just above the base of Bed 9 (Hsü *et al.* 1985).

Similar objections were also raised at the Moscow meeting of the IUGS/IGCP P€–€ Boundary Working Group in 1984; Cowie reporting the views of other members (1985). These noted that the Meishucun section 'is composed of dolomites and phosphorites, which are known everywhere as deposits that have many breaks, and certain horizons with faunas can be missing. Some assemblages in these kinds of deposits can also be of mixed origin, leading to difficulty in correlation. Faunas from the Meishucun section commonly are badly preserved. This leads to faulty determinations and sometimes to the creation of a large number of taxa that show only the character of the preservation and not the true taxonomic identity, thus giving an illusion of correlation.' It should, in fairness, be pointed out that the *sunnaginicus* Zone is also thin: Ulukhan-–Sulugur (4.4–6.4 m), Dvortsy (9 m), Isit (10.5–15 m), and mildly phosphatic (Rozanov & Sokolov 1984).

5. 'In addition, there is today evidence that certain Tommotian forms which have short vertical distribu-

tion on the Siberian Platform in fact reach into younger deposits elsewhere (e.g. Mongolia and the Avalon Platform), and this may also turn out to be the case here.' In fact, the long ranges of certain Tommotian small shelly fossils in Mongolia (Voronin *et al.* 1982) are probably due to stratigraphical admixture at unconformities, as discussed above. The long ranges of species on the Avalon Platform of maritime Canada are also questioned by Brasier (1986*a*). The problem of long ranges may be diminished if first appearances of new taxa (i.e. immigration bio-events) can be employed, as discussed further below.

6. 'Luo's correlation . . . of the base of Bed 7 at Meishucun with the base of Bed 8 at the Ulukhan Sulugur section of the Aldan, Siberia, said to be based on shelly fossil assemblages, appears unsupported in that a comparison of the faunal lists produced in documentation of the candidate sections shows that the two beds do not have any stratigraphically significant species in common'. This difficulty is diminished, however, if *Bemella jacutica* spp. is accepted as part of the *Latouchella korobkovi* plexus, since the former appears in Bed 8 of the Ulakhan Sulugur (Rozanov & Sokolov 1984). *Obtusoconus* sp. and other 'Chinese' forms are now known from the basal Pestrotsvet Formation, Selinda River, Eastern Siberia (V. V. Khomentovsky, pers. comm. 1987).

7. Other objections to the Meishucun section raised at the Moscow meeting (Cowie 1985) were as follows: 'The age of Zone II at Meishucun is not established, and palaeobiological and magnetostratigraphical correlations suggest that it may not be at the desired earliest level. It may be younger than the Tommotian stage and equivalent in age to the immediately overlying Atdabanian Stage, both in the Early Cambrian of Siberia.' As mentioned above, the Atdabanian correlation has not been proven biostratigraphically. The magnetostratigraphical correlation must also be questionable, since the technique relies on calibration against a tested biostratigraphical scale, usually in deep-sea deposits with faunal control on hiatuses and accurately known rates of deposition. This is not yet possible for these boundary sections. Obviously, the potential of magnetostratigraphy for intrabasinal correlation must be proven first.

'Conservation of the Meishucun section depends on the cooperation and exploitation policy of the phosphate mining authorities.' This is a matter of concern, since a stratotype must be inviolate.

'Moreover the radioisotope dating is suspect and grossly inaccurate.' These results are reviewed by Cowie

& Johnson (1985) and Cowie (1985) and are not pursued further here, save to note that they are unusual in being based on illites and conflict with results obtained from 'Pan-African' igneous events from the Middle East, Africa, and the Avalon terrane.

'Finally, the global distribution of trace fossils near the Precambrian–Cambrian boundary is still poorly known and zonation by trace fossils is not a viable method of stratigraphy.' This refers to the significant occurrence of *Rusophycus* and *Cruziana* traces shortly above Marker B. It is now too harsh a verdict, and overlooks the possibility that the boundary may, in practice, have to be guided by trace-fossil changes in the majority of outcrops from North America, South America, Australia, Europe, and much of Asia.

3.11.3. Correlating Marker C (basal Zone IV)

The appearance of redlichiid trilobites of *Parabadiella* sp. at Marker C heralds the Qiongzhusian Stage and provides another possible datum for the boundary. Here, however, there are major problems of correlation and much careful taxonomic work is required.

Rozanov (1984) and Rozanov & Sokolov (1984) correlate Marker C with about the base of the *Judomia* Zone (i.e. low Upper Atdabanian) of Siberia. This view may be supported by the presence of typical Upper Atdabanian, and higher, small shelly fossils such as *Lapworthella cornu*, *Tannuolina multiforata*, *Botsfordia caelata*, and *Pelagiella* sp. in high Zone III to Zone V rocks at Meishucun, Maidiping and elsewhere. The first redlichiid trilobites of Siberia also appear in the *Judomia* Zone, or just below.

Luo *et al.* (1984) and Xing & Luo (1984) correlate Marker C with the base of the Atdabanian in Siberia. This would suggest that *Profallotaspis* and *Parabadiella* were nearly synchronous, although they regard the latter as the world's oldest trilobite.

Zhang (1984) considers that Marker C falls below the American and Moroccan *Fallotaspis* Zones. This is based on a supposed close relationship between *Eoredlichia* and *Lemdadella* from the *Eofallotaspis* Zone of Morocco, as well as with *Pararedlichia* from the *Fallotaspis tazemmourtensis* Zone of Morocco (e.g. Sdzuy 1978). Qian (1984) implied a correlation of Marker C with the middle, *regularis* Zone of the Siberian Tommotian, on the basis of the appearance of lipped hyoliths and *Lapworthella*. This correlation, would, however, imply that *Parabadiella* and *Eoredlichia* are Tommotian trilobites. The record of a putative trilobite cephalon from the basal Tommotian (Fedorov *et al.* 1979) has suggested to Zhang (op. cit.) a correlation between Marker C and the basal Tommo-

tian of Siberia. However, the supposed cephalon is now discredited as a poorly preserved *Tumulduria* shell fragment (Bengston *et al.* 1986).

3.12. THE BIO-EVENT APPROACH TO CORRELATION

The potential of the bio-event approach outlined on pp. 156–7 is demonstrated in Fig. 3.12. Shown here are are broadly homotaxial points of first appearance in selected taxa through each of the main areas discussed above. Also shown are sedimentological events and the tops of sedimentary cycles traceable across relatively large distances. Note how the new sections from India and Iran largely support the evidence for homotaxial bio-events. This indicates that broad correlation is possible along this belt.

3.13. THE PALAEOGEOGRAPHICAL APPROACH TO CORRELATION

It is obviously easier to correlate *within* facies belts, or within palaeogeographical provinces, than it is to correlate *across* them. Thus, correlation is easier along broadly parallel palaeogeographical belts of the Tethyan region than across the strike from Meishucun to Aldan.

Yin & Qian (1986) have recognized eight palaeogeographical divisions for the earliest Cambrian small shelly faunas of China. When traced beyond China, broader concepts may have to apply, however, and three main belts are distinguished here. The Inner Palaeotethyan Belt is found in Yunnan, Sichuan, Guizhou, Shaanxi, the eastern Yangtze Gorges of western Hubei, the Lesser Himalayas, parts of northern Pakistan, and northern Iran. These areas possibly comprised portions along the ?northern marginal terrane of Gondwana. Condensed phosphatic sediments are found here in Chinese Zone I, while Zone III–IV sediments are typically thicker siliciclastics with thin carbonates, best characterized by chancelloriids such as *Allonnia tripodophora* and *Archiasterella pentactina*, with *Pelagiella lorenzi* group gastropods in Zone V. Endemic redlichiid trilobites predominate from Zone IV onwards in this belt, suggesting that habitats had little access to the open ocean.

The Outer Palaeotethyan belt has a more northerly distribution, from the microcontinents and island arcs of southern Kazakhstan, through the Tianshan–Hinggan Basin and Tarim Platform of western Xinjiang, to the northern parts of the Yangtze Platform in Fangxian

County of Hubei. The relatively emergent North China Platform may also belong here. Major phosphatic sediments in Kazakhstan appear to be of Zone I–II age and those of Xinjiang may be of similar age. In Kazakhstan and Fangxian, a disconformity occurs between dolomites of presumed Zone II age and dolomites of Qiongzhusian age. The latter bear *Allonnia tripodophora* but with *Cambroclavus* spp., *Microcornus parvulus*, *Adyshevitheca adyshevi*, and *Rhombocorniculum cancellatum*, of presumed Zone III–IV age. Cambroclavid elements are particularly characteristic. The upper parts of Zone V in Maly Karatau, Xinjiang, and Fangxian are indicated by pelagic trilobites *Hupeidiscus*, *Schizudiscus*, and *Tsunyidiscus* respectively. Later trilobite zones have a mixed Chinese and pelagic, cosmopolitan aspect. Periodic connections with Siberia and the Avalon–Baltic region are indicated.

The Outer Mongolian Belt lay even further away, probably along the subsiding ?eastern margin of the Siberian Platform, continuous with the Altai–Sayan and Tannuola Mountains of the Soviet Union. Here, putative Zone I assemblages are followed by Zone II assemblages plus successive Tommotian taxa that arguably allow recognition of the *regularis* to *lenaicus* and higher zones of the Siberian Platform interior.

Since correlation along these various belts is more readily achieved than across them, attempts should first be made to establish the detailed pattern of events *along* the belts. Patterns of migration and diachronism across the belts may then be calibrated against a regional event-stratigraphy. Interestingly, Chinese Markers A and B can be recognized (arguably) across a very wide part of Asia, from West Anabar and the Olenek Uplift, to western Mongolia and the Inner and Outer Palaeotethyan belts. It is mainly along the Lena, Aldan, Uchur–Maya, and Yudoma River sections that faunas differ, with more tommotiids, hyoliths and archaeocyathans in the early Tommotian. Correlations become contentious here, even across the Siberian Platform (e.g. Zhuravleva 1975; Missarzhevsky 1983; Khomentovsky 1986).

3.14. EVENTS AND CORRELATION

Below is given an outline of the sequences of biological and geological events across the Precambrian–Cambrian boundary in the Palaeotethyan Belt, with special reference to the Inner Palaeotethyan Belt. This outline is developed from Brasier (1986*b*, 1987) and can be modified as further evidence becomes available.

Zones 0 to V are discussed with reference to sections in southern China; potential markers and their correlation are then discussed for each zone.

3.14.1. Zone 0

This is the informal name of a transitional interval between the Vendian and Zone I assemblages of presumed Cambrian type. It includes equivalents of the Manykayan (Missarzhevsky 1983) or Nemakit–Daldynian (Sokolov & Fedonkin 1984) chronostratigraphical unit at the top of the Vendian. Datum points with potential for widespread correlation are as follows.

1. Grand Cycle top of the Donglangtan Member, Wangjiawan, East Yunnan. This marks an abrupt change from peritidal dolomites to subtidal shales of the Jiucheng Member, and the start of a new 'clearing-upwards' cycle that contains Marker A near the top. Such cycle tops can be widely correlated in the Cambrian of North America (e.g. Aitken 1981; Palmer 1981) and in the Yudomian/Vendian of the Siberian Platform (e.g. Khomentovsky 1986). Their relationship to faunal and floral occurrences deserves careful study. It is notable, however, that the first 'shelly fossil' small assemblages appear above such a cycle top in east Yunnan, Lesser Himalaya of India, northern Pakistan, northern Iran, southern Kazakhstan, and northern Mongolia (see above) and across the Siberian Platform (e.g. Khomentovsky 1986).

2. Disappearance of the soft-bodied *Charnia* fauna. A late Vendian (Kotlin) decline in size is suggested to have taken place in the Ediacarian or *Charnia* fauna (Sokolov & Fedonkin 1986). Medusoids survived at least into the lower part of the *Anabarites trisulcatus* Zone in Siberia (Khomentovsky 1986).

3. Increased preservation of macroflora, including *Chuaria* and vendotaeniid algae in the Upper Vendian (Sokolov & Fedonkin 1986). Such assemblages occur between Ediacarian macrofaunas and Zone I macrofaunas in the Yangtze region (Xing & Ding 1985; Xing *et al.* 1985), and at comparable horizons in other areas. Clastic strata, bearing large algae of the *Chuaria*, *Shouhsenia*, and *Vendotaenia* groups plus sapropelic material, occur low down in the successions of Yunnan and northern Iran, the latter occurrence being associated with early remains of *Protohertzina* sp., *Hyolithellus* sp., and molluscs. At the western limit of the Redlichiid realm in Spain, a similar algal assemblage was found associated with the first arthropod traces of *Monomorphichnus* and diminutive, complex deposit-feeding traces of 'Cambrian-type' (Brasier *et al.* 1979), although Crimes (Chapter 8) suggests a younger age.

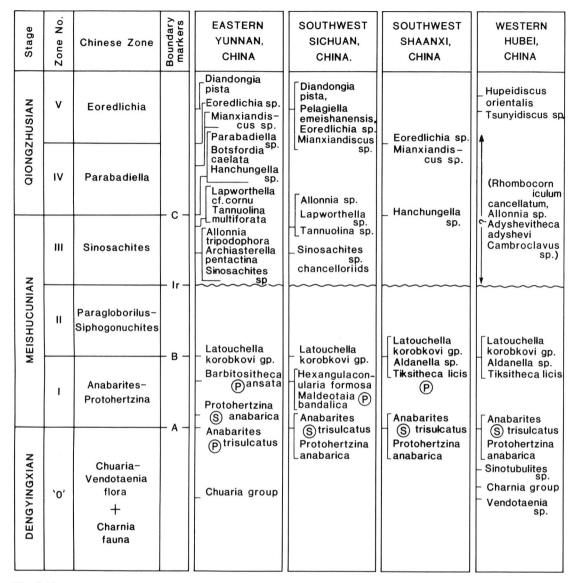

Fig. 3.12. Correlation of first-appearance datum points in the Meishucun section of China with homotaxial horizons in adjacent wavy lines represent disconformities or unconformities; dashed lines represent suspected disconformities.

4. First appearance of protoconodonts of the *Protohertzina anabarica* group, plus tubes of *Anabarites trisulcatus* and *Cambrotubulus decurvatus*. This assemblage typically appears within dolomites in the upper part of a grand cycle (beneath Zone I) and can be compared with the *Anabarites trisulcatus* Zone of Missarzhevsky (1983) and Khomentovsky (1986). It is distinctly present in northern Iran, west Anabar, and the Olenek Uplift of Siberia and may correlate with strata bearing *Sinotubulites* in the Dengying Formation of Hubei (e.g. Xing & Ding 1985). The order of appearance of taxa within this zone may prove useful, though these currently differ between localities and authors.

3.14.2. Zone I

The base of Zone I is here taken at Marker A of Meishucun. Zone I skeletal assemblages of low diversity

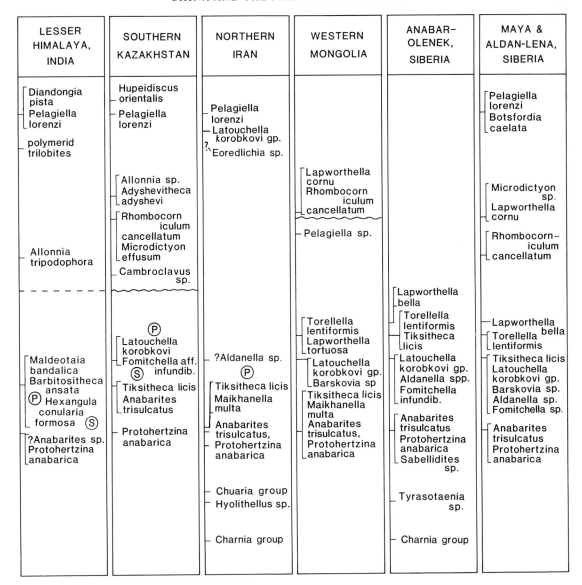

LESSER HIMALAYA, INDIA	SOUTHERN KAZAKHSTAN	NORTHERN IRAN	WESTERN MONGOLIA	ANABAR-OLENEK, SIBERIA	MAYA & ALDAN-LENA, SIBERIA
Diandongia pista Pelagiella lorenzi	Hupeidiscus orientalis Pelagiella lorenzi	Pelagiella lorenzi Latouchella korobkovi gp. ? Eoredlichia sp.			Pelagiella lorenzi Botsfordia caelata
polymerid trilobites			Lapworthella cornu Rhombocorniculum cancellatum		
	Allonnia sp. Adyshevitheca adyshevi				Microdictyon sp. Lapworthella cornu
	Rhombocorniculum cancellatum Microdictyon effusum		Pelagiella sp.		Rhombocorniculum cancellatum
Allonnia tripodophora	Cambroclavus sp.				
				Lapworthella bella	
			Torellella lentiformis Lapworthella tortuosa	Torellella lentiformis Tiksitheca licis	Lapworthella bella Torellella lentiformis
Maldeotaia bandalica Barbitositheca ansata Ⓟ Hexangula conularia formosa Ⓢ	Ⓟ Latouchella korobkovi Fomitchella aff. Ⓢ infundib. Tiksitheca licis Anabarites trisulcatus	?Aldanella sp. Ⓟ Tiksitheca licis Maikhanella multa	Latouchella korobkovi gp. Barskovia sp Tiksitheca licis Maikhanella multa Anabarites trisulcatus, Protohertzina anabarica	Latouchella korobkovi gp. Aldanella spp. Fomitchella infundib. Anabarites trisulcatus Protohertzina anabarica Sabellidites sp.	Tiksitheca licis Latouchella korobkovi gp. Barskovia sp. Aldanella sp. Fomitchella sp. Anabarites trisulcatus Protohertzina anabarica
?Anabarites sp. Protohertzina anabarica	Protohertzina anabarica	Anabarites trisulcatus, Protohertzina anabarica Chuaria group Hyolithellus sp.		Tyrasotaenia sp.	
		Charnia group		Charnia group	

regions. Modified from Brasier & P. Singh (1987). 'P' refers to major phosphogenic intervals; 'S' to major siliceous (chert) intervals;

and simple skeletal plan often appear towards the top of widespread dolomites, and usually range through phosphatic–clastic strata deposited during a major phosphogenic event. Assemblages of this type occur widely across Asia and into the Middle East (Fig. 3.12). Some events with potential for correlation are as follows.

1. First appearance of the spinose-walled ?mollusc

Cassidina pristinis–Maikhanella multa group; simple hyoliths of the 'Circotheca' or 'Conotheca' group, *Spinulitheca billingsi* group, *Turcutheca craseocochlia* group; tubes of the *Tiksitheca licis* group; *Fomitchella infundibuliformis* and other trumpet-shaped elements.

2. Cycle top of the Xaiowaitoushan Member, Meishucun, east Yunnan: an irregular, ?karstic discontinuity surface occurs between dolomites and phosphorites of

Zone I in east Yunnan and south-west Sichuan, suggesting the emergence at the top of a depositional cycle. Elsewhere across the Yangtze Platform, Lesser Himalaya of India, northern Pakistan, northern Iran, and southern Kazakhstan there is a rapid to abrupt change in facies from dolomites (often cherty and supratidal) to phosphatic limestones, dolomites, or cherts. In western Mongolia, an abrupt change is also recorded between dolomites and limestones, while significant downcutting intervened between the top of the Yudoma dolomites and limestones, and the transgressive, often argillaceous, limestones of the Lower Tommotian Pestrotsvet Formation of Siberia (Khomentovsky 1986).

3. Negative δ^{13}C isotope trend: carbon isotopes across the disconformity in two sections at Meishucun show a slight decrease (–1.8 to –1.9, and –2.3 δ^{13}C, Wu Xiche & Oyang Ling, pers. comm. 1986). It is interesting to compare these results with the carbon isotope curve for the Dvortsy section, Aldan River of Siberia (Magaritz *et al.* 1986). There, a similar trend towards negative values is seen across the Yudoma/Petrotsvet Formation boundary (i.e. the Nemakit–Daldynian/Tommotian boundary according to the concept of Khomentovsky 1986).

4. The following taxa first appear above this break at Meishucun or correlated sections: *Salanacus* sp., *Barbitositheca ansata*, *Hexangulaconularia formosa* group, *Maldeotaia bandalica–Ganloudina symmetrica* group, *Allatheca*, and related hyoliths with rounded triangular cross-sections.

5. Major phosphogenic event: this followed at some point within Zone I of the Palaeotethyan Belt, perhaps in response to a transgressive pulse and oceanic overturn of stratified waters (Cook & Shergold 1986). The phosphorites of Meishucun, Lesser Himalaya, and northern Mongolia show distinctive, strongly negative δ^{13}C values (e.g. Xu *et al.* 1985; Banerjee 1986). These may have been enhanced by diagenesis of organic matter, but the results invite comparison with a negative spike recorded from the *sunnaginicus* limestones of Dvortsy, Siberia (cf. Magaritz *et al.* 1986).

6. Rising δ^{13}C values: carbon isotope values rise from a low of c.–8 δ^{13}C in the middle of Bed 6 phosphorites at Meishucun, to about –1 δ^{13}C near its top, and continue into Bed 8 (cf. Xu *et al.* 1985). This rising trend invites comparison with that from the base to middle part of the *regularis* Zone of Dvortsy, Siberia (cf. Magaritz *et al.* 1986).

3.14.3. Zone II

Coiled, sculptured molluscs of the *Latouchella korobkovi* and *Aldanella attleborensis* groups appeared suddenly and widely at Marker B in Meishucun and correlated levels elsewhere. This event suggests invasion from a neighbouring source, perhaps during palaeogeographical changes. The fauna occurs in clastics, carbonates, and phosphorites of varying ecological and palaeoclimatic settings, and may indicate a 'bloom' of opportunistic forms. Some notable events are as follows.

1. Abrupt change in facies: at Meishucun and Maidiping, Marker B seems to coincide with the beginning of regressive sedimentation; this needs to be investigated elsewhere.

2. Mass appearance of the *Latouchella korobkovi* and *Aldanella attleborensis* groups and abrupt decline of the Zone I assemblage. The molluscan assemblage cannot be clearly recognized, however, in the more offshore sequences of Guizhou, Shaanxi, and Fangxian of Hubei, nor in northern Sichuan and the Lesser Himalyas of India. Their first appearance may therefore have depended upon shallow lagoonal facies. Neither are *Siphogonuchites triangulatus* nor *Zhijinites* sp. perfect indices, since the former occurs at a lower position in Shizhonggou, Shaanxi (and possibly in northern Iran), while the latter ranges low at Gezhongwu, Guizhou, Even so, the change from Zone I to Zone II assemblages is very sharp across much of the Yangtze Platform and suggests a migratory or ecostratigraphical event. Correlation is suggested with the *Latouchella korobkovi* group first-appearance datum of the Aldan–Lena region and western Mongolia, interpreted by some as of *regularis* Zone age (Voronin *et al.* 1982; Rozanov & Sokolov 1984).

3. Shoaling and provincialism: returning limestones and dolomites become more fenestral and cherty upwards at Meishucun. Columnar stromatolites complete the phosphogenic cycle in India (P. Kalia, pers. comm. 1986). In Maly Karatau of Kazakhstan, *Torellella* cf. *biconvexa* occurs in the upper part of a comparable lithological cycle, suggesting a range into late Tommotian or early Atdabanian times. The age of the upper part of this zone in the other sections is less certain.

Faunas above the base of Zone II are notably provincial, even across Yunnan (e.g. the curious *Scoponodus* fauna of Huize County). Slow deposition may also have resulted in condensation and hiatuses. Homotaxial correlation in this part of the column is therefore difficult.

4. Cycle top and Dahai ferruginous event: shoaling culminated in Fe–Mn or limonitic crusts and karst-like discontinuity surfaces across the Yangtze Platform (e.g. Dahai Member of Meishucun), suggesting a period of stillstand and emergence (Xiang *et al.* 1981). Fe–Mn enrichment also occurs in stromatolites at the top of the cycle in southern Kazakhstan, while the stromatolites in India are pyritic. Faunas are poor and evidence for synchroneity at the top of this cycle is needed.

3.14.3. Zone III

Faunas of Zone III appear some time after the break at the base of the Qiongzhusi Formation. A change from calcareous to clastic (often carbonaceous) sedimentation is widespread in the Inner Palaeotethyan Belt. Some important events are as follows.

1. Iridium and negative $\delta^{13}C$ event: geochemical anomalies are found in black shales of the Qiongzhusi Formation, just above the discontinuity surface in the Yangtze Gorges and Yunnan (e.g. Hsü *et al.* 1985). They are also vanadium enriched (Xiang *et al.* 1981).

2. The chancelloriids *Allonnia tripodophora* and *Archiasterella pentactina* and the tommotiid *Tannuolina* sp. first appear. Similar assemblages also occur in India and Pakistan and may be traced through the redlichiid faunal province to France and Spain. In France, *Allonnia tripodophora* first occurs below the trilobite *Bigotina* (e.g. Doré & Reid 1965) and in Spain with lower to middle Atdabanian archaeocyathans below the trilobite *Lemdadella* (e.g. Liñan 1978). *Archiasterella pentactina* first appears in Spain with trilobites of high *Callavia* Zone aspect (Sdzuy 1969). A roughly mid-Atdabanian correlation for the *Allonnia* fauna is consistent with the presence of *Rhombocorniculum cancellatum* with *Cambroclavus* spp. (and *Allonnia* at higher levels) above the ferromanganese-enriched discontinuity in Kazakhstan (Missarzhevsky & Mambetov 1981) and in correlated beds in Fangxian of Hubei (Qian & Zhang

1983), although Qian (1984) infers a Zone IV–V age for these in present terminology. It may be coincidental that the top of the *R. insolutum* beds in England (top of Home Farm Member) and comparable strata in south-east Newfoundland (near the top of the Smith Point Limestone) are also ferromanganese-enriched carbonates, at the top of a major depositional cycle.

3. First appearances of the *Lapworthella cornu* group, as in the lower part of Bed 13, Meishucun. Further taxonomic work is needed, but their appearance suggests comparison with the Upper Atdabanian of Siberia and the *Serrodiscus bellimarginatus* interval of the Avalon terrane.

3.14.4. Zone IV

Calcitic trilobite skeletons began to appear during Zones IV (with *Parabadiella* and *Mianxiandiscus*). This interval contains the brachiopod *Botsfordia caelata*, which first appears in the Lower Botomian of Siberia and the *Protolenus* fauna of the Avalon terrane. Successive events within this zone are not yet clear.

3.14.5. Zone V

Pelagic eodiscid trilobites became significant and widespread in Zone V (*Eoredlichia*) times in China (e.g. *Tsunyidiscus* and *Hupeidiscus*). Gastropods of the *Pelagiella lorenzi* group appeared widely at about this time (Sichuan, Lesser Himalaya, southern Kazakhstan, northern Iran, and Siberia), although the supposition requires more careful analysis. The brachiopod *Diandongia pista* is also reported from Yunnan, Sichuan, and similar levels in the Lesser Himalaya.

3.14.6. Conclusion

A framework of biological and geological events is therefore emerging along the Palaeotethyan Belt from China to Iran. This belt clearly has the potential to provide standard successions for fine-scale calibration of late Precambrian and early Cambrian times.

REFERENCES

Ahluwalia, A. D. & Gupta, V. J. (1984). Tal Phosphorite—viewed unhurriedly at a stratigraphic crossroad. *Bull. Ind. geol. Assoc.*, **17**, 29–38.

Aitken, J. D. (1981). Generalizations about Grand Cycles. *US geol. Surv., Open File Rep.*, **18–743**, 8–14.

Assereto, R. (1966). Geological map of upper Djadjerud and Lar valleys (Central Elburz, Iran), scale 1:50 000, with explanatory notes. *Inst. Geol. Univ. Milano, Ser. G., Publ.*, **232**, 86 pp.

Azmi, R. J. (1983). Microfauna and age of the Lower Tal Phosphorite of Mussoorie Syncline, Garwhal Lesser Himalaya, India. *Himalayan Geol.*, **11**, 373–409.

Azmi, R. J. (1985). Skepticism and discrepancies in Krol-Tal: a synthesis. *Bull. Ind. geol. Assoc.*, **18**, 29–41.

Azmi, R. J. & Joshi, M. N. (1983). Conodont and other biostratigraphic evidences on the age and evolution of the Krol Belt. *Himalayan Geol.*, **11**, 198–223.

Azmi, R. J. & Pancholi, V. P. (1983). Early Cambrian (Tommotian) conodonts and other shelly microfauna from the Upper Krol of Mussoorie Syncline, Garwhal Lesser Himalaya with remarks on the Precambrian–Cambrian boundary. *Himalayan Geol.*, **11**, 360–72.

Azmi, R. J., Joshi, M. N., & Juyal, K. P. (1981). Discovery of the Cambro-Ordovician conodonts from the Mussoorie Tal Phosphorite: its significance in correlation of the Lesser Himalaya. *Contemporary Geoscient. Res. Himalaya*, **1**, 245–50.

Banerjee, D. M. (1986). Proterozoic and Cambrian phosphorites—Regional review: Indian subcontinent. In *Proterozoic and Cambrian phosphorites*, (eds P. J. Cook & J. H. Shergold), pp. 70–90. Cambridge University Press.

Bengtson, S., Missarzhevsky, V. V., & Rozanov, A. Yu. (1984). The Precambrian–Cambrian boundary: a plea for caution. *IUGS/IGCP Working Group on Precambrian–Cambrian Boundary Circular Newsletter*, June 1984, 14, 15 (unpublished; limited circulation).

Bengtson, S., Fedorov, A. B., Missarzhevsky, V. V., Rozanov, A. Yu., Zhegallo, E. A., and Zhuravlev, A. Yu. (1986). *Tumulduria incomperta* and the case for Tommotian trilobites. *Lethaia* **20**, 361–70.

Bhatt, D. K., Mamgain, V. D., Misra, R. S., & Srivastava, J. P. (1983). Shelly microfossils of Tommotian age (Lower Cambrian) from the Chert–Phosphorite Member of Lower Tal Formation, Maldeota, Dehra Dun District, Uttar Pradesh. *Geophytology*, **13**, 116–23.

Bhatt, D. K., Mamgain, V. D., & Misra, R. S. (1985). Small shelly fossils of early Cambrian (Tommotian) age from Chert–Phosphorite Member, Tal Formation, Mussoorie Syncline, Lesser Himalaya, India and their chronostratigraphic evaluation. *J. palaeont. Soc. India*, **30**, 92–102.

Brasier, M. D. (1986a). The succession of small shelly fossils (especially conoidal microfossils) from English Precambrian–Cambrian boundary beds. *Geol. Mag.*, **123**, 237–56.

Brasier, M. D. (1986b). Precambrian–Cambrian boundary biotas and events. In *Global bio-events*, (ed. O. Walliser), Lecture Notes in Earth Sciences, **8**, pp. 109–20. Springer-Verlag, Berlin.

Brasier, M. D. (1987). 'Inner Tethyan' Precambrian–Cambrian boundary sequences from China, India, Pakistan and Iran. *Abstracts of the International Symposium on Terminal Precambrian and Cambrian Geology, Yichang*, pp. 1–2.

Brasier, M. D. (ed.) (in press). *Stratigraphy of China, Volume 4: the Cambrian System of China*. Scottish Academic Press, Edinburgh.

Brasier, M. D. & Singh, P. (1987). Microfossils and Precambrian–Cambrian boundary stratigraphy at Maldeota, Lesser Himalaya. *Geol. Mag.*, **124**, 323–45.

Brasier, M. D., Perejon, A., & de San José, M. A. (1979). Discovery of an important fossiliferous Precambrian–Cambrian sequence in Spain. *Estudios Geol.*, **35**, 379–83.

Bushinsky, G. I. (1966). Old phosphorites of Asia and their genesis. *Akad. Nauk. SSSR, Trudy Geol. Inst.*, **149**. [In Russian. English translation by Israel Programme for Scientific Translations, 266pp. Jerusalem, 1969.]

Chang Wentang (1984). Precambrian–Cambrian boundary problems. In *Developments in geoscience, Academia Sinica*, pp. 41–50. Science Press, Beijing.

Chen Meng'e, Chen Yiyuan, & Qian Yi (1981). [Some tubular fossils from Sinian–Lower Cambrian boundary sequences, Yangtze Gorges.] *Bull. Tianjin Inst. Geol. M. R., Chinese Acad. geol. Sci.*, **3**, 117–124. [In Chinese.]

Chen Ping (1984). [Discovery of Lower Cambrian small shelly fossils from Jijiapo, Yichang, West Hubei and its significance.] *Prof. Pap. Strat. Palaeont.*, **13**, 49–64. [In Chinese.]

Cook, P. J. & Shergold, J. H. (1986). Proterozoic and Cambrian phosphorites—nature and origin. In *Proterozoic and Cambrian phosphorites*, (eds P. J. Cook & J. H. Shergold), pp. 367–86. Cambridge University Press.

Cowie, J. W. (1978). IUGS/IGCP Project 29 Precambrian–Cambrian Boundary Working Group in Cambridge, 1978. *Geol. Mag.*, **115**, 151–2.

Cowie, J. W. (1984). Introduction to papers on the Precambrian–Cambrian boundary. *Geol. Mag.*, **121**, 137–8.

Cowie, J. W. (1985). Continuing work on the Precambrian–Cambrian boundary. *Episodes*, **8**, 93–7.

Cowie, J. W. & Johnson, M. R. W. (1985). Late Precambrian and Cambrian geological time-scale. In *The *chronology of the geological record*, (ed. N. J. Snelling), Mem. geol. Soc. Lond., **10**, pp. 47–64. Blackwell Scientific, Oxford.

Doré, F. & Reid, R. E. (1965). *Allonnia tripodophora* nov. gen., nov. sp., nouvelle éponge du Cambrien inférieur de Carteret (Manche). *C. R. Soc. géol. Fr.*, 1965, 20–1.

Eganov, E. A. & Sovetov, Yu. K. (1979). Karatau—a model for phosphorite deposition. *Akad. Nauk SSSR, Sib. Otdel., Trudy Inst. Geol. Geofis.*, **427**, 1–190. [In Russian.]

Ergaliev, G. K. & Pokrovskaya, N. V. (1977). *Lower Cambrian trilobites of Lesser Karatau (southern Kazakhstan).* Akademiya Nauk Kazakhskoy SSR, Ordena Trudovogo Krasnogo Zhameni Inst. Geol. Nauk K. I. Satpaeva, 98 pp. Izdat. 'Nauka' Kazakhskoy SSR, Alma-Ata. [In Russian.]

Fedorov, A. B., Egoreva, L. I., & Savitsky, V. E. (1979). The first find of ancient trilobites in the lower part of the stratotype Tommot tier of Lower Cambrian (River Aldan). *Dokl. Akad. Nauk SSSR*, **249**, (5), 1188–90. [In Russian.]

Fuchs, G. & Mostler, H. (1972). Der erste Nachweis von Fossilien (Kambrischen Alters) in der Hazira Formation, Pakistan. *Mitt. Geol. Palaeont. Gesell., Innsbruck*, **2**, 1–12.

Gansser, A. (1974). The Himalayan Tethys. *Riv. Ital. di Paleont. Stratigr. Mem.*, XIV, 393–411.

Glaessner, M. F. (1984). *The dawn of animal life.* Cambridge University Press.

Hamdi, B. & Golshani, F. (1983). Preliminary investigation on the Lower Cambrian and Cambrian–Precambrian boundary of North Iran. *Internal Report of the Tehran Geological Survey of Iran*, 1–10.

Hamdi, B., Brasier, M. D., & Jiang Zhiwen (in prep.). A succession of early skeletal microfaunas from Precambrian–Cambrian boundary strata, Elburz Mountains, Iran. *Geol. Mag.*, submitted.

Hasan, M. T. (1986). Proterozoic and Cambrian phosphorites—deposits: Hazara, Pakistan. In *Proterozoic and Cambrian phosphorites*, (eds P. J. Cook & J. H. Shergold), pp. 190–201. Cambridge University Press.

He Ting–gui (1984). [Discovery of *Lapworthella bella* assemblage from Lower Cambrian Meishucun Stage in Niuniuzhai, Leibo County, Sichuan Province.] *Prof. Pap. Stratigr. Palaeont.*, **13**, 23–34. [In Chinese.]

He Ting–gui, Pei Fang, & Fu Guanghong (1984). Some small shelly fossils from the Lower Cambrian Xinji Formation in Fangcheng County, Henan Province. *Acta Palaeont. Sinica*, **23**, 350–7.

Hsü, K. J., Oberhansli, H., Gao, J. Y., Sun Shu, Chen Haihong, & Krahenbühl, U. (1985). 'Strangelove ocean' before the Cambrian explosion. *Nature*, **316**, 809–11.

Jiang Zhiwen (1980). [The Meishucun Stage and fauna of the Jinning County, Yunnan.] *Bull. Chinese Acad. geol. Sci., Ser. I*, **2**, 75–92. [In Chinese.]

Jiang Zhiwen (1985). Evolution of shelly fossils and the end of the late Precambrian. *Precambr. Res.*, **29**, 45–52.

Khomentovsky, V. V. (1986). The Vendian System of Siberia and a standard stratigraphic scale. *Geol. Mag.*, **123**, 333–48.

Kirschvink, J. L. & Rozanov, A. Yu. (1984). Magnetostratigraphy of Lower Cambrian strata from the Siberian Platform. A palaeomagnetic pole and a preliminary polarity time-scale. *Geol. Mag.*, **121**, 189–203.

Kumar, G., Raina, B. K., Bhatt, D. K., & Jangpangi, S. (1983). Lower Cambrian body- and trace-fossils from the Tal Formation, Garwhal Synform, Uttar Pradesh, India. *J. palaeont. Soc. Ind.*, **28**, 106–11.

Kumar, G., Raina, B. K., Bhargava, O. N., Maithy, P. K., & Babu, R. (1984). The Precambrian–Cambrian boundary problem and its prospects, Northwest Himalaya, India. *Geol. Mag.*, **121**, 211–9.

Kumar, G., Bhatt, D. K., & Raina, B. K. (1987). Skeletal microfauna of Meishucunian and Qiongzhusian (Precambrian–Cambrian boundary) age from the Gauga Valley, Lesser Himalaya, India. *Geol. Mag.*, **124**, 167–71.

Landing, E. & Brett, C. E. (1982). Lower Cambrian of eastern Massachusetts: microfaunal sequence and the oldest known borings. *Geol. Soc. Am., Abstr. Prog.*, **14**, p. 33.

Liñan, E. G. (1978). *Bioestratigrafía de la Sierra de Cordoba.* Tesis Doctorales de la Universidad de Granada. Departamento de Paleontologia, Facultad de Ciencias.

Luo Huilin (1981). [Trilobites from the Chiungchussu Formation (Lower Cambrian) in Meishucun of Jinning, Yunnan Province.] *Acta Palaeont. Sinica*, **20**, 331–40. [In Chinese.]

Luo Huilin, Jiang Zhiwen, Xu Zhongjiu, Song Xueliang, & Xue Xiaofeng (1980) [On the Sinian–Cambrian boundary of Meishucun and Wangjiawan, Jinning County, Yunnan.] *Acta geol. Sinica*, **2**, 95–111. [In Chinese.]

Luo Huilin, Jiang Zhiwen, Wu Xiche, Song Xueliang, & Ouyang Lin (1982). [*The Sinian–Cambrian boundary in Eastern China.*] People's Publishing House, Beijing. [In Chinese.]

Luo Huilin *et al.*, (1984). *Sinian–Cambrian boundary stratotype section at Meishucun, Jinning, Yunnan, China.* People's Publishing House, Yunnan.

Magaritz, M., Holser, W. T., & Kirschvink, J. L. (1986). Carbon-isotope events across the Precambrian/Cambrian boundary on the Siberian Platform. *Nature*, **320**, 258–9.

Missarzhevsky, V. V. (1973). [Conodont-shaped organisms from the Precambrian and Cambrian boundary strata of the Siberian Platform and Kazakhstan. In *Palaeontological and biostratigraphical problems in the Lower Cambrian of Siberia and the Far East*, (ed. I. T. Zhuravleva)], pp. 53–7, pls 9–10. Nauka Publishing House, Novosibirsk. [In Russian.]

Missarzhevsky, V. V. (1982). [Subdivision and correlation of the Precambrian–Cambrian boundary beds using some groups of the oldest skeletal organisms.] *Byulleten' Moskovskogo Obshchestva Ispytatelei Prirody, Otdelenie Geologii*, **57**, (5), 52–67. [In Russian.]

Missarzhevsky, V. V. (1983). [Stratigraphy of oldest Phanerozoic deposits of Anabar Massif.] *Stratigr. & Palaeogeogr. Soviet Geol.*, **9**, 62–73. [In Russian.]

Missarzhevsky, V. V. & Mambetov, A. J. (1981). Stratigraphy and fauna of Cambrian and Precambrian boundary beds of Maly Karatau. *Trudy Akad. Nauk. SSSR, Moscow* **326**, pp. 1–90, pls I–XVI. [In Russian.]

Palmer, A. J. (1981). On the correlation of Grand Cycle tops. *US geol. Surv., Open File Rep.*, **81–743**, 156–9.

Parrish, J. T., Ziegler, A. M., Scotese, C. R., Humphreville, R. G., & Kirschvink, J. L. (1986). Early Cambrian palaeogeography, palaeoceanography and phosphorites. In *Proterozoic and Cambrian phosphorites*, (eds P. Cook & J. H. Shergold), pp. 280–94. Cambridge University Press.

Qian Jianxian & Xiao Bing (1984). [An early Cambrian small shelly fauna from Aksu–Wushi region, Xinjiang.] *Prof. Pap. Stratigr. Palaeont.*, **13**, 65–90. [In Chinese.]

Qian Yi (1977). [Hyolitha and some problematica from the Lower Cambrian Meishucun Stage in Central and SW China.] *Acta palaeont. Sinica*, **16**, 255–78. [In Chinese.]

Qian Yi (1983). [Sinian–Cambrian boundary in China. In *Studies on stratigraphic boundaries in China*], pp. 1–11. [In Chinese.]

Qian Yi (1984). Early Cambrian–Late Precambrian small shelly faunal assemblage with a discussion on Cambrian–Precambrian boundary in China. In *Developments in geoscience*, Academic Sinica, pp. 9–20. Science Press, Beijing.

Qian Yi & Yin Gongzheng (1984*a*). [Small shelly fossils from the lowest Cambrian in Guizhou.] *Prof. Pap. Stratigr. Palaeont.*, **13**, 91–124. [In Chinese.]

Qian Yi & Yin Gongzheng (1984*b*). [Zhijinitidae and its stratigraphical significance.] *Acta palaeont. Sinica*, **23**, 215–23. [In Chinese.]

Qian Yi & Zhang Shi-ben (1983). [Small shelly fossils from the Xihaoping Member of the Tongying Formation in Fangxian County of Hubei Province and their stratigraphical significance.] *Acta palaeont. Sinica*, **22**, 82–93. [In Chinese.]

Qian Yi, Chen Meng'e & Chen Yi-Yuan (1979). [Hyolithids and other small shelly fossils from the Lower Cambrian Huangshangdong Formation in the eastern part of the Yangtze Gorge.] *Acta palaeont. Sinica*, **18**, 207–32. [In Chinese.]

Qian Yi, Yu Wen, Liu Diyong, & Wang Zongzhe (1985). [Restudy of the Precambrian–Cambrian boundary section at Meishucun of Jinning, Yunnan.] *Kexue Tongbao*, **30**, 1086–90. [In Chinese.]

Rai, V. & Singh, I. B. (1983). Discovery of trilobite impressions in the Arenaceous Member of Tal Formation, Mussoorie area, India. *J. palaeont. Soc. India*, **28**, 114–7.

Riding, R. & Voronova, L. (1984). Assemblages of calcareous algae near the Precambrian/Cambrian boundary in Siberia and Mongolia. *Geol. Mag.*, **121**, 205–10.

Rozanov, A. Yu. (1984). The Precambrian–Cambrian Boundary in Siberia. *Episodes*, **7**, 20–4.

Rozanov, A. Yu. & Sokolov, B. S. (1982). Precambrian–Cambrian boundary: recent state of knowledge. *Precambr. Res.*, **17**, 125–31.

Rozanov, A. Yu. & Sokolov, B. S. (eds) (1984). [*Lower Cambrian stage subdivision. Stratigraphy.*] Akad. Nauk SSSR, Iztdatelstvo 'Nauka', Moscow. [In Russian.]

Rozanov, A. Yu. *et al.* (1969). [The Tommotian Stage and the Cambrian lower boundary problem.] *Trudy Geol. Inst., Nauka, Moscow*, **206**. [In Russian; English translation by US Dep. of the Interior, 1981.]

Rushton, A. W. A. (1973). Cambrian fossils from the Hazira Shale, Pakistan. *Nature, Phys. Sci.* **243**, (130), p. 142.

Sdzuy, K. (1969) Unter- und mittelkambrische Porifera (Chancelloriida und Hexactinellida). *Paläont. Z.*, **43**, 115–47.

Sdzuy, K. (1978). The Precambrian–Cambrian boundary beds in Morocco. *Geol. Mag.*, **115**, 83–94.

Seger, F. E. (1977). Zur Geologie des Nord-Alamut Gebeites (Zentral-Elburz, Iran): Eidgenossische Technische Hochschule, Zurich, Unpublished Thesis No. 6093, 161 pp., 92 figs, 3 pls.

Shah, S. K. (1982). Cambrian stratigraphy of Kashmir and its boundary problems. *Precambr. Res.*, **17**, 87–98.

Shah, S. M. J. (ed.) (1977). Stratigraphy of Pakistan. *Mem. geol. Surv. Pakistan*, **12**, 138 pp.

Shergold, J. H. & Brasier, M. D. (1986). Biochronology of Proterozoic and Cambrian phosphorites. In *Proterozoic and Cambrian phosphorites*, (eds P. J. Cook & J. H. Shergold), pp. 295–326. Cambridge University Press.

Singh, I. B. (1983). A note on the nature of the stromatolites of Krol sediments, Nainital, Kumaun Himalaya with special reference to *Conophyton*. *Geophytology*, **13**, 111–15.

Singh, I. B. & Rai, V. (1983). Fauna and biogenic structures in Krol-Tal succession (Vendian–Early Cambrian), Lesser Himalaya: their biostratigraphic and palaeoecological significance. *J. palaeont. Soc. India*, **28**, 67–90.

Singh, I. B., Shukla, V., Rai, V., & Kapoor, P. K. (1984). Ichnogenus *Skolithos* in the Tal Formation of Mussoorie area. *J. geol. Soc. India*, **25**, 102–7.

Singh, P. & Shukla, D. S. (1981). Fossils from the Lower Tal: their age and its bearing on the stratigraphy of Lesser Himalaya. *Geosci. J.*, **II**, 157–76.

Sokolov, B. S. & Fedonkin, M. A. (1984). The Vendian as the terminal system of the Precambrian. *Episodes*, **7**, 12–19.

Sokolov, B. S. & Fedonkin, M. A. (1986). Global biological events in the late Precambrian. In *Global bio-events*, (ed. O. Walliser), Lecture Notes in Earth Sci., 8, pp. 105–8. Springer-Verlag, Berlin.

Stocklin, J., Ruttner, A., & Nabavi, M. (1964). New data on the Lower Paleozoic and Precambrian of North Iran. *Rep. geol. Surv. Iran*, **1**.

Tewari, V. C. (1984). Discovery of Lower Cambrian stromatolites from Mussoorie Tal Phosphorites, India. *Current Sci.*, [*India*], **53**, 319–21.

Tewari, V. C. (1985). Some new observations on the Krol-Tal problem of Lesser Himalaya, India. *Bull. Indian geol. Assoc.*, **18**, 43–50.

Tripathi, G., Jangpangi, B. S., Bhatt, D. K., Kumar, G., & Raina, B. K. (1984). Early Cambrian brachiopods from 'Upper Tal', Mussoorie Syncline, Dehradun District, Uttar Pradesh, India. *Geophytology*, **14**, 221–7.

Voronin, Yu. I. *et al.* (1982). [*The Precambrian/Cambrian boundary in the geosynclinal areas* (*the reference section of Salany-Gol, MPR*).] Transactions of the Joint Soviet–Mongolian palaeontological expedition, **18**. Nauka Izdatelstvo, Moscow. [In Russian.]

Wang Yangeng *et al.* (1984). *The Upper* [*Precambrian and Sinian–Cambrian boundary in Guizhou.*] The People's Publishing House of Guizhou. [In Chinese.]

Wolfart, R. (1981). Lower Palaeozoic rocks of the Middle East. In *Lower Palaeozoic rocks of the Middle East, Eastern and Southern Africa and Antarctica*, (ed. C. H. Holland), pp. 5–130. John Wiley, Chichester.

Xiang Liwen, Wang Jingbin, Cheng Shoude, & Zhang Tairong (1985). [*Stratigraphy and trilobite faunas of the Cambrian in the western part of northern Tianshan, Xinjiang.*] Geol. Mem., Ser. 2, No. 4. Geological Publishing House, Beijing. [In Chinese.]

Xiang Liwen *et al.* (1981). [*Stratigraphy of China. No. 4. The Cambrian System of China.*] Geological Publishing House, Beijing. [In Chinese.]

Xiao Ligong & Zhou Benhe (1984). [Early Cambrian Hyolitha from Huainan and Huoqiu County in Anhui

Xing Yusheng & Ding Qixiu (1985). [Basic features of biota. In *Biostratigraphy of Yangtze Gorge area, Volume 1, Sinian* (eds Zhao-Ziqiang & Zheng Shusen)], pp. 21–30, 128. Geological Publishing House, Beijing. [In Chinese.]

Xing Yusheng, Ding Qixiu, Luo Hulin, He Tinggui, & Wang Yangeng (1983). [The Sinian–Cambrian boundary of China.] *Bull. Inst. Geol., Chinese Acad. geol. Sci.*, **10**, 1–206, pls 1–29. [In Chinese.]

Xing Yusheng, Ding Qixiu, Luo Hulin, He Tinggui, & Wang Yangeng (1984). The Sinian–Cambrian boundary of China and its related problems. *Geol. Mag.*, **121**, 155–70.

Xing Yusheng *et al.* (1985) *Late Precambrian palaeontology of China*. Geol. Mem., Ser. 2, No. 2. Geological Publishing House, Beijing. [In Chinese.]

Xing Yusheng, & Luo Huilin (1984). Precambrian–Cambrian boundary candidate, Meishucun, Jinning, Yunnan, China. *Geol. Mag.*, **121**, 143–54.

Xu Dao-Yi, Zhang Qin-Wen, Sun Yi-Ying, & Yan Zheng (1985). Three main mass extinctions—significant indicators of major natural divisions of geological history in the Phanerozoic. *Modern Geol.*, **9**, 1–11.

Yang Xianhe & He Tinggui (1984). [New small shelly fossils from Lower Cambrian Meishucun Stage of Nanjiang area, northern Sichuan.] *Prof. Pap. Stratigr. Palaeont.*, **13**, 35–47. [In Chinese.]

Yang Xianhe, He Yuanxiang, & Deng Shouhe (1983). [On the Sinian–Cambrian boundary and the small shelly fossil assemblages in Nanjiang area, Sichuan.] *Bull. Chengdu Inst. Geol. Min. Res., Chinese Acad. geol. Sci.*, **4,** 91–105. [In Chinese.]

Yang Zunyi, Chen Yuqi, & Wang Hongzhen (1986). *The Geology of China.* Oxford Monogr. Geol. Geophys., No. 3. Clarendon Press, Oxford.

Yin Gongzheng & Qian Yi (1986). Biogeographical divisions of earliest Cambrian small fossils in China. *Acta Paleont. Sinica*, **25,** 338–44.

Yin Jicheng, Ding Lianfang, He Tinggui, & Lin Changbao (1980*a*). [On the Sinian–Cambrian boundary in Emei County, Sichuan.] *Bull. Chinese Acad. geol. Sci., Ser. I,* **2,** 59–74. [In Chinese.]

Yin Jicheng, Ding Lianfang, He Tinggui, Li Shilin, & Shen Lijuan (1980*b*). [*The palaeontology and sedimentary environment of the Sinian System in Emei–Ganluo area, Sichuan.*] People's Publishing House, Sichuan. [In Chinese.]

Yue Zhao (1987). [The discovery of *Tannuolina* and *Lapworthella* from Lower Cambrian in Meishucun (Yunnan) and Maidiping (Sichuan) sections.] *Prof. Pap. Stratigr. Palaeont.*, **16,** 173–80. [In Chinese.]

Zhou Benhe & Xiao Ligong (1984). [Early Cambrian monoplacophorans and gastropods from Huainan and Huoqiu Counties, Anhui Province.] *Prof. Pap. Stratigr. Palaeont.*, **13,** 125–40. [In Chinese.]

Zhuravleva, I. T. (1975). [Description of the palaeontological characteristics of the Nemakit–Daldyn Horizon and its possible equivalents in the territory of the Siberian platform. In *Equivalents of the Vendian Complex in Siberia*, (ed. B. S. Sokolov and V. V. Khomentovsky)], pp. 62–100. *Trudy Inst. Geol. Geofis., Akad. Nauk, SSSR, Moscow*, **232,** 240 pp. [In Russian.]

4

Siberia and Eastern Europe

J. W. Cowie

4.1. SIBERIA

In the period from 1950 to the present day, the vast Siberian sector of the USSR (Fig. 4.1) has seen substantial developments in research on the rocks and fossils which occur below the earliest known trilobites. The main descriptive regional stratigraphy and palaeobiology was carried out by many eminent Soviet workers in the 1950s and 1960s, but the research has continued into the 1980s, with important contributions up to the present. An early paper on the topic, in English, was written by Rozanov (1967), but the basic reference in this field is the volume by Rozanov *et al.*

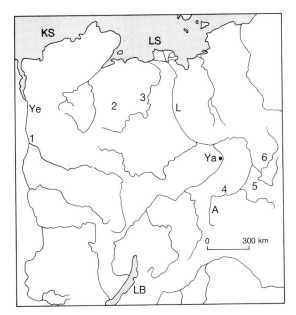

Fig. 4.1. East Siberian localities: KS, Kara Sea; LS, Laptev Sea; Ye—Yenisei River; L, Lena River; Ya, Yakutsk; A—Aldan River; LB—Lake Baikal; 1, Sukharikha River: Turakhan Uplift; 2, Anabar Massif; 3, Olenek Uplift; 4, Dvortsy–Ulakhan-Sulugur localities; 5, Uchur–Maya district; 6, Yudoma–Maya district.

(1969): *The Tommotian Stage and the Cambrian lower boundary problem*, which was translated into English in its entirety in 1981, with funds from the USA. The chapter by Voronova & Rozanov (1969, pp. 1–13) on the history of the Cambrian and Palaeozoic lower boundary summarizes the Soviet work quite well. The bibliography is now out of date, but even in 1969 comprised about 30 pages (approximately 400 entries), and, although it has a global range, it also shows that the Soviet literature is very extensive. Some of this information is difficult to obtain outside the USSR. Amongst the most dynamic and productive workers in this field are and were A. Yu. Rozanov, B. S. Sokolov, V. V. Khomentovsky, and the late V. E. Savitsky. The geological world is greatly indebted also to the rest of the large cadre of dedicated Soviet geologists who have forwarded research on the Precambrian–Cambrian boundary in the USSR and globally.

In this brief account it is impossible to give more than the main outlines of the current state of research, as known to the author, while directing attention to the key publications available, especially those in English; see Chapter 7 for biostratigraphical interpretations.

The most recent review summary in English—Rozanov (1984), divides the Siberian Precambrian–Cambrian boundary outcrops (Fig. 4.2) between

1. the South-eastern Siberian Platform;

2. the Northern Platform of Siberia.

4.1.1. South-eastern Siberian Platform

At Ulakhan–Sulugur on the Aldan River the lithostratigraphy is as follows (Fig. 4.2: oldest first; see Rozanov 1984, pp. 22–3).

Yudoma Formation (predominantly dolomites)

Bed 1. Yellowish and light-grey, find-grained, laminated dolomites with lenses of oncolitic limestones. Thickness 1.2 m.

Bed 2. Medium-grained, laminated, light-grey dolo-

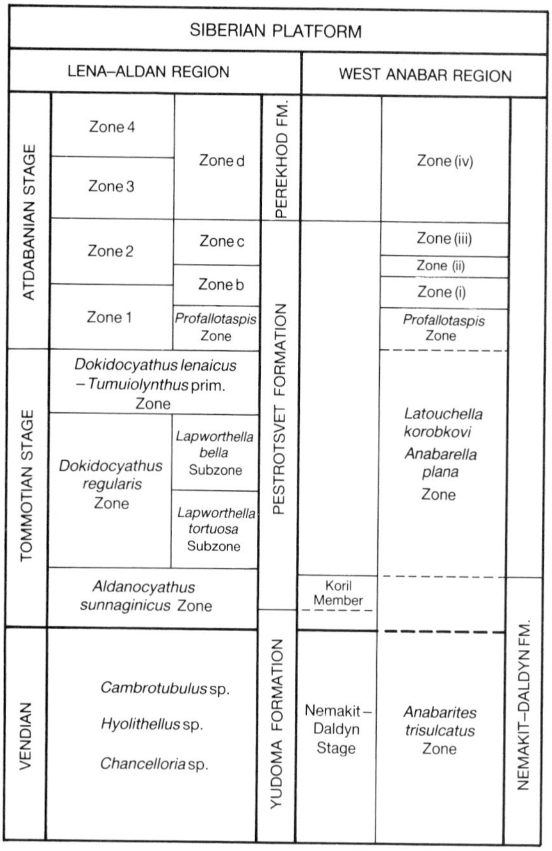

Fig. 4.2. Correlation of the Precambrian–Cambrian boundary deposits of the southern and northern Siberian Platform (after Rozanov 1984).

mites, brecciated in the lower half, oolitic in the upper part. Thickness 0.85 m.

Bed 3. Fine-grained, light-grey oolitic dolomites with a thin layer of brecciated dolomite in the upper part. Thickness 0.9 m.

Bed 4. Fine-grained, thick- to thin-bedded light-grey dolomites. Thickness 0.8 m.

Bed 5. Yellowish-grey dolomitic breccias, with oolites. Thickness 0.3 m.

Bed 6. Dolomite. Thickness 0.7 m.

Bed 7. Dolomitic breccia with oolites and/or oncolites. Thickness 0.3 m.

In these Beds 1–7, fossils are occasionally found e.g. *Chancelloria*, probably *Cambrotubulus*, and the microphytolite *Nubecularites*. In the type section of the Tommotian Stage at Dvortsy, some kilometres upstream from Ulakhan–Sulugur, older beds, 20 m below the putative equivalent of Bed 8, yield *Chancelloria*, *Hyolithellus*, and *Cambrotubulus*. The base of Bed 8 is proposed by Rozanov to define the base of the Cambrian System, although there are indications at that horizon of a disconformity or at least a considerable diastem. The top of Bed 7 is overlain by Bed 8, with traces of slight reworking. Bed 8 is a laterally impersistent layer of glauconitic sandstones with an admixture of quartzose material. Abundant fossils are found here for the first time in this succession (working upwards). Bed 8 is 0.1 m thick.

Bed 9. Light-grey laminated dolomites and sandy brecciated dolomites. Thickness 0.7 m.

Bed 10. Light-grey, coarse-bedded, sugary dolomites and sandy brecciated dolomites with *Turcutheca*, fragments of brachiopods, and *Nubecularites*. Thickness 0.4 m.

Bed 11. Light-grey, sugary, cross-bedded dolomites, intercalated with thin layers of sandy and brecciated dolomites. Thickness 0.3 m.

Pestrotsvet Formation (predominantly limestones)

Bed 12. Grey limestones, sometimes tinged with green or pink, highly glauconitic. At the base of the bed there is much clastic material and the top of the underlying Yudoma Formation is uneven and pocketed. Small bioherms with archaeocyathans occur throughout the entire bed. Fossils are abundant, with archaeocyathans, hyoliths, molluscs, and skeletal problematica. Thickness 3.5 m.

Bed 13. Argillaceous red and cherry-red, bedded limestones with an admixture of glauconite in the lower part of the bed. Abundant fossils include hyoliths, molluscs, and skeletal problematica. Thickness 17 m.

Bed 14. Red argillaceous bedded limestones, with rare thin layers of grey and greenish grey limestones. Numerous bioherms with archaeocyathans occur throughout the entire bed.

The Aldan River sections have been most intensively studied over more than twenty years, and there are excellent cliff sections which are accessible up gullies and near the river shores. The sections extend over a hundred kilometres in flat-lying, near-horizontal bedded strata. The section at Ulakhan–Sulugur is at present adopted by the USSR National Stratigraphical Committee as the Precambrian–Cambrian boundary Type for the USSR, and as a candidate for the Global Stratotype Section and Point (GSSP). The Soviet choice is based on:

(1) the priority discovery of the rich upper Yudoma Formation fauna there;

(2) the lowest part of the outcrop of the putative Precambrian–Cambrian boundary beds can be observed there continuously over a lateral distance of 1300 m.

4.1.2. Northern Platform of Siberia

This is a term which includes (Fig 4.1) the Anabar Massif, the Olenek Uplift, and the Igarka–Norilsk dislocation along the Sukharika River (Turakhan Uplift). Sections in these relatively remote and inaccessible regions have been considered as locations for a possible Precambrian–Cambrian (PЄ–Є) stratotype (Rozanov 1984, p. 21), especially in Anabar, because it is here that the type section of the Nemakit–Daldyn Stage is located (Fig. 4.3). Arguments in favour of this northern platform region as a site for a candidate PЄ–Є stratotype section arise partly because the lithological sequence is more monofacial than in the south-eastern platform (briefly described above) and probably has a rich skeletalized fauna older than the faunas of the Aldan region and of pre-Tommotian age.

Rozanov (1984, pp. 21–2) discussed problems which have been current since the 1960s and do not, as yet, seem to have been resolved to general satisfaction. These discussions are listed below:

1. The sections on the Sukharika River are of special interest because they contain zones of archaeocyathans similar to those in the Lena River–Aldan River region, and therefore an analogous succession of zones can be established. Transitional beds from the Nemakit–Daldyn Stage to the Tommotian Stage are only slightly dolomitized. Rozanov (1984, p. 21) suggests that the Sukharika River successions enable an analogy to be drawn between the limestones there and the dolomites of the Aldan River, with correlation being possible over the approximately 1400 km distance between the two regions of Siberia.

2. The Olenek Uplift (Fig. 4.1) exposes the Precambrian to Cambrian transition in a terrigenous–carbonate facies, with faunas readily establishing correlation with the lower zone of the Tommotian Stage. These rocks also contain acritarchs typical of the Lontova level of the East European Platform (Sokolov & Fedonkin 1984) and lower down, Kotlin acritarchs, vendotaeniids, and a rich Vendian–Ediacarian fauna have been discovered (Rozanov & Sokolov 1982). It is to be hoped that this region, with its rich Precambrian–Cambrian fossils correlating with the Aldan River section and the East European Platform (intracontinental correlations) can be further investigated, and in the future may be visited by non-Soviet geologists. It has been inaccessible to the PЄ–Є Working Group to-date and it is, for various reasons, difficult to reach.

3. Long-term controversy regarding the correlation and stratigraphical schemes for the Siberian Precambrian–Cambrian boundary successions is still causing debate. Rozanov (1984, p. 21–2) introduces a short summary of current discussion but does not, perforce, include work from the second half of the 1980s which, it is hoped, will emerge for a global audience in 1988. Areas of particular interest may be the Uchur River–Maya River and Yudoma–Maya districts (Fig. 4.1) and other areas currently being investigated.

As mentioned in Chapter 1 of this book, the International Working Group on the Precambrian--Cambrian Boundary of IUGS has twice visited the Aldan–
Lena outcrops in Yakutia, east Siberia: in 1973 (Cowie & Rozanov 1974) with a large party (28 foreign and 60 Soviet geologists), and in 1981 (Cowie & Rozanov 1983) with a much smaller group (4 foreign and 11 Soviet geologists).

In 1973, discussions were held in the field area (Cowie & Rozanov 1974) which are in many cases still relevant to consideration of a GSSP candidature in Siberia; a few brief abstracts are given here:

1. '. . . the base of the Pestrotsvet . . . slightly disconformable contact on the underlying Yudoma . . . suggests a time gap of unknown length', op cit. p. 238.

2. '. . . a thin stratum (Bed 8) of glauconite–carbonate sandstones which is impersistent laterally, rests with slight washout on brecciated dolomites and contains fragments of archaeocyathids, hyolithids, hyolithelminths and other fossils. Investigations and discussion developed the opinion by many that Bed 8 accumulated at the time of deposition of Bed 12 by downward movement of glauconitic sediment and fossil fragments along solution enlarged fractures and bedding joints in

already lithified dolomite of the Yudoma Formation. Connecting fissure-fillings joining up the two beds were convincingly pointed out. Others believed, however, that the movement was in the reverse direction, from Bed 8 upwards to Bed 12'. op cit., p. 239.

3. 'The boundary suggested ... the base of the *Aldanocyathus sunnaginicus–Tiksitheca licis* zone, on the Aldan at Ulakhan–Sulugur, seems to be associated there with facies changes, unconformities of unknown time gap (possibly small) and is presumably fundamentally based on the first known occurrence of shelly fossils'. op cit., p. 249.

These 1973 discussions were considered again at length during the visit to the Aldan River in 1981 (Cowie & Rozanov 1983). Fieldwork in 1981 was on *Dvortsy* which displays all but the lowest 10 m of the Yudoma Formation, the whole of the Pestrotsvet formation for which it is the type section and part of the overlying Tumuldur Formation; *Byukteleekh* which assists correlation between Dvortsy and *Ulakhan–Sulugur*. These three critical outcrops each have a horizontal strike length of 2–3 km with almost complete exposure. A number of points emerged from discussions (Cowie & Rozanov 1983):

1. Sediment 'leakage', disconformities, and facies changes mentioned in 1973 with strong discussion of Bed 8 (Cowie & Rozanov 1974) with alternatives:

(a) the origin of Bed 8 and its fauna was *in situ* between Beds 7 and 9;

(b) that stratigraphical leakage had brought 'Bed 8' down along 'neptunean dykes' and karst cavities from its original *in situ* position with Bed 12 above.

For several years after 1973 intensive and detailed research on a large scale (exceptional for any region) was carried out. The data from the study of glauconite and sedimentary structures showed that (a) above was probably correct.

Between Bed 8 and Bed 12 in the matrix of the bedrock are found fragments of brachiopods, chancellorids, hyolithelminthes and hyoliths which can be seen in thin-section, and these fossils represent a Tommotian fauna, albeit somewhat impoverished. The sedimentary types and structures from the lower part of the Yudoma Formation (180 m thick) display a similarity to bedding types seen near the base of the Tommotian Stage which lies in the uppermost Bed 11 (1.5 m thick) of the Yudoma Formation. Glauconite enrichment is notable, however, from Bed 8 upwards. The green colour of Bed 8

makes it a good marker horizon throughout the 1.5 km length of cliff sections of Ulakhan–Sulugur. The fissure material seen between Beds 8 and 12 is more fine grained and a different shade of green from Beds 8 and 12. All members of the investigating team agreed, *in 1981*, that the first position (a) above was correct and there is now little or no support for stratigraphical leakage as in (b) above. If Bed 8 is actually *in situ* then the interruption between the Yudoma Formation and Pestrotsvet Formation could be very short and located within the *sunnaginicus* Zone.

2. Everywhere that the P$\in$–$\in$ boundary is studied, there are difficulties with facies, and nowhere in the world has a location with a completely monofacial boundary succession been discovered at this level. In the Aldan region there is a change of facies near the suggested boundary point, but the change may not be radical. It is from shallower water (?intertidal) to shallow water (near the tidal limit) with carbonate (dolomite and limestone) sedimentation in both cases. The palaeontological change is from algal stromatolitic biofacies to algal–archaeocyathan biofacies, but the first appearance of rich Phanerozoic-type faunal assemblages occurs first in the algal stromatolitic biofacies.

3. Within the Tommotian Stage, zones and horizons are only correlatable with the Siberian Platform. The zonation of the Tommotian Stage cannot be recognized elsewhere. The zones are based on archaeocyathan and small shelly fossils. Tommotian archaeocyathans are present only in the Siberian Platform, and zonation according to small shelly fossils is different in each region where they are found (e.g. China; People's Republic of Mongolia; Karatau in Kazakhstan, USSR; and the East European/Russian Platform) and are not zonally correlatable with Siberia but only by common assignment to the Tommotian Stage. Therefore international–inter-continental correlations demand the use of the base of the Tommotian Stage as a feasible and practicable boundary level.

4. Everywhere that there are changes of assemblages in the Phanerozoic Eon, whether general or specific (as in the Pestrotsvet Formation), there are usually changes in facies in the broadest meaning of the term. It is necessary, therefore, to assess how great an effect the change of facies has on the change of assemblages, although it is impossible to assess this completely. Change of lithology is only one feature of changes of facies. Facies changes can be, or may not be, recorded in the lithology. The same lithology may involve different facies aspects, while different lithologies may have more-

or-less the same biofacies. For example, salinity may remain the same, even though there are changes in the sea-floor sedimentary environment, so that a group of planktonic organisms can still prosper at the surface of the sea without detectable change. The downward sedimentation of this planktonic group will continue unabated on to the changed sea-floor lithofacies. In such cases, only changes of salinity or temperature can be of sufficient importance to generate a change in the planktonic element of the final lithology.

Kirschvink & Rozanov (1979, 1984) have co-operated to collect and interpret magnetostratigraphical data from the Aldan River and the Lena River Precambrian–Cambrian sections. Correlation by remanent magnetic techniques has been established over the approximately 200 kilometre distance between the outcrops at a number of stratigraphical levels (Nemakit–Daldyn, Tommotian, and Atdabanian Stages). Further work here and elsewhere is approaching publication.

The base of Bed 8 at Ulakhan–Sulugur (GSSP candidate) has shown an increased percentage of iridium (Nazarov *et al.* 1983). Like similar geophysical-geochemical results (e.g. carbon isotopes or iridium) from other stratigraphical levels in Precambrian–Cambrian successions in Morocco and China, the data are still too few and far between (in time and space) to postulate the global synchronous events which are hoped for by stratigraphers.

In a plenary session of the Working Group on the Precambrian–Cambrian boundary in Bristol, UK, in May 1983, a majority of all members presented voted to put the Ulakhan–Sulugur stratotype candidate to a postal ballot of all the Voting Members. The result of the postal vote was a rejection of the Siberian candidate by 9 against, 7 in favour, with 3 abstentions. The reasons given by members were that:

(a) the section was too inaccessible;

(b) the section was unsuitable for isotopic dating;

(c) the section was neither continuous nor monofacial;

(d) the section has disconformities and facies changes close to the candidate boundary point horizon;

(e) doubts are raised that the succession of the candidate stratotype on the Aldan River is complete. Horizons of skeletalized faunal assemblages of earliest Cambrian age may be missing here due to the disconformities and facies change.

4.1.3. Accessibility of Ulakhan–Sulugur and the Aldan River

Scheduled routine flights by the Soviet airline 'Aeroflot' are readily available from Moscow to Yakutsk at a low cost by world standards. There are no special restrictions concerning visas.

From Yakutsk to the Aldan River is best travelled by helicopter with a two-hour flight at a reasonable cost (assuming that twenty persons share). Flights are possible by fixed-wing aircraft to Chagda town or Aldan town at small cost. It is necessary to use either a rigid boat on the Aldan River (USSR Academy of Sciences can provide boats) or rubber boats. There are plenty of good camp-sites with water. Fuel oil can be obtained from barges moored locally on the Aldan and navigation signs, markers, and buoys are very plentiful, indicating the constantly surveyed, dredged, and monitored channels used by the regular river traffic.

In summer it is possible to use the road from Yakutsk to Tommot town (500 km distance). A railway is now being built towards the Soviet Far East, terminating at the coast of the Sea of Okhotsk, and when completed could be used to reach Tommot on the Aldan River (about 600 km upstream from Ulakhan–Sulugur). The best time for fieldwork is from the end of June to the beginning of August when the days are warm to hot (25–32°C) and the nights are cool (5–10°C). In Yakutsk, hotel accommodation is available, while in the Aldan River region a tented field-camp or a river boat with bunks should be used. Conservation is assured in this remote region with no industrialization or urban pressure.

4.2. EAST EUROPEAN PLATFORM

This unit is taken here to include the Russian Platform which is recognized from Western Podolia (Ukraine) in the south to the White Sea (Arctic coast) in the north and the Precambrian–Cambrian strata which are also known in Poland and the Baltic states of the USSR. This eastern European region is the type area of the Vendian ('terminal system of the Precambrian') which is the oldest stratigraphical division of the sedimentary cover to the crystalline basement and Proterozoic aulacogens immediately below the Cambrian System.

Deep drilling has played a very important part in studies, but exposures at the surface are important in addition to borehole cores (Sokolov & Fedonkin 1984). A long-term programme has been carried out by a Polish–Soviet Working Group, resulting in many publications on the late Precambrian and Lower Cambrian strata and fossils (Rozanov & Sokolov 1982; Fedonkin et al. 1983).

The Vendian of the Russian Platform is terminated by the Rovnian stage (or Rovno horizon) (Fig. 4.3) characterized by acritarchs, sabellitids, rare *Platysolenites*, *Aldanella*, and, most importantly in the context of present interest, trace fossils of 'Palaeozoic' type: *Phycodes*, *Treptichnus*, and *Gyrolithes*. Soviet geologists compare the Rovno horizon with the Nemakit–Daldyn horizon (Nemakit–Daldynian) on the grounds of stratigraphical position and palaeobiology.

The earliest Cambrian deposits are the Zbrutch Formation in the Ukraine (south-west slope) and the Lontova Horizon in the Moscow and Baltic (south-east slope) region which are equated with the Tommotian Stage. These strata contain *Platysolenites*, *Yanichevskyites*, *Sabellidites*, *Aldanella*, *Onuphionella*, and abundant trace fossils, metaphyta, and abundant acritarchs (Fig. 4.3).

The East European Platform Precambrian–Cambrian boundary successions are therefore of great importance for providing good correlation with regions elsewhere which also have abundant trace fossils at this level. The region is one of terrigenous sedimentation, and the lithological characteristics and included succession of fossils are important in global comparisons to deduce whether changes were due to:

(1) evolution of the biosphere; and/or

(2) environmental changes.

In the north-western part of the East European Platform (Mens & Pirrus 1986), the Lontova Stage is widely distributed (except in the far north) and contains soft-bodied metazoa, platysolenitids, hyoliths?, gastropods, vendotaenids, and acritarchs. Mens & Pirrus (1986) place the Precambrian–Cambrian boundary at a lower level, at the base of the Baltic Series (Rovno *and* Lontova) on the basis of their estimate of the importance of certain changes in the organic world, than Sokolov & Fedonkin (1984) who select the base of the Lontova as their PЄ–Є boundary. In this Mens & Pirrus gain support also from Makhnach et al. (1986).

In the Lublin area of south-east Poland the succession shows sedimentary continuity from Vendian to Cambrian strata environmentally interpreted as the tidal zone of a shallow epicontinental sea. The Upper Vendian Lublin Formation (*Vendotaenia* Zone) is overlain by the Lower Cambrian Mazowsze (Masovian) Formation (divided into a *Sabellidites* lower zone and a *Platysolenites* upper zone). The Vendian–Cambrian boundary, palaeobiologically based, is placed at the boundary between the *Vendotaenia* and *Sabellidites* zones (Lendzion 1986)—a further variation from the opinion of Sokolov & Fendonkin (1984).

Although there is disagreement among workers on the East European Platform Precambrian–Cambrian boundary strata as to the best stratigraphical level to choose, there is no doubt that this great region will play an important part in future discussions, especially in contributions concerning the biostratigraphical role of trace fossils.

SYSTEM	STAGE	UKRAINIAN SHIELD (SW SLOPE) PODOLIA – VOLHYN		MOSCOW SYNECLISE BALTIC SHIELD (SE SLOPE)	
CAMBRIAN	TOMMOTIAN	BALTIC SERIES	Zbrutch Formation	BALTIC STAGE / BALTIC SERIES	Lontova Horizon
VENDIAN	ROVNIAN = NEMAKIT –DALDYNIAN		Khmeinitski Formation		Rovno Horizon
VENDIAN	KOTLINIAN = POVAROVIAN	KANILOV SERIES	Four formations	VALDAY SUPERSERIES / KOTLIN = POVARONO SERIES	Reshma Formation
VENDIAN	KOTLINIAN = POVAROVIAN	KANILOV SERIES	Four formations	VALDAY SUPERSERIES / KOTLIN = POVARONO SERIES	Lyubim Formation

Fig. 4.3. Correlation of the Precambrian–Cambrian deposits of the East European Platform (Russian Platform) (after Sokolov & Fedonkin 1984).

REFERENCES

Cowie, J. W. & Rozanov, A. Yu. (1974). IUGS Precambrian–Cambrian Boundary Working Group in Siberia. *Geol. Mag.*, **111**, (3), 237–52.

Cowie, J. W. & Rozanov, A. Yu. (1983). Precambrian–Cambrian boundary candidate, Aldan River, Yakutia, U.S.S.R. *Geol. Mag.*, **120**, (2), 129–39.

Fedonkin, M. A. *et al.* (1983). *Upper Precambrian and Cambrian palaeontology of the East-European Platform.* Wydawnictwa Geologiczne, Warszawa, 1–158.

Kirschvink, J. L. & Rozanov, A. Yu. (1979). Palaeomagnetism of fossiliferous Lower Cambrian sediments: new results from the Tommotian stage of the Siberian platform. *Abstracts of IUGG interdisciplinary symposia.* December 1979, Canberra, Australia No. 17, p. 508.

Kirschvink, J. L. & Rozanov, A. Yu. (1984). Magnetostratigraphy of Lower Cambrian strata from the Siberian Platform, a palaeomagnetic pole and a preliminary polarity time scale. *Geol. Mag.*, **121**, 189–203.

Lendzion, K. (1986). Sedimentation of the Vendian–Cambrian marine sequence, Poland. *Geol. Mag.* **123**, (4), 361–5.

Makhnach, A. S., Veretennikov, N. V., & Shkuratov, V. I. (1986). Vendian rocks of the western part of the East European Platform: stratotypic range, boundaries and principles of their establishment. *Geol. Mag.*, **123**, (4), 349–56.

Mens, K. & Pirrus, E. (1986). Stratigraphical characteristics and development of Vendian–Cambrian boundary beds on the East European Platform. *Geol. Mag.*, **123**, (4), 357–60.

Nazarov, M. A., Barsukova, L. D., Kolesov, G. M., & Alekseyev, A. S. (1983). Iridium abundances in the Precambrian–Cambrian boundary deposits and sedimentary rocks of the Russian Platform. *Abstr. 14th Lunar Planet. Sci. Conf.* pt. 2, 546–47.

Rozanov, A. Yu. (1967). The Cambrian lower boundary problem. *Geol. Mag.*, **104**, (5), 415–34.

Rozanov, A. Yu. (1984). The Precambrian–Cambrian boundary in Siberia. *Episodes*, **7**, (1), 20–24.

Rozanov, A. Yu. & Sokolov, B. S. (1982). Precambrian–Cambrian boundary: recent state of knowledge. *Precambr. Res.* **17**, (2), 125–31.

Rozanov, A. Yu. *et al.* (1969). *The Tommotian Stage and the Cambrian lower boundary problem.* pp. 1–380. Trudy Geol. Inst. Akad. Nauk SSSR. 206. [English edition, 1981, Amerind Publishing Co., New Delhi.]

Sokolov, B. S. & Fedonkin, M. A. (1984). The Vendian as the terminal system of the Precambrian. *Episodes*, **7**, (1), 12–19.

Voronova, L. & Rozanov, A. Yu. (1969). The Cambrian lower boundary problem. In *The Tommotian Stage and the Cambrian lower boundary problem* (Rozanov *et al.*), pp. 5–18. Trudy Geol. Inst. Akad. Nauk SSSR, 206. [English edition, 1981, Amerind Publishing Co., New Delhi.]

5

Sections in England and their correlation

M. D. Brasier

5.1. INTRODUCTION

English sections spanning the Precambrian–Cambrian boundary are richly fossiliferous, providing successions of considerable interest, for several reasons:

1. They are of historical importance, being close to the original 'type area' of the Cambrian System of Sedgwick & Murchison (1835), while the palaeontological nature of the Nuneaton section was first outlined by Lapworth (1882). The work of Cobbold (e.g. 1919, 1921) is also seminal.

2. The first-described and demonstrably Precambrian *Charnia* fauna (Ford 1958) comes from this region, at Charnwood Forest, Leicestershire, beneath fossiliferous Cambrian strata.

3. The late Precambrian strata are intruded by igneous events that provide isotopic dates (e.g. Cribb 1975; Odin *et al.* 1983, 1985), though their 'young' ages suggest that a cautious approach should be adopted for the present (e.g. Cowie & Johnson 1985).

4. Overlying Cambrian rocks are richly fossiliferous at several levels, notably in the older *Hyolithes* Limestone of Nuneaton in Warwickshire and the younger Comley Limestone of Comley and Rushton, Shropshire. These faunas are commonly cited in description and correlation of Tommotian and Meishucunian assemblages, while the region provides the type localities for important taxa such as *Rhombocorniculum* Walliser 1958, *Latouchella* Cobbold 1921, and *Lapworthella* Cobbold, 1921.

5. The Nuneaton section has the most complete succession of Cambrian faunas in the British Isles (e.g. Rushton 1974) and is one of the most fossiliferous in Europe.

6. The lithological succession of Charnwood–Nuneaton compares closely with that of the Bonavista and Avalon peninsulas, south-east Newfoundland.

7. Most of the faunal elements are pandemic, 'outer detrital belt' forms with good potential for international correlation. The assemblage shares many taxa with maritime Canada, Balto-Scandinavia, and the Siberian Platform.

8. The successions are readily accessible, close to motorways, and within a few hours' drive of London or Manchester airports.

These are points in its favour, but there are problems too. Exposures are ephemeral, restricted by the whims of quarrying activity and official protection. Also, much of the data reside in a few exposures of highly condensed limestones. Although no strong case has been made for placing the Precambrian–Cambrian boundary stratotype in England, the successions provide important and independent evidence for the sequence of faunal and geological events, and they could provide both auxiliary stratotypes and zonal stratotypes in the future.

5.2. GEOLOGICAL SETTING

Precambrian igneous rocks are unconformably overlain by Lower Cambrian sediments in inliers around the margins of the English Midlands (Fig. 5.1) at Malvern (Worcestershire), Comley to the Wrekin and the Rushton district, (Shropshire), Lickey (Warwickshire), and Nuneaton to Charnwood (Warwickshire to Leicestershire, the latter beyond the top right hand edge of Fig. 5.1). The general lithology, palaeontology, and stratigraphy of English and Welsh Cambrian rocks have been reviewed by Cowie *et al.* (1972) and Rushton (1974), while the palaeogeographical setting and facies development were reviewed by Brasier (1980). In each case, the sequence consists of late Precambrian igneous rocks, usually including lavas, tuffs, and intrusives, unconformably overlain by shallow-water sandstones, becoming more glauconitic and phosphatic upwards. These transgressive sedimentary cycles culminate in thin, reddish pink nodular limestones at Nuneaton (the

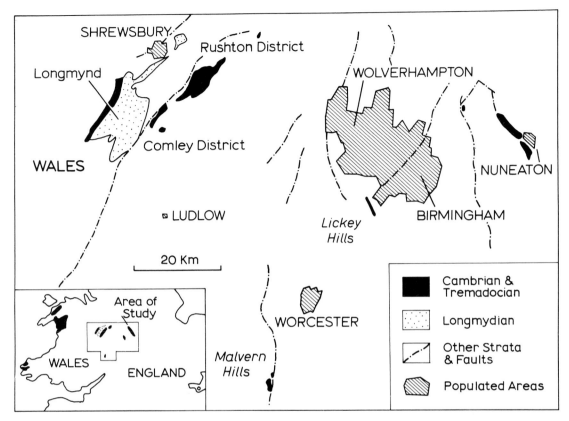

Fig. 5.1. Localities with late Precambrian and Cambrian strata at the surface in the English Midlands and the Welsh Borderland (after Brasier & Hewitt 1979).

Hyolithes Limestone, which is the upper part of the Home Farm Member), at Lickey near Birmingham (where the *Hyolithes* Limestone was later removed into Permian breccias), and in Shropshire (the Comley Limestone). Phosphorite conglomerates, authigenic glauconite, carbonate hardgrounds, Fe–Mn crusts, and a rapid succession of faunal assemblages show that these limestones are the condensed equivalents of thicker clastic sequences elsewhere (Brasier *et al.* 1978; Brasier & Hewitt 1979, 1981; Brasier 1980; Northolt & Brasier 1986; Shergold & Brasier 1986; Brasier 1986).

The setting seems to have been one of late Precambrian island arcs (e.g. Thorpe *et al.* 1984) marginal to continental blocks that were later incorporated into Gondwanaland, stretching from Massachusetts to the English Midlands (e.g. Fig. 5.2 I–L, Brasier 1980). Volcanicity, intrusion, and deformation came to a climax before the main Cambrian transgressions, when the island arcs were effectively converted into an 'Avalon

Platform' terrane, locally known as the English Midland Platform. Differential faulting of the essentially igneous basement had an important control over sedimentation across the English part of the Avalon Platform (e.g. Fig. 5.3), as suggested by the proximity of outcrops to major faults and the near-reciprocation of thickness changes between eastern and western margins of the English Midland Platform. More profound faults may have controlled development of the Welsh Basin to the west, which received a thick and largely unfossiliferous accumulation of Lower Cambrian sediments (for a fuller account of these see Rushton 1974).

The palaeontology and stratigraphy of the Comley Limestone of Shropshire owes much to the painstaking early work of E. S. Cobbold (e.g. 1921; Cobbold & Pocock 1934). Further micropalaeontology of the phosphatic fossils by Walliser (1958), Matthews (1973), Matthews & Missarzhevsky (1975), and Hinz (1983, 1987) has added considerably to our knowledge.

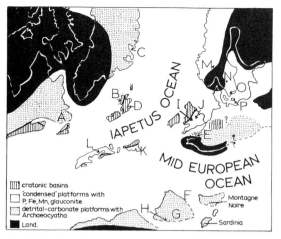

Fig. 5.2. The palaeogeographical reconstruction for English Precambrian–Cambrian boundary strata suggested by Brasier (1980). The Avalon Platform (L to J), English sections (I and J), adjacent to Baltic Platform (M–P), separated from North America (A–D) and Gondwana (F–H).

Cobbold (1919) also made the first important description of the *Hyolithes* Limestone fauna at Nuneaton, and listed taxa found in derived *Hyolithes* Limestone clasts of the Nechells Breccia (Cobbold *in* Boulton 1924). Preliminary information on phosphatic microfossils at Nuneaton was added by Missarzhevsky (*in* Cowie *et al.* 1972) and Matthews & Missarzhevsky (1975). The stratigraphical succession of fossils and sediments was later worked out by Brasier *et al.* (1978), followed by work on trace fossils and facies analysis (Brasier & Hewitt 1979), and then by an outline scheme for the faunal succession (Brasier & Hewitt 1981). More recently there have been systematic studies of small shelly fossils, including molluscs, from the *Hyolithes* Limestone (Brasier 1984), and of conoidal phosphatic microfossils from this and the Comley Limestone (Brasier 1986). The latter work showed successions of species of small shelly fossils which, taken with other evidence, confirms a similar sequence of faunal events in Newfoundland, England, and Siberia.

These two key areas of Charnwood–Nuneaton and Shropshire will now be discussed in further detail.

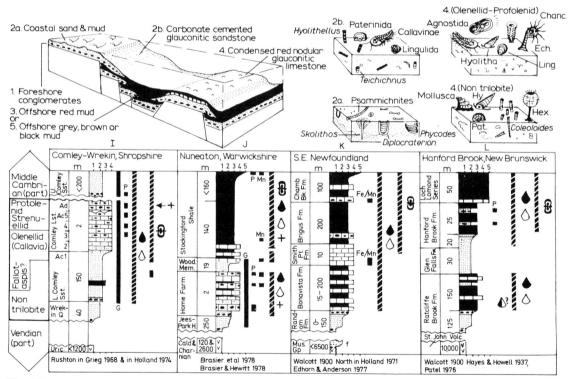

Fig. 5.3. Lower to Middle Cambrian facies models for the Avalon Platform (after Brasier 1980 where references are fully cited). Illustrates reciprocating rates of sedimentation in response to block movements (described later in text). Occurrences of glauconite grains (G), phosphate grains (P), manganiferous levels (Mn), ferruginous levels (Fe); and ranges of hexactinellid sponge spicules (+), phosphate shells (◇), calcareous shells (◆), agglutinated shells (△), and open-ocean trilobites (⚮) shown. All confirm the influence of oceanic conditions on the Avalon Platform.

5.2.1. Charnwood–Nuneaton area

Precambrian–Cambrian rocks of the Charnwood–Nuneaton area (Fig. 5.4) have a typically 'Charnioid' north-west–south-east trend. Of these, the Precambrian Charnian Group is best seen to the north of Leicester, where the rocks protrude beneath a cover of Triassic alluvium, exposing a suite of volcanics and volcaniclastics that form the basement of much of the East Midlands and East Anglia. The Charnian Group disappears beneath Triassic cover to the south of Charnwood Forest, although 'inselbergs' of south Leicestershire diorite intrusions appear at scattered intervals through their cover. Some 15–20 km south-west of Charnwood, however, these rocks are uplifted along a north-west–south-east trending fault that bounds the Warwickshire Coalfield, again exposing Precambrian volcanics (the Caldecote Volcanic Formation) and a thick succession of overlying Cambrian sediments, stretching from Nuneaton to Atherstone and dipping south-west.

The palaeogeographical setting is incompletely known, but the Charnian volcanics appear to be part of an island-arc suite that accreted on to the edge of Gondwanaland in late Precambrian times. The Cambrian succession over this craton was tilted but largely undeformed, whereas the rocks to the north-east of Charnwood are tectonized and probably formed part of a Palaeozoic Caledonian basin (e.g. Whittaker & Chadwick 1984).

Charnian Group

A soft-bodied metazoan fauna with petalonamean *Charnia masoni* and basal disc *Charniodiscus concentricus* Ford, medusoid ?*Cyclomedusa* sp., and ?arthropod *Pseudovendia charnwoodensis* Boynton & Ford, occur in volcanic tuffs of the Woodhouse Beds, at the top of the Maplewell Group of Charnwood Forest, *c*.15–20 km north-east of Nuneaton (Boynton 1978; Boynton & Ford 1979). This fauna includes some remains of giant *Charnia* (estimated length of *c*.1 m) and compound associations of ?*Charnia* (the 'water lily' fossil of Boynton 1978; Brasier 1985). Possible worm burrows of *Planolites* type also occur in these beds. *Cyclomedusa* sp. has been found at higher levels, in overlying Brand Series tuffs, and possible medusoids occur in tuffs of the Blackbrook Series, beneath the Maplewell Series (Boynton 1978). Whole-rock K–Ar dates of 684 ± 29 Ma obtained from dacitic 'porphyroid' intrusives are considered to post-date the Woodhouse beds (Meneisy & Miller 1963), but may come from an interval which consists mainly of tuff and does not form a closed isotopic system (Jenkins 1984). K–Ar ages obtained by Meneisy & Miller (1963) from the porphyroid mass on the north side of Bardon Hill (604 ± 26 Ma) and from a similar porphyroid on High Sharpley (583 ± 25 Ma) are more consistent with dates in south-east Newfoundland and elsewhere in the Avalon–Armorican region.

Geophysical studies show the subsurface continuity of the tilted Charnian rocks beneath the cover of Triassic rocks to the upfaulted Caldecote Volcanic Formation of Nuneaton (Whitcombe & McGuire 1981). The Woodhouse Beds may therefore lie up to a maximum of about ten kilometres below the first skeletal faunas, separated by conglomerates and shales of the Brand Group and volcanics of the Caldecote Formation. This assumes neither folding nor repetition of strata but it does compare with south-east Newfoundland, where a similar thickness of late Precambrian sediments may be present (Hofmann *et al.* 1979; Anderson & Conway Morris 1982). The strata concealed here might possibly compare with the Longmyndian strata of Shropshire.

Caldecote intrusive events

The Charnian and Caldecote rocks were broadly folded, faulted, intruded, and eroded prior to the trangression of the Hartshill Formation (Brasier *et al.* 1978). Thus it would seem possible to obtain reasonably precise Rb–Sr whole-rock isochron dates to constrain the age of the base of the transgressive sediments. Dates of 540 ± 57 Ma have been obtained from the distinctive Charnwood South Diorites (markfieldites) intrusive into the Charnian (Thorpe *et al.* 1984) and exposed beneath the basal Hartshill Formation, though no dates have yet been obtained from diorite dykes beneath the unconformity in this area.

Hartshill Formation

The overlying Hartshill Formation consists of *c*.270 m of coastal sandstones and mudstones, with minor limestones, that clearly represent a variety of successive nearshore to offshore facies (e.g. Fig. 5.5). Trace fossils are abundant through the section. *Arenicolites* sp., *Didymaulichnus* sp., *Neonereites uniserialis* Seilacher, and ?*Psammichnites* sp. appear in the first few tens of metres of the Park Hill Member (30–61 m, Brasier *et al.* 1978; Brasier & Hewitt 1979, 1981; Brasier 1984); to which may now be added ?*Bergaueria* sp., *Hormiscioidea canadensis* Crimes & Anderson, and *Monocraterion* sp. The overlying Tuttle Hill Member (160–180 m) also has *Gordia*, while the arthropod trace of *Isopodichnus* appears in the Jees Member (6 m), a few metres below the first skeletal faunas. Intense bioturbation also appears for the first time near the base of the Jees Member, *c*.6 m below the first skeletal faunas.

The lithology, stratigraphical position, sedimento-

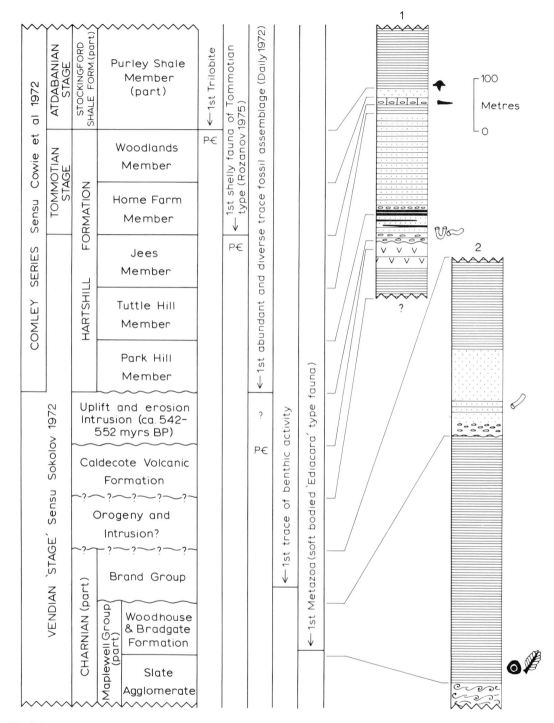

Fig. 5.4. The sequence at Nuneaton (1) and the upper part of the Charnian sequence (2), showing the successive datum points of Brasier *et al.* (1978). The Tommotian–Atdabanian boundary is now thought to lie *within* the Home Farm Member (Brasier 1986), or within the disconformity at its base. The Precambrian–Cambrian boundary currently lies within the unconformity at the base of the Park Hill Member (in the sense of China A or B, or the *Phycodes pedum* ichnofossil Zone of Newfoundland). At right are symbols for the first occurrences of the Ediacaran fauna, benthic trace fossils, abundant and diverse trace fossils, shelly fauna including trilobites.

LITHOFACIES

BIOFACIES

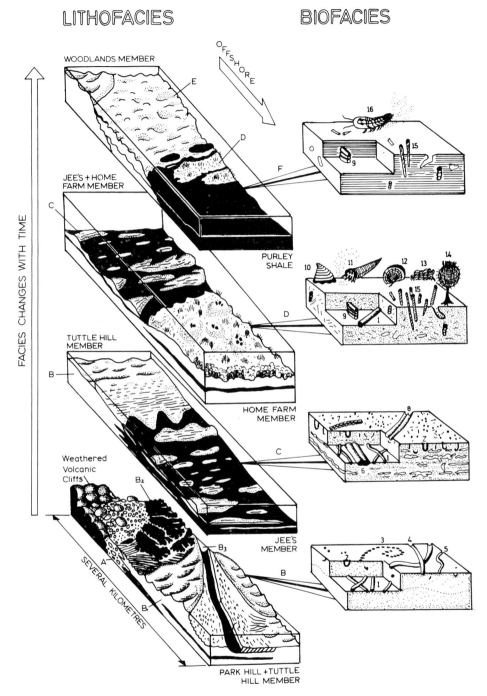

Fig. 5.5. Block diagrams of lithofacies and biofacies in the Hartshill Formation and Purley Shale, illustrating their lateral relationships and the changes with time. A, littoral conglomerates; B, coastal sands and muds; C, interlaminated sands and muds; D, condensed pink nodular limestones; E, sheet sandstones; F, offshore shales. 1, *Planolites*; 2, *Arenicolites*; 3, *Hormisiroidea*; 4, cf. *Psammichnites*; 5, *Gordia*; 6, *Taphrhelminthopsis*; 7, *Isopodichnus*; 8, *Didymaulichnus*; 9, *Teichichnus*; 10, *Bemella* (mollusc); 11, hyolith; 12, paterinid brachiopod; 13, ferromanganese stromatolites; 14, protospongiid; 15, *Coleoloides*; 16, olenellid trilobite. A, beach facies; B, coastal sands and muds (B_1, sand flats; B_2, mud flats; B_3, channel sands); C, interlaminated sands and muds; D, shelly limestones; E, sheet sands; F, offshore muds. (After Brasier & Hewitt 1979.)

logy, and trace-fossil assemblage from the base of the Hartshill Formation to the base of the Home Farm Member, compare closely with those of the Random Formation sandstones of south-eastern Newfoundland (cf. Hiscott 1982; Crimes & Anderson 1985).

The base of the Home Farm Member is marked by 0.15 to 0.45 m of the Quartzose Conglomerate Bed, which compares with that found widely at the top of the Random Formation throughout south-east Newfoundland (e.g. Walcott 1900). There, the beds beneath the conglomerate appear to have suffered local erosion (Dr T. P. Fletcher, pers. comm. 1986) and the beds above may be of varying ages (Bengtson & Fletcher 1983).

Home Farm Member

The lithological and faunal successions through the sandstones, shales and limestones of the Home Farm Member (*c*.2 m) are shown in Fig. 5.6. Because of the extreme condensation of the succession, the beds will be described in detail.

Quartzose Conglomerate   This is a tabular cross-bedded unit (0.15 to 0.45 m), with concave foresets; the grain size varies from sandy mudstone to polygenetic, inequigranular conglomerate. Clasts include red and grey shale intraclasts, vein quartz, sheared quartzite, plagioclase, glauconite, and volcanic rock. The cement varies from syntaxial quartz overgowths to poikilitic calcite. The trace fossil *Planolites* occurs in shale partings, and thorough breakdown of the rock has revealed a few fragments of the tommotiid *Sunnaginia neoimbricata* Brasier, large *Micromitra phillipsi* Holl brachiopods, tubes of *Hyolithellus* cf. *micans* Billings morphotype A, and single fragments of the hyolith ?*Turcutheca* sp. and brachiopod ?'*Obolus*' *groomi* (Matley).

Sandstone Bed   This unit (0.1–0.4 m) consists of thin, laminar-bedded conglomerates, sandstones, and shales passing up into relatively massive maroon micaceous sandstones with some poikilitic calcite cement, and indications of cross-bedding. These yield a scarce fauna of the brachiopod *Micromitra phillipsi*, the tube *Hyolithellus* cf. *micans*, and poorly preserved internal casts believed to be of the tube *Coleoloides typicalis* Walcott.

Phosphatized Limestone Conglomerate, Beds 1i–iii   This unit (0.12 m) comprises part of Bed 1 of the *Hyolithes* Limestone, marking the beginning of calcareous deposition in the sequence. Bed 1 is widely divisible into a lower unit of sandy limestone (Bed 1i); a complex middle bed of phosphatic limestone conglomerate intraclasts in a limestone matrix, topped by a phosphatized limestone hardground (Bed 1ii–iii); a thin overlying unit of

glauconitic course-sandy limestone (1iv) and a finer sandy limestone at the top (1v). The latter two lie above a discontinuity surface and are described separately.

A rich assemblage of small shelly fossils appears in Beds 1ii–iii, many of which shows signs of phosphatization and abrasion. This unit is characterized by the tommotiid *Sunnaginia neoimbricata* and also includes tubes *Hyolithellus* cf. *micans*, *Coleoloides typicalis*, and *Torellella lentiformis* Missarzhevsky; protoconodont *Hertzina elongata* Müller; tommotiids *Camenella baltica* Bengtson, *Eccentrotheca grandis* Brasier; wiwaxiid-like *Halkieria* sp.; hyolith *Allatheca* sp., and brachiopod *Micromitra phillipsi*. The presence of a phosphorite hardground at the top of this unit suggests that the phosphatization was *in situ* and that the winnowed fossils need not have travelled far. There is a possibility, however, that the Phosphatized Limestone Conglomerate is highly condensed and admixes some older and younger faunal elements.

It should be noted that calcareous forms are rare in this bed, the fauna being mainly of more-resistant phosphatic elements. This may be one explanation for the absence of gastropod *Aldanella attleborensis* (Shaler & Foerste) which occurs in the lower levels in the calcareous facies of Newfoundland.

Glauconitic Sandy Limestones, Beds 1iv–v   Above the discontinuity marked by the phosphatized hardground, in units 1iv–v (0.04 m), the fauna is sparser but has greater affinity with the overlying *Coleoloides* Limestone Beds. These thin units contain tubes of *Coleoloides typicalis*, *Hyolithellus* cf. *micans* type B, and *Torellella lentiformis*; tommotiid *Camenella baltica*; brachiopod *Micromitra phillipsi*; and hyolith *Allatheca* sp., as in the underlying units. Two calcareous forms also make their first appearance here: the lipped-hyolith *Burithes alatus* (Cobbold) and the monoplacophoran mollusc *Bemella pauper* (Billings).

Coleoloides Limestones, Siltstone, and Shales, Beds 2–10ii   Overlying strata comprise about 1 m of maroon or dark-grey sandy limestones intercalcated with siltstones and thin shales. The lower part of this sequence (0.56 m thick) is of nodular limestones with clastic intercalations (Beds 2–7), passing up into a more massive limestone with several discontinuities and variable microfacies (Beds 8–12). There is a gradual upward change in the faunal character of these beds, with a major change at the base of a discontinuity in unit 10iii, which is therefore discussed separately. Below this discontinuity, assemblages are dominated by calcareous forms, with 'meadows' of *Coleoloides typicalis* and pentactine spicules of hexactinellid sponges. More

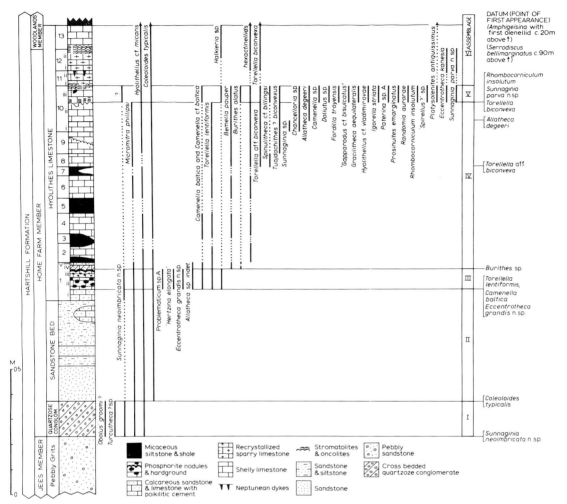

Fig. 5.6. Composite range chart of microfossils and small shelly fossils through the Home Farm Member at Jees and Woodlands Quarries, Nuneaton. Solid line—occurrence; dotted line—inferred. (After Brasier 1986.)

occasional specimens of the tommotiid *Camenella baltica*, tube *Torellella lentiformis*, monoplacophoran *Bemella pauper* and hyolith *Burithes alatus* also occur.

From about Bed 8 upwards, the clastic content is much reduced, and new faunal elements begin to appear: tubes of *Torellella* aff. *biconvexa* Miss. and hyolith *Spinulitheca* cf. *billingsi* (Syssoiev) in Bed 8; hyolith *Crestjahitus*? (= *Tuojdachithes*?) *biconvexus* (Cobbold) in Bed 9; chancelloriid *Chancelloria* sp. in Bed 10i; and hyolith *Allatheca degeeri* (Holm) in Bed 10ii.

Hyolith Shell Bed, Bed 10iii This richly fossiliferous unit (0.07 m) overlies an iron-impregnated and stroma-

tolite-encrusted discontinuity surface within the massive limestone of Bed 10. Large grains of authigenic glauconite are also present. *Torellella lentiformis* and *Micromitra phillipsi* do not range up into this bed, and the fauna contains distinct species of *Torellella*, *Sunnaginia*, *Eccentrotheca*, and paterinid brachiopods plus many molluscs not seen in the underlying beds. Most important is the appearance of the index protoconodont, *Rhombocorniculum insolutum* within this assemblage.

The fauna of Bed 10iii is as follows: tubes of *Coleoloides typicalis*, *Hyolithellus* cf. *micans* Billings morphotype B, *H. vladimirovae* Missarzhevsky, *Torellella biconvexa* Missarzhevsky, *Platysolenites antiquissimus* Eichwald; tommotiids *Camenella* cf. *baltica* and

Camenella sp., *Eccentrotheca kanesia* Landing, Nowlan, & Fletcher, and *Sunnaginia parva* Brasier; chancelloriid *Chancelloria* sp.; wiwaxiid-like *Halkieria* sp.; protoconodont *Rhombocorniculum insolutum* Missarzhevsky; hyoliths *Allatheca degeeri*, *Burithes alatus*, *Doliutus* sp., *Crestjahitus?* (= *Tuojdachithes?*) *biconvexus* (Cobbold), and *Gracilitheca aequilateralis* (Cobbold); monoplacophorans *Bemella pauper*, *Igorella striata* (Cobbold), *Prosinuites emarginatus* (Cobbold), *Randomia aurorae* Matthew; bivalves *Fordilla* sp. and *Paterina* sp. A, and hexactinellid spong spicules. The records of *Gapparodus* cf. *bisulcatus* (Müller) and *Spirellus* sp. (Brasier 1984) now seem doubtful, while *?Hyperammina* and *Yanischevskyites* are placed in *Platysolenites antiquissimus*.

Sparry and Algal Limestones, Beds 11 and 12   Overlying limestones (0.15 m) have suffered patchy recrystallization to sparry calcite, possibly during emergence at the end of the Home Farm Member times. They contain 'meadows' of *Coleoloides* alternating with discontinuity surfaces encrusted with domal and columnar ferromanganese stromatolites plus oncolites. Glauconite and limonite grains and infills, and spar-filled vugs and calcite sheet cracks also occur. The sparse fauna includes tubes of *C. typicalis*, *H.* cf. *micans* morphotypes B and C, *Halkieria* sp., and *Torellella biconvexa*; tommotiids *Eccentrotheca kanesia* and *Sunnaginia parva*, inarticulate brachiopod fragments and hexactinellid sponge spicules.

The top surface of Bed 12 of the Home Farm member is a ferromanganese crust, draped with macroscopic flakes of detrital mica from the base of Woodlands Member, marking the base of the next sedimentary cycle.

Woodlands Member

This unit (19 m) comprises 'regressive' sandstones overlying the preceding limestones. The base is actually Bed 13 of the *Hyolithes* Limestone, being 0.09 m of glauconitic quartzose sandstone with a poikilitic calcite cement. This contains some reworked fossils from the underlying Bed 12. This bed passes up into maroon shales, siltstones, and sandstones with lunate megaripples, confirming a return to coastal sand-bar sedimentation. Above this lowest metre-or-so of the Woodlands Member, sediments consist of massive, glauconitic, feldspathic, sheet sandstone units, probably formed by storm surges. This part of the member is devoid of trace and shelly fossils. The uppermost 0.9 m consists of calcareous sandstones, hematitic sandstones, and grey–maroon, mottled *Coleoloides*-bearing limestone lenses. Bioturbation is present here, and fossil remains include tubes of *C. typicalis* and *Torellella* sp. This calcareous facies appears to be transitional between the underlying sheet-sandstone facies and the conformably overlying maroon shales of the Purley Shale Formation. Thus it seems that the sandstones of the Woodlands Member were a coarse clastic base to the Purley Shale transgressive cycle.

Purley Shale Formation

This is a 210 m-thick unit of relatively uniform maroon shales, providing a thick fossiliferous Lower to low Middle Cambrian sequence. Exposure is poor and much of the data rely on trenching (e.g. Rushton 1966).

One metre above the base of the Purley Shales are found two closely spaced calcareous nodule bands. These yield trilobite remains, possibly of *Callavia* sp., plus tubes of *Coleoloides typicalis*, sponge spicules, unidentified conodonts (cf. *Amphigeisina*), *Platysolenites antiquissimus* tubes, and *Teichichnus* burrows (Brasier *et al.* 1978; Brasier 1986). *Callavia* sp. also occurs about 12 m higher in the section (Smith & White 1963).

Locality 1A of Rushton (1966) is a green mudstone band, yielding *Serrodiscus bellimarginatus* (Shaler & Foërste), about 58 m above the base of the Purley Shales. Locality 1b is a second green bed, about 67 m above the base of the formation, yielding trilobites *S. bellimarginatus*, *Serrodiscus* cf. *serratus* R. & E. Richter, *Ladadiscus llarenai* (R. & E. Richter), *Strenuella* (*Myopsostrenua*) *sabulosa* Rushton; gastropod *Pelagiella* cf. *primaeva* (Shaler & Foërste); tubes of *Coleoloides typicalis*, a hyolith, and an eoorthid brachiopod.

Locality 2A of Rushton (1966) lies about 137 m above the base of the Purley Shales. These are fossiliferous green mudstones, yielding trilobites *Serrodiscus ctenoa* Rushton, *Acidiscus theristes* Rushton, *Tannudiscus balanus* Rushton, *Chelediscus acifer* Rushton, *Acimetopus?* sp., *Condylopyge amitina* Rushton, *Ellipsostrenua heyi* Rushton, and *Atops?* sp.; gastropod *Pelagiella* sp.; tubes of *Salterella?* sp. and *Coleoloides typicalis*; brachiopods *Lingulella* cf. *westergaardi* Kautsky, *Botsfordia* cf. *pulchra* (Matthew), *Botsfordia* cf. *granulata* (Redlich), *Botsfordia?* sp., and worm tracks. A similar brachiopod assemblage was found at locality 2B, about 11 m above 2A, while tubes of *Coleoloides typicalis* and inarticulate brachiopods occurred at locality 2C, about 155 m above the base of the formation.

The base of the Middle Cambrian lies within the Purley Shale Formation, below an assemblage with trilobites *Ptychagnostus* cf. *praecurrens* (Westergaard) and *Paradoxides sedgwickii porphyrus* Rushton. This fauna is late within the *Paradoxides pinus* Zone and occurs at locality 2D, some 198 m above the base of the Purley Shales.

5.3. CORRELATION WITH NEWFOUNDLAND

The Charnwood–Nuneaton succession shows remarkable lithological and palaeontological similarities with the succession in south-east Newfoundland (e.g. Landing *et al.* 1988; Landing & Benus 1988). Further aspects of correlation will be discussed below, but the following general correlations can be made.

The tuffs and slates of the Charnian Group, bearing the *Charnia masoni* fauna, correspond with similar lithologies and faunas in the Conception Group of south-east Newfoundland. Conglomerates and red shales of the Brand Series, at the top of the Charnian in Charnwood Forest, may compare with the dark-red shales of the Mistaken Point Formation, at the top of the Conception Group. The dates of 604 ± 26 and 583 ± 25 Ma obtained from Charnian porphyroids are thought to be contemporaneous with the volcanicity. In south-east Newfoundland, the Conception Group overlies volcanics of the Harbour Main Group, with precise U–Pb zircon dates of $606.3^{+3.7}_{-2.9}$ Ma, intruded by rhyolite dykes dated at $585^{+3.4}_{-2.4}$ Ma (Krogh *et al.* 1983). Volcanics from beneath the Rencontre ($=$ Doten Cove) Formation of the Burin Peninsula may be of this age; they yield U–Pb zircon dates of $623^{+1.9}_{-1.7}$ to $606^{+3.7}_{-2.9}$ Ma (Cowie & Johnson 1985). The Conception Group itself is thought to be intruded by the Holyrood Granite, dated at 585 ± 15 Ma (Cowie & Johnson 1985) or 594 Ma (Glaessner 1984). An age of about 566 ± 5 Ma now seems likely for the *Charnia masoni* fauna, on the basis of U–Pb zircon dates from the lower Mistaken Point Formation (Benus in press) and the above correlation.

The Caldecote Volcanic Formation of Nuneaton signals a local return to volcanic conditions, culminating in subaerial deposition followed by intrusive events dated at 540 ± 57 Ma. The Caldecote volcanics may therefore correlate with the Eastern Uriconian Volcanics of Shropshire, dated at 558 ± 16 Ma (Thorpe *et al.* 1984) and with Arvonian volcanics that lie conformably beneath and intercalate with thick basal conglomerates in North Wales (Wood 1969). The Sarn granite of nearby Llêyn may be related and yields an intrusive Rb–Sr age of 549 ± 17 Ma (Thorpe *et al.* 1984). Volcanic rocks and tuffs also occur above the *Charnia masoni* fauna in the lower part of the Musgravetown Group (Bull Arm Volcanics and Big Head Formation), Hodgewater Group, and St John's Group of south-eastern Newfoundland (e.g. Hofmann *et al.* 1979).

The Chapel Island formation of south-east Newfoundland marks the initial stages of the Cambrian transgression. Similar deposits are not present at Nuneaton (unless within the base of the thick Hartshill Formation). The Chapel Island Fauna suggests that the succeeding Random Formation of Newfoundland was deposited within the range of *Aldanella attleborensis* (Bengtson & Fletcher 1983).

Sandstones of the Hartshill formation beneath the Home Farm Member compare in lithology and lithofacies with the Random Formation (e.g. Brasier & Hewitt 1979; Brasier 1980; Hiscott 1982), although the Hartshill Formation is thicker and *Diplocraterion* and *Cruziana* of the Random Formation (Crimes & Anderson 1985) are not yet known from Nuneaton.

The Quartzose Conglomerate Bed at the base of the Home Farm Member compares with a similar conglomerate at the base of the Bonavista Formation (cf. Walcott 1900; Landing & Benus 1988). The *Aldanella attleborensis* fauna of the basal Bonavista Formation (Bengtson & Fletcher 1983) is not definitely present at Nuneaton; this may be because of different facies, or because of a hiatus. Such a hiatus and unconformity is widely suspected in parts of south-eastern Newfoundland (Hiscott 1982; Bengtson & Fletcher 1982; Benus & Landing 1984; Landing & Benus 1988). The change from thick coastal sandstones to condensed shelf limestones at Nuneaton is striking, comparing with similar local changes in Newfoundland (Bengtson & Fletcher 1983) and suggesting active block movements at this time.

The alternating clastics and limestones from the Sandstone Bed to Bed 7 of the Home Farm Member compare with the Bonavista Formation, though they are extremely condensed in comparison with the 275 m seen on the northern Bonavista Peninsula. However, the latter also thin to a feather edge in the eastern localities of St Mary's and Conception Bays (Benus & Landing 1984; Landing & Benus 1988). The appearance of conspicuous nodular and bedded limestones within the middle to upper parts of the Bonavista Formation marks a biofacies change from *Aldanella* to *Coeloloides* faunas (Benus & Landing 1984). Thus the *Sunnaginia neoimbricata* fauna from the Quartzose Conglomerate to Bed 1 may correlate with lower or middle parts of the Bonavista Formation (or reworked therefrom); the *Coeloloides* limestones and clastics from Beds 2 to 7, compare with the middle to upper Bonavista Formation, especially the 'Cuslett Formation' of Landing & Benus (1988), here relegated to the status of a member.

The amalgamated limestones of Beds 8–12 may compare better with the lower to middle Smith Point Limestone Formation ('Fosters Point Formation' of Landing & Benus 1988). The fauna from Bed 10iii upwards undoubtedly shares the following with the middle Smith Point Limestone: *Coleoloides typicalis*, *Allatheca degeeri*, *Doliutus* sp., *Randomia aurorae*, *Fordilla* sp., *Sunnaginia parva*, and *Eccentrotheca kanesia*. Specimens of *S. parva* obtained about 1.8 m (6 feet) below the trilobites in the Smith Point Limestone (original concept) appear to be 'more developed', however, than those in Bed 10iii. The ferruginous crusts, stromatolites, and oncilites from Beds 10iii to 12 nevertheless compare directly with similar facies in the upper part of the Smith Point Limestone (e.g. Landing & Benus 1984). Shoaling and emergence are thought to have taken place during and after deposition of these ferromanganese stromatolites.

A widespread discontinuity is believed to separate the 'non-trilobite' faunas of the Fosters Cove Member of the Smith Point Limestone Formation, from *Callavia* faunas found in the very top of the limestone (Bengtson & Fletcher 1983; Landing & Benus 1988). This discontinuity is tentatively correlated with the discontinuity between the Home Farm Member and Woodlands Member sandstones, of which the upper ferruginous calcareous lenses may compare with the very top of the Smith Point Limestone (i.e. basal Brigus Formation). Thus the composite Smith Point Limestone Formation may correlate with the Home Farm and Woodlands Members together. Landing & Benus (1984) note herring-bone cross-bedding at the base of the Brigus Formation, which would confirm correlation with the latter.

The change to maroon Purley Shales compares directly with that in the Brigus Formation of south-east Newfoundland, and the faunal horizons within can be matched closely between the two areas, with similar successions of *Callavia* sp., *Serrodiscus bellimarginatus*, *Strenuella sabulosa*, and *Acidiscus theristes* faunas (e.g. Rushton 1966; Bengtson & Fletcher 1983). This similarity between Nuneaton and south-east Newfoundland is the more remarkable when contrasted with the Shropshire sections.

5.4. SHROPSHIRE

The Precambrian–Cambrian outcrops of Shropshire lie along the north-western margin of the Midland craton, some 75 km due west of Nuneaton (Fig. 5.1). These rocks have a north-east–south-west 'Caledonian' trend, lying either side of the major Church Stretton Fault line. This fault, and the Pontesford–Linley Fault further west, roughly delimits the incomplete, condensed, and undeformed deposits of the Midland craton from the thick and deformed deposits of the Welsh Basin of early Palaeozoic times (see following comments).

The standard interpretation for this sequence begins

with rocks to the west of the Church Stretton Fault, where the Precambrian Longmynd Group is uplifted as a horst. According to James (1956), the Longmynd is a syncline with an overturned western limb, with older 'Uriconian' volcanics passing up into sedimentary and volcaniclastic Longmyndian rocks. This block was not definitely overstepped until early Arenig times. To the east of the Church Stretton Fault are found uplifted slices of Precambrian Uriconian volcanics (e.g. the hills known as the Lawley, Caer Caradoc, and the Wrekin) overstepped by Lower Cambrian sediments, passing up into Middle Cambrian to Tremadoc rocks. The main contention here is that type Uriconian volcanics are contemporaneous with volcanic basement to the west of the Longmynd (James 1956). This would mean that the whole, or greater part, of the great thickness of Longmyndian sediments is omitted from the eastern sequence. But this bold interpretation is uncertain, and Barber (1985) has briefly suggested that major transcurrent movements along the Church Stretton and other faults have juxtaposed allochthonous terranes. Cowie & Johnson (1985, p. 59) have noted that a similar Rushton Schists date (536 ± 8 Ma) (probably reset) occurs either side of the Wrekin (=Church Stretton) Fault in the north of the area, and therefore major transcurrent displacements may have ceased by 533 ± 12 Ma when the Ercall granophyre was intruded. For the present, the two areas will be discussed separately.

5.4.1. The western succession

Precambrian rocks to the west of the Church Stretton Fault are disposed in an isoclinally folded syncline, overturned towards the east. The oldest rocks are 'Uriconian' volcanics exposed along the western margin of the outcrop. The succeeding sediments of the Longmyndian comprise a thick pile of Eastern Longmyndian marine or lacustrine sediments, including tuffs (Stretton Series), unconformably overlain by conglomeratic red beds of the Western Longmyndian (Wentnor Series).

The Stretton Series contain filamentous microfossils not inconsistent with a late Precambrian age (Peat 1984) and well-preserved remains of the soft-bodied problematical macrofossil *Arumberia*, whose remains also occur in the latest Precambrian St John's Group of south-east Newfoundland (Bland 1984). Illitic shales from the Stretton Series yielded isochrons of 452 ± 31 Ma and 529 ± 6 Ma, interpreted as dewatering dates, extrapolated backwards to depositional ages of 600 Ma (see Bath 1974). Deformation is thought to have taken place soon after deposition.

Outcrops of purple sandstones beneath the Wrekin

Quartzite were at one time thought to be evidence for Western Longmydian strata on the east side of the Church Stretton Fault. That interpretation is questioned by Cowie & Johnson (1985) and herein. Although no longer seen, such rocks are as likely to be feldspathic facies at the base of the Wrekin Quartzite, widely present in the Lickey Hills and Nuneaton. These could have been reverse-faulted out in local 'overthrusting' of the Wrekin Quartzite over the Uriconian.

5.4.2. The eastern succession:

Precambrian rocks and the boundary

The succession of events is best seen on the flanks of the Wrekin Hill, to the north of the outcrop. Much of the interest here has focused on the potential for isotopic dating (e.g. Odin *et al.* 1983; Thorpe *et al.* 1984), though some of these remain controversial (e.g. Cowie & Johnson 1985).

Present evidence indicates that the Ercall granophyre lies above the Uriconian volcanics and is unconformably overlain by the Wrekin Quartzites (Cope & Gibbons 1987). This implies that the thick succession of Longmyndian sediments known further south is here omitted. Interpolation of these units, and reference to the isotopic dates discussed in Thorpe *et al.* (1984), Cowie & Johnson (1985), and Odin *et al.* (1985) allow a tentative succession of events to be assembled for the Church Stretton–Longmynd–Wrekin area:

1. Metamorphism of the Rushton Schists (667 ± 20 Ma; Thorpe *et al.* 1984).

2. Uriconian volcanicity (558 ± 16 Ma), at a time of normal to mixed polarity, followed by reversed polarity and dolerite intrusion (< 558, < 533? Ma; Thorpe *et al.* 1984).

3. Deposition of the transitional Helmeth Grits and the Stretton Series of the Longmyndian, with intermittent volcanic eruptions. The Stretton Series resembles that described in Section 5.4.1.

4. Emergence and deposition of the Western Longmyndian alluvial conglomerates and sandstones.

5. Deformation of the Longmyndian–Uriconian synclinal pile, with overturning in the west (James 1956; Grieg *et al.* 1968).

6. Intrusion of the Ercall granophyre (533 ± 13 Ma, Thorpe *et al.* 1984; 531 ± 5 Ma, Compston *et al. in* Cowie & Johnson 1985) and local contact metamorphism of the Rushton Schists (536 ± 8 Ma, Thorpe *et al.* 1984). Geochemical data indicate that the Ercall granophyre is not simply an intrusive equivalent of Uriconian rhyolites (Beckinsale *et al. in* Cowie & Johnson 1985) and recent quarrying confirms the unconformable relationship between the granophyre and Wrekin Quartzite (Cope & Gibbons 1987).

7. Faulting and erosion, leading to removal of Longmyndian strata and exposure of Rushton Schists in the Wrekin area.

8. Coastal sand deposits of the Wrekin Quartzite transgressively overstepped the faulted and eroded Uriconian and Rushton Schists (Cobbold & Pocock 1934) and possibly the Western Longmyndian of the Cwms area (Greig *et al.* 1968). The quartzite bears abundant *Diplocraterion uriconiensis* (Calloway) near the top (Brasier & Hewitt 1979) and unidentifiable inarticulate brachiopod fragments (Cowie & Johnson 1985). Correlation with the Random Formation is suggested (Anderson 1981).

9. A sharp (?disconformable) change from quartzites to siltstones and shales of the Lower Comley Sandstones is accompanied by the appearance of the first biostratigraphically useful assemblages of small shelly fossils, including *Mobergella* cf. *turgida* Bengtson 10 cm above their base (Rushton 1972; Bengtson 1977). *Platysolenites antiquissimus* Eichwald, *Camenella baltica* Bengtson, and *Coleoloides typicalis* Walcott occur *c.*7–10 m above the base (new data).

The Ercall granophyre therefore lies within a complex succession of volcanic, block-tectonic and sea-level events whose significance for wider correlation is uncertain. Strong similarity between the Nuneaton and Avalon (Newfoundland) successions, however, suggests that tectonic phases may prove useful for event stratigraphy. The biostratigraphical setting of the overlying Comley Series is of greater importance, however.

Wrekin Quartzite Formation

Pebbly beds at the base of this unit (< 50 m) include Uriconian and Ercall granophyre pebbles, but the basal contact often seems to be faulted and arkosic sediments may be missing (see above). The base rests on Ercall granophyre, Uriconian lavas, or Rushton Schists. The formation is a typical coastal sandstone, though mainly of orthoquartzite, with large- and small-scale ripples and a general coarsening upwards in grain size. Trace fossils are scarce but include *Planolites* and *Diplocraterion*. The latter was found close to the locality of

Arenicolites uriconiensis Callaway (1878). These are only seen on one bedding plane, as '*Bifungites*'-like cross-sections with clay-filled paired holes and interlinking spreite. The only verified skeletal fossil found is an inarticulate brachiopod fragment (Cowie & Johnson 1985).

Lower Comley Sandstone

The junction between the Wrekin Quartzite and Lower Comley Sandstone (150 m) is sharp in the Ercall Quarry and may be disconformable. The Lower Comley Sandstone is distinguished throughout by abundant glauconite grains but there are distinct faunal and lithological beds (e.g. Cobbold 1921; Cobbold & Pocock 1934; Rushton 1974).

The base of the Lower Comley Sandstone comprises the *Obolella groomi* Beds (Ab1), of greenish grey sandstones and silty micaceous shales with local conglomeratic beds and decalcified rottenstones, together *c.*8 m thick. Further work needs to be done on this assemblage because it may be highly condensed, like the Home Farm Member of Nuneaton, and faunal lists from the 8 m thickness may be misleading: small shelly fossils *Mobergella* cf. *turgida* Bengtson, *Camenella baltica* Bengtson, *Platysolenites antiquissimus* Eichwald, *Coleoloides typicalis* Walcott, *Torellella* cf. *lentiformis* Miss., *Hyolithellus* cf. *micans* Billings; hyolith *Burithes strettonensis* (Cobbold); brachiopods '*Obolus*' *groomi* (Matley), *Micromitra phillipsi* (Holl), *Walcottina lapworthi* Cobbold, and *W. elevata* Cobbold.

The Ab2 Beds are overlying muddy sandstones with some glauconitic beds. They contain hyolith *Allatheca degeeri*(?) (Holm) and tube *Hyolithellus micans*. In the Rushton area the *Acrothele prima* Shale comprises buff-weathering shales with sandy seams, lying some 30 m above the local base of the Lower Comley Sandstone. These shales bear *Hyolithellus micans*, brachiopods *Lingulella* sp., *Acrothele prima* (Matthew), *Acrotreta gemmula* Matthew, and bradoriid crustaceans *Bradoria robusta* (Matthew), *B. benepuncta* (Matthew), *B.* cf. *obesa* Matthew, and questionable *Beyrichona tinea* Matthew (Cobbold & Pocock 1934).

The Ab3 Beds comprise green and reddish sandstones with some calcareous beds. They contain olenellid fragments, doubtfully referred to *Holmia* (Cobbold 1921), *Fallotaspis* (Hupé 1952) or *Kjerulfia* (Bergström 1973) plus *Hyolithellus micans*.

The Ab4 Beds consist of green sandstones with calcareous nodules. These contain rare bradoriid crustaceans. The middle to upper Lower Comley Sandstone of the Shoot Rough Road borehole yielded bradoriids *Bradoria* cf. *nitida* (Wiman), *Indiana exigua* (Cobbold), and *Indiana minima* Wiman plus tubes of *Hyolithellus micans* and *Torellella laevigata* (Rushton *in* Greig *et al.*

1968). Assemblages from uncertain levels above the *Acrothele prima* Shale have also yielded the brachiopod *Kutorgina*? *anglica* Cobbold and bradoriid crustaceans *Beyrichona* cf. *rotundata* Matthew, *Aluta rotundata salopiensis* Cobbold, *A. ulrichi* Cobbold, and *Indiana exigua* Cobbold (Cobbold & Pocock 1934).

Callavia *Sandstone*

The Callavia Sandstone Bed (Ac1) of the Lower Comley Sandstone is a transitional unit of fine-grained green sandstones (*c.*1.3 m) with quite abundant fossils (Fig. 5.7). These include trilobites: *Callavia broeggeri* Walcott, *Kjerulfia*? *granulata* Raw, *Judomia*? *pennapyga* (Raw), *Dipharus attleborensis* (Shaler & Foerste); brachiopods *Paterina* cf. *labradorica* (Billings) and *Obolella atlantica* Walcott var. *comleyensis* Cobbold; and mollusc *Helcionella rugosa* Hall. Reworked fragments in the basal Middle Cambrian that appear to come from Ac1 include *Strenuella calceola* Cobbold (Cobbold 1931).

Comley Limestones

The Comley Limestone Formation (*c.*1.8 m) comprises five members, each lithologically and faunally distinct, separated from one another by disconformities. There is a general upward progression from sandy limestone to pure limestone with phosphorite clasts, returning to sandy limestone with phosphorite clasts in Bed Ad. The succession is interpreted as a major transgression over a sediment-starved bank, ending with a regression in Ad times and probable emergence at the end of early Cambrian times. The succession with major faunal ranges is shown in the composite range chart of Fig. 5.7; new data from Hinz (1987) are included below.

The Red Callavia Sandstone Bed (Ac2) is a nodular calcareous sandstone up to 0.75 m thick, bearing abundant glauconite plus quartz and other minerals in a calcareous matrix. The fauna is similar to but richer than that of the underlying green sandstone of Ac1: trilobites *Callavia broeggeri* (Walcott), *Nevadia* (or *Nevadella*) *cartlandi* Raw, *Micmacca protolenoides* Cobbold, *Strenuella spinosa* Cobbold, *S. calceola* Cobbold, *Ellipsocephalus* sp., *Triangulaspis vigilans* (Matthew) and *Dipharus attleborensis* (Shaler & Foerste); brachiopods: *Paterina* cf. *labradorica* (Billings), *P. minor* Cobbold, *Lingulella viridis* Cobbold, *Obolus parvulus* Cobbold and *Obolella atlantica* Walcott; small shelly fossils include *Rhombocorniculum cancellatum* (Cobbold), *Eccentrotheca kanesia* Landing, Nowlan & Fletcher, *Torellella biconvexa* Miss., *Hyolithellus micans;* hyolith '*Hyolithes*' *compressus* Cobbold; molluscs: *Latouchella costata* Cobbold, *Helcionella rugosa* Hall, and *Scenella elevata* Cobbold; and various brador-

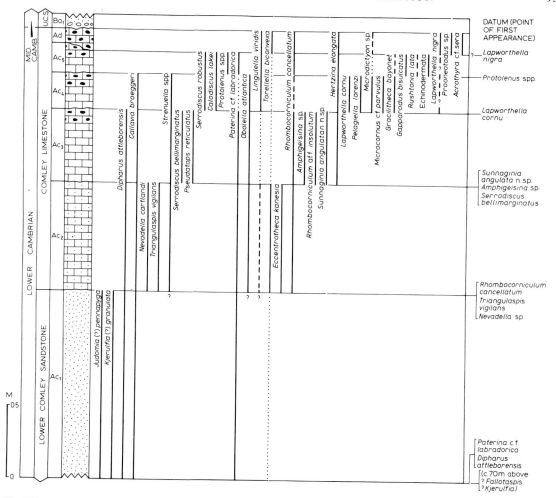

Fig. 5.7. Composite range chart of selected trilobites, microfossils, and small shelly fossils through the upper Lower Comley Sandstone and Comley Limestone of Comley and Rushton, Shropshire. Taxa shown are selected forms with biostratigraphical potential. For microfossils and small shelly fossils: solid line—occurrence at Comley; dotted line—inferred; dashed line—occurrence at Rushton only. (After Brasier 1986.)

iid crustaceans (Cobbold 1921; Cobbold & Pocock 1934).

The undifferentiated *Callavia* Limestone microfauna of Hinz (1987) may belong here or in Bed Ac3. This notably contains: *Coleoloides typicalis*, *Cowiella reticulata* Hinz, *Eccentrotheca kanesia*, *Sunnaginia imbricata* group, *Lapworthella* sp., *Amphigeisina danica* (Poulsen), cf. *Protohertzina cultrata* Missarzhevsky, and *Chancelloria lenaica* Zhuravleva & Korde.

The *Serrodiscus bellimarginatus* Limestone Bed (Ac3) is a pale-grey and pinkish limestone up to about half-a-metre thick, containing patches with glauconite and sandy material and much phosphatic matter in the upper part. The fauna includes trilobites *Callavia broeggeri*, *Strenuella pustulata* (Cobbold), *Pseudatops reticulatus* (Walcott), *Dipharus attleborensis*, and *Serrodiscus bellimarginatus* (Shaler & Foerste); brachiopods *Paterina* cf. *labradorica*, *P. minor*, *Obolus parvulus* Cobbold, and *Obolella atlantica;* small shelly fossils *Rhombocorniculum cancellatum*, *R. spinosus* (Hinz). *Amphigeisina* sp., *Hertzina elongata* Müller, *Sunnaginia angulata* Brasier, *Hyolithellus micans;* mollusc *Helcionella cingulata* Cobbold (?).

The Strenuella Limestone Bed (Ac4) is a reddish purple to grey sandy limestone about 0.2 m thick at Comley and 0.4 m thick at Rushton. It contains glauconite and phosphatic matter but the detrital fraction is finer-grained and less abundant than in underlying beds. The fauna includes trilobites: *Callavia broeggeri*, *Strenuella strenua* (Billings), *S. cobboldi* (Richter), *S. (Comluella) platycephala* Cobbold, *Calodiscus lakei* Rasetti (= *lobatus* of Lake; Cobbold; and Ahlberg), *Serrodiscus robustus* (Kobayashi) (= *speciosus* of Lake; Cobbold), *Cobboldites comleyensis* (Cobbold); brachiopods *Lingulella viridis* (Cobbold), *Paterina minor*(?), *Obolus parvulus*, *Obolella atlantica;* hyoliths *Gracilitheca bayonet* (Matthew), *Microcornus* cf. *parvulus* Mambetov, *Burithes* sp., *Lenatheca* sp.; small shelly fossils *Hadimopanella apicata* Wrona, *Lapworthella cornu cornu* (Wiman) or *L. cornu nigra* Cobbold, *Cowiella reticulata* Hinz, *Rhombocorniculum cancellatum*, *Hertzina elongata* Müller, *Amphigeisina danica* (Poulsen), *Hyolithellus micans*, *Chancelloria lenaica*; molluscs *Latouchella* sp., *Yochelcionella* sp., *Anabarella indecora* Miss., cf. *Stenotheca rugosa abrupta* Shaler & Foerste, *Pelagiella lorenzi* Kobayashi; sponge *Calcihexactina* sp. and several ostracodes and hyoliths.

The *Protolenus* Limestone Bed (Ac5) is a dark- to pale-grey or brownish phosphatic limestone 0.15 m thick and very fossiliferous. Echinoderm calcite ossicles are abundant, glauconite is present, and detrital minerals are rare. The fauna includes *Protolenus (Latoucheia) latouchei* Cobbold, *P. morpheus* Cobbold, *Mohicana clavata* Cobbold, *Strettonia comleyensis* Cobbold, *Calodiscus lakei*, *Serrodiscus robustus*, *Cobboldites comleyensis* (Cobbold), *Runcinodiscus index* Rushton; brachiopods *Paterina labradorica*, *Obolus parvulus*, and *Lingulella viridis*; small shelly fossils *Lapworthella cornu*, *Rhombocorniculum cancellatum*, *Hertzina elongata*, *Gapparodus bisulcatus* (Müller), *Amphigeisina danica*, *Rushtonia lata* Cobbold, *Hyolithellus* cf. *micans*, *Microdictyon* cf. *effusum* Bengtson, Matthews & Missarzhevsky (? = *Archaeocyathus*(?) sp. of Cobbold, 1931), *Chancelloria lenaica*, *Lenastella*? sp.; hyoliths *Allatheca degeeri*, *Gracilitheca bayonet*, *Microcornus* cf. *parvulus*, '*Hyolithes*' *crassus* Cobbold, '*H.*' cf. *lenticularis* Holm, '*H.*' *holmi* Cobbold, '*H.*' *pallidus* Cobbold, '*H.*' *similis* Walcott, '*H.*' *sculptilis* Cobbold; molluscs *Helcionella cingulata*, *Latouchella costata* Cobbold, *Igorella striata* (Cobbold); cf. *Stenotheca rugosa abrupta*, *Pelagiella* sp.; and cystoid echinoderm ossicles.

The *Lapworthella* Limestone (Ad) ranges from 0 to 0.15 m, locally transgressing underlying members and coming to rest on the Lower Comley Sandstone. It clearly post-dates a phase of local erosion, but precedes the regression prior to the basal Middle Cambrian Quarry Ridge Grits. It comprises a white, pinkish, or black calcareous bed with phosphate granules, manganiferous nodules, abundant glauconite and pockets of coarse and fine detritus. The fauna lacks trilobites and comprises phosphatic remains of brachiopods *Acrothyra sera* (Matthew), *A. comleyensis* Cobbold, *Acrotreta*(?) sp., *Lingulella viridis*; small shelly fossils *Rhombocorniculum cancellatum*, *R. spinosus*, *Hertzina elongata*, *Prooneotodus* sp., *Lapworthella cornu cornu*, *L. cornu nigra*, *Eccentrothera kanesia*, *Cowiella granuolata* Hinz, *Torellella biconvexa*, *Hyolithellus* cf. *micans*, possible *Microdictyon* sp.(?) (= polyzoan of Cobbold 1931; Cobbold & Pocock 1934); bradoriid crustaceans *Indiana obtusa* (Cobbold), *Hipponicharion* sp.; mollusc *Pelagiella* sp., and sponge spicules. This assemblage shows clear signs of abrasion and differential preservation; the smooth form of *Lapworthella cornu nigra* may owe something to this abrasion.

The Ad Beds are overlain by a thin phosphatic skin, followed by conglomeratic beds of the Quarry Ridge Grits, bearing clasts of Lower Comley Sandstone, Lower Comley Limestone, and possible Precambrian. This assemblage bears the Middle Cambrian trilobite *Paradoxides groomi* Lapworth.

5.4.3. Correlation

The Precambrian succession of Shropshire broadly compares with that of south-east Newfoundland (e.g. Anderson 1972, 1981; Landing *et al.* 1988). The Shropshire Cambrian is very arenaceous, however, contrasting with the argillaceous post-Random strata of Nuneaton and New England–Newfoundland. The lithostratigraphy is more comparable with that on the margins of the Baltic Shield at Mjøsa in Norway, or in Scania and Bornholm, though the fauna is more directly comparable with that of the Avalon region.

On the western side of the Church Stretton Fault, the western 'Uriconian' may compare with the Bull Arm volcanics of south-east Newfoundland. The succeeding tuffaceous and argillaceous Stretton Series of the Eastern Longmyndian may then compare with the Gibbet Hill Formation, Signal Group, and Bighead Formation, Musgravetown Group, which also contain *Arumberia* (Bland 1984). The succeeding molasse facies of the Eastern Longmyndian compares with upper parts of the Signal Hill and Musgravetown Groups.

On the eastern side of the Church Stretton Fault, the Rushton Schists metamorphism is dated at 667 ± 20 Ma, which compares with others in England and Wales: Stanner–Hunter metasediments (702 ± 8), Charnian porphyroids (684 ± 29 Ma), Malvernian gneisses (670 ± 10 Ma), Rosslare Complex (*c*.650 Ma), and Johnston Complex (*c*.643 Ma; see Thorpe *et al.* 1984). Comparison may also be made with the Coldbrook volcanics of New Brunswick (795 ± 80 Ma,

Anderson 1972). These events fall within Cadomian Phase I of the Avalon and Armorican terranes, dated at between 690 and 650 Ma (Jenkins 1984).

The eastern Uriconian volcanics, dated at 558 ± 16 Ma, compare with igneous events dated at 570–530 Ma throughout England and Wales (Thorpe *et al.* 1984; Cowie & Johnson 1985) including the Sarn Complex intrusions of Llêyn (549 ± 17 Ma) and possibly contemporaneous Arvonian volcanics and intrusives of Caernarfon, and the post-Charnian diorites (540 ± 57 Ma) cutting the Caldecote Volcanic Formation. Volcanicity notably continued through the earliest stages of thick conglomerate accumulation in Caernarfon (Wood 1969); the latter conglomerates may correlate with the molasse of the Rencontre (= Doten Cove) Formations, and the upper Musgravetown, Hodgewater, and Signal Hill Groups of south-east Newfoundland. Further south, in the Harlech Dome area of North Wales, transgressive sandstones just above the volcanics contain *Platysolenites antiquissimus*, a rough index for a Tommotian age (Rushton 1978).

This major igneous and metamorphic 'event' at *c.*570–530 Ma is also found below the first skeletal faunas or transgressive marine sandstones in north-west France, southern Spain, Morocco, Sardinia, Israel, Arabia, Massachusetts, New Brunswick, Nova Scotia and south-east Newfoundland (e.g. Thorpe *et al.* 1984; Jenkins 1984; Cowie & Johnson 1985; Odin *et al.* 1985). These igneous rocks and their covering Cambrian deposits may not be everywhere of the same age, but the Hoppin Hill Granodiorite of eastern Massachusetts, dated at 553 ± 10 Ma (Gale *in* Cowie & Johnson 1985) was clearly prior to the classic *Aldanella attleborensis* fauna (e.g. Theokritoff 1968; Landing & Brett 1982). These dates contrast with the 'older' dates obtained from glauconites in the Soviet Union and illites in China (e.g. Cowie & Johnson 1985). It may be argued that the latter incorporate materials reworked into black shales, limestones, and phosphatic sediments. However, the Rb–Sr and U–Pb dates from the Shuijingtuo Formation of Hubei, China, are consistent at about 570 Ma (Cowie 1985). The fauna of the Shuijingtuo Formation, however, is at least of *Rhombocorniculum cancellatum* age (e.g. Qian & Zhang 1983) and the date is suspect.

It has been suggested that the igneous events outlined here resulted from massive island-arc accretion at a time of rapid sea-floor spreading (Thorpe *et al.* 1984) culminating in the compressive Avalonian Orogeny (Landing *et al.* 1988). This may in turn have been connected with the subsequent transgression and phosphogenic events, favouring the evolutionary explosion of skeletal fossils (Cook & Shergold 1984; Brasier 1985).

Lithostratigraphical correlation of the Wrekin Quartzite has been made with the Random Formation of south-east Newfoundland (Anderson 1981) but not usually with the Hartshill Formation of Nuneaton (Brasier & Hewitt 1979).

The sparse remains of inarticulate brachiopods (Odin *et al.* 1985) suggest that the unit is unlikely to be older than the Tommotian Stage of Siberia or the Meishucunian Stage of China. Paterinid brachiopods typically appear at the base of the Tommotian Stage in Siberia (e.g. Rozanov & Sokolov 1984) and in Zone II of China (e.g. Xing *et al.* 1983).

Diplocraterion is a distinctive trace fossil with clear stratigraphical potential. It occurs in the Chapel Island and Random Formations, stratigraphically close to the lowest *Aldanella attleborensis* fauna of the Burin Peninsula, but below that of the Bonavista and Avalon Peninsulas of south-east Newfoundland (e.g. Crimes & Anderson 1985; Narbonne *et al.* 1987). On the East European Platform it appears in the Lontova Formation and equivalents, close to the first *Platysolenites antiquissimus*, *A. polonica*, and *Aldanella kunda* and below the levels with *Mobergella* (Urbanek & Rozanov 1983; Fedonkin 1985). At Mjøsa in Norway and in Scania it appears near the top of the Ringsaker and Hardeberga quartzites respectively, below silty and glauconitic beds of the *Schmidtiellus mickwitzii* Zone (e.g. Martinsson 1974; Bergström 1981). In Siberia, it appears in the Tommotian Stage (Fedonkin 1988).

Thus the appearance of *Diplocraterion* near the top of the Wrekin Quartzite resembles the stratigraphical situation at Mjøsa and in Scania. The unit may tentatively be correlated with the Hardeberga Quartzite. This implies a pre-Lukati age (cf. Vidal 1981) and confirms a probable Tommotian correlation (Rozanov & Sokolov 1984).

The Ab1 assemblage with *Camenella baltica* and *Mobergella* cf. *turgida* clearly compares with the *Mobergella* fauna from a similar stratigraphical setting at Småland in Sweden (Bengtson 1977), which may form the base to the *Schmidtiellus mickwitzi* Zone of Scandinavia (Bergström, 1981). Since that contains *Rhombocorniculum insolutum* and *Mobergella*? sp. in Bornholm, a correlation of the *Mobergella* fauna of Ab1 with Bed 10iii of the Home Farm Member of Nuneaton and the upper Bonavista to Smith Point Limestone Formations of south-eastern Newfoundland seems likely. The previously suggested correlation with the basal Home Farm Member (Brasier & Hewitt 1981) was based on the co-occurrence of '*Obolus*' *groomi*, *Micromitra phillipsi*, and *Hyolithellus micans*.

The poorly fossiliferous beds Ab2 to Ab4 are insecurely correlated but the presence of bradoriid crustaceans and (?)*Kjerulfia* suggests a low *Callavia* Zone age, perhaps correlative with the Woodlands Member to Lower Purley Shale of Nuneaton, top Smith Point to low Brigus Formations of south-east Newfoundland, Rispbjerg Sandstone of Scania and Bornholm (on stratigraphical position), and Brastad Shale of Mjøsa, Norway.

The fossiliferous beds from Ac1 to Ac5 are more readily correlated with strata in south-eastern Newfoundland. Beds Ac1 to Ac4 fall within the *Callavia* Zone of Hutchinson (1962), while the presence of *Dipharus attleborensis* from Ac1 to Ac3 compares with the middle part of that Zone and the lower part of the Brigus Formation. *Serodiscus bellimarginatus* and *Calodiscus lakei*, combined with *Callavia broeggeri* in Beds Ac3 and Ac4, indicate the upper part of the *Callavia* Zone, at about the level of the Redlands Limestone in the Brigus Formation. The fauna with *Serrodiscus speciosus* and *Strenuella sabulosa* known from Nuneaton and Newfoundland, does not appear to be present. The *Protolenus* fauna in Ac5 compares with the *Protolenus* Zone of Newfoundland, containing forms correlated with the Toyonian of Siberia (Dr T. P. Fletcher, pers. comm. 1986). Similar assemblages of small shelly fossils are also found in Shropshire and New England–Newfoundland (see Chapter 2) and these assist with wider correlation. *Rhombocorniculum cancellatum* in Ac1 to Ac2 compares with the *Judomia* Zone. *R. cancellatum* plus *Lapworthella cornu*, *Microdictyon* sp., *Hadimopanella aspicata*, *Pelagiella lorenzi*, and *Gracilitheca bayonet* in Ac4 suggests correlation with the latest Atdabanian–Botomian assemblages of Siberia, the Lower Botomian of Salanygol, Mongolia, the Shabakty Formation of Southern Kazakhstan, and the Qiongzhusi Formation of the Yangtze Platform. This assemblage also compares with that of the condensed Gislöv Formation of Scania and Bornholm, which likewise passes up into a phosphatic limestone conglomerate with reworked microfossils, resembling Bed Ad.

5.5. SUGGESTED SEQUENCE OF EVENTS

There is no single 'complete' succession through late Precambrian–early Cambrian events in England and Wales but the complementary sections of Charnwood–Nuneaton and Shropshire provide fossil assemblages through a series of events closely similar to those of south-east Newfoundland. These faunas and events may now be briefly summarized into a composite succession as follows.

1. Cadomian Phase I, *c*.700–650 Ma BP, may be represented by the Rosslare Complex of Eire, the Malvern Complex of England, the Mona Complex sediments of North Wales, the Johnston Complex to Pebidian volcanics of South Wales, the Stanner–Hunter Group of Wales, and the Rushton Schists of Shropshire. These may be correlated with the Gneiss de Brest, Brittany, the Ouarzate Group Schists of the Anti-Atlas, and the Coldbrook volcanics of New Brunswick.

2. Cadomian Phase II, *c*.610–590 Ma BP, may be represented by Mona Complex metamorphism and intrusion in North Wales, Dimetian intrusions of South Wales, and ?Western Uriconian volcanics of Shropshire. These igneous events may be correlated with the Long Cove, Rocky Harbour, and Harbour Main volcanics and Holyrood Granite of south-east Newfoundland, the Vire-Carolles granite of Normandy and the Taolecht granite of Morocco. The Laplandian glaciation subsequently deposited the Gaskiers Formation in Newfoundland; this level is not seen in England.

3. The *Charnia masoni* fauna is preserved in the Blackbrook to Brand Series of the Charnian Group and the Conception Group of the Avalon Peninsula, south-east Newfoundland. The suggested age is about 566 Ma BP.

4. Caldecote–late Uriconian–Arvonian igneous events, *c*.560–540 Ma BP, are represented by the Caldecote volcanics of Nuneaton, the Markfield diorites of Charnwood and Nuneaton, ?the Barnt Green volcanics of Birmingham, the Eastern Uriconian volcanics of Shropshire, the Arvonian volcanics of North Wales, the Sarn granite and metamorphism of the Llêyn Peninsula. These may be correlated with the Cobo adamellite of Guernsey, various Moroccan granites, the Bull Arm volcanics of south-east Newfoundland and the Hoppin Hill granodiorite of Massachusetts.

5. The *Arumberia* fauna is preserved in the Stretton Series of Shropshire and the Signal Hill/Hodgewater Groups of south-east Newfoundland, associated with argillites and water-lain tuffs. Igneous events continued and horst-and-graben (or listric fault) tectonics developed at about this time.

6. Molasse and non-marine facies are developed in the Wentnor Series of Shropshire and the Tryfan to Glog grits and conglomerates of North Wales; the latter are contemporaneous with volcanicity. These may be correlated with the upper Signal Hill/Hodgewater/Musgravetown Groups, and the Doten Cove/Rencontre Formations of south-east Newfoundland. These facies changes are apparently associated with the Avalonian

Orogeny and the development of horst-and-graben (or listric fault) tectonics, during which the North Wales Basin, the Warwickshire Trough, and the 'Chapel Island' Basin began to subside. Plutonic intrusions were also uplifted and exposed. Sediments in North Wales and Newfoundland passed upwards from non-marine ?fanglomerates and red beds to shallow-marine environments.

7. The initial stages in this Cambrian transgression show a similar succession of biotas in the Burin Peninsula, East European Platform, and northern Scandinavia: replacement of a vendotaeniid algal flora —*Harlaniella podolica* trace-fossil assemblage by an assemblage with *Phycodes pedum* and *Sabellidites*; appearance of trace fossils: *Rusophycus* sp., *Arenicolites* sp., *Diplocraterion* sp., tube *Platysolenites antiquissimus*, plus gastropods of the *Aldanella attleborensis* group and simple hyoliths. A major rise in sea-level during the latter interval probably laid down the following units: Dolwen Grit (with *P. antiquissimus*) and Glog Grit in North Wales; Wrekin Quartzite and Hartshill 'quartzites' in England; Random Formation in south-east Newfoundland; Ringsaker and Hardeberga Quartzite in Norway and Sweden, and the Lontova Formation of the East European Platform (e.g. Føyn & Glaessner 1979; Urbanek & Rozanov 1983; Narbonne *et al.* 1987).

8. Block movements shifted the centres of deposition away from Nuneaton (resulting in the condensed Home Farm Member) and away from the Burin Peninsula, south-east Newfoundland (causing non-deposition or erosion). Breaks at this level occur throughout Avalonia, and Baltica and Spitsbergen.

9. Deposition resumed, bearing the *Aldanella attleborensis* fauna of the lower Bonavista Formation, south-east Newfoundland, lower Hoppin Slate and Weymouth Formation, Massachusetts; condensation and non-deposition were suspected at this level throughout most of England and Baltica, although evidence for deposition is possibly available from boreholes beneath Oxfordshire (Dr. A. W. A. Rushton *et al.* unpublished data).

10. A change towards more nodular limestone sedimentation was accompanied by a change to a *Camenella–Coleoloides* fauna in the middle–upper Bonavista Formation, south-east Newfoundland and middle Home Farm Member, Nuneaton, with onlapping deposits in Baltica.

11. Thicker limestones bear the *Rhombocorniculum*

insolutum and *Randomia* fauna of the upper Home Farm Member, Nuneaton, correlated with the lower to middle Smith Point Limestone of south-east Newfoundland; both have several Fe–Mn enriched and dolomitized discontinuity surfaces. These are correlated with the Ab1 glauconitic sandstones in Shropshire and clastic facies in the Malvern Hills and Baltica that bear a *Mobergella* fauna. Of these, the Green Shales of Bornholm yield K–Ar glauconite dates of *c.*538 Ma BP (Dr V. Berg-Madsen, pers. comm. 1986). Early olenellids made their appearance on platforms at about this time, such as *Profallotaspis* and *Fallotaspis* in Siberia and *Schmidtiellus* in the Balto-Scandinavian region.

12. Further block movements followed local emergence, bringing about thick argillaceous sedimentation at Nuneaton and western Avalon but condensed or non-deposition in Bonavista and eastern Avalon, south-east Newfoundland. Glauconitic sandstone deposition continued in Shropshire.

13. The *Callavia* fauna appeared, as in the upper Smith Point Limestone–basal Brigus Formation of south-east Newfoundland and the lower Purley Shale Formation of Nuneaton. *Rhombocorniculum cancellatum* appeared at about this time in Newfoundland and Siberia.

14. The *Dipharus attleborensis* fauna appeared, as in the lower Brigus Formation of south-east Newfoundland and the Ac1–Ac2 Beds of Shropshire. Condensed deposition in Shropshire began at this time, perhaps in response to the foundering of adjacent blocks. *Microdictyon effusum* appeared widely with *Rhombocorniculum cancellatum* at about this level, as in Newfoundland and Kazakhstan.

15. The *Serrodiscus bellimarginatus* fauna appeared, as in the Brigus Formation of south-east Newfoundland, the Purley Shale Formation of Nuneaton, and the Ac3–Ac4 Beds of Shropshire. Widespread small shelly fossils include *R. cancellatum*, *Lapworthella cornu*, *Microdictyon effusum*, *Hadimopanella apicata*, *Pelagiella lorenzi*, and *Gracilitheca bayonet*. Eustatic deepening is suggested by widespread pagetiids, and blind eodiscids and conocoryphids.

16. The *Strenuella sabulosa* fauna of the Brigus Formation and the Purley Shale Formation may be represented by a hiatus between Ac4 and Ac5 beds in Shropshire. This appears at the time of a major faunal turnover, with extinction of *Callavia* plus the Serrodiscus bellimarginatus fauna.

M. D. BRASIER

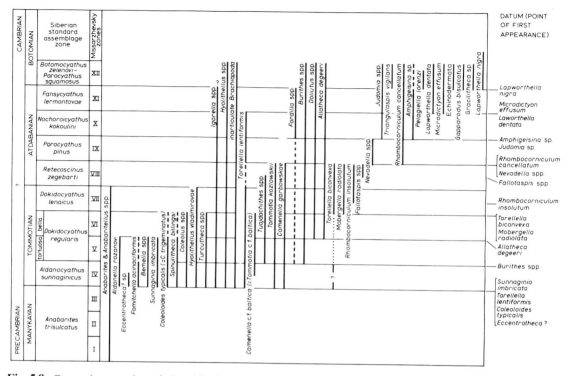

Fig. 5.8. Composite range chart of selected fossils through the Precambrian–Cambrian boundary beds of the Siberian Platform. Taxa selected are useful for biostratigraphical correlation·with England. Solid line—common occurrence; dashed line—restricted occurrence and uncertain range; dotted line, not reported. (Modified from Brasier 1986.)

17. The *Protolenus* fauna appeared, as in the Brigus formation and the Ac5 Beds of Shropshire. The brachiopod *Botsfordia caelata* group appeared widely at about this time.

18. The *Acidiscus theristes–Serrodiscus ctenoa* fauna appeared, as in the Brigus and Purley Shale Formations.

19. Block movements, possibly involving movements on the Church Stretton Fault and uplift of the Longmynd Horst, led to partly reworked *Lapworthella cornu nigra* faunas of the Ad Beds, locally unconformable over older beds in Shropshire. Manganese salts formed in Shropshire, south-east Newfoundland, and North Wales, the latter spanning turbiditic sedimentation. Continued movements resulted in renewal of thicker, clastic deposition in Shropshire. Movements along the Tornquist Line in Scania and Bornholm led to

hiatuses and highly condensed Middle Cambrian strata while on the margins of the North American Craton, the early Middle Cambrian is often missing (Palmer 1981).

20. Appearance of the *Paradoxides bennetti–P. oelandicus* faunas of south-east Newfoundland, Nuneaton and Shropshire, marks the start of mid-Cambrian times.

This sequence of biological and geological events allows some preliminary correlations between the sections in England and south-eastern Newfoundland. The writer has also drawn attention to the similar sequence of first-appearance datum points in the faunas of England and Siberia (Fig. 5.8, Brasier 1986). Further work is needed, but it does indicate that biostratigraphical correlation between these areas should be possible, once the nature of these bio-events has been carefully studied and further taxonomic work undertaken.

REFERENCES

Anderson, M. M. (1972). A possible time span for the late Precambrian of the Avalon peninsula, southeastern Newfoundland, in the light of worldwide correlation of fossils, tillites, and rock units within the succession. *Can. J. Earth Sci.*, **9**, 1710–26.

Anderson, M. M. (1981). The Random Formation of southeastern Newfoundland: a discussion aimed at establishing its age and relationship to bounding formations. *Am. J. Sci.*, **281**, 807–30.

Anderson, M. M. & Conway Morris, S. (1982). A review, with descriptions of four unusual forms, of the soft-bodied fauna of the Conception and St. John's Groups (Late Precambrian), Avalon Peninsula, Newfoundland. *Proc. Third N. Am. Paleont. Conv.*, **1**, 1–8.

Barber, A. (1985). A new concept of mountain building. *Geol. Today*, **1**, 116–121.

Bath, A. H. (1974). New isotopic age data on rocks from the Long Mynd, Shropshire. *J. geol. Soc. Lond.*, **130**, 567–74.

Bengtson, S. (1977). Aspects of problematic fossils in the early Palaeozoic. *Acta Univ. Uppsalliensis*, **415**, 1–71.

Bengtson, S. & Fletcher, T. P. (1983). The oldest sequence of skeletal fossils in the Lower Cambrian of southeastern Newfoundland. *Can. J. Earth Sci.*, **20**, 525–36.

Benus, A. P. (in press). Sedimentological context of a deep-water Ediacaran fauna (Mistaken Point Formation, Avalon Zone, Eastern Newfoundland). In *Trace fossils, small shelly fossils and the Precambrian–Cambrian boundary—Proceedings* (eds E. Landing & G. M. Narbonne). Bull. N.Y. State Mus., **463**.

Benus, A. P. & Landing, E. (1984). Depositional environment and biofacies of the Bonavista Formation (early Cambrian [Tommotian–lower Atdabanian]) eastern Newfoundland. *Geol. Soc. Am. Abstr. Prog.*, **16**, p. 3.

Bergström, J. (1973). Classification of olenellid trilobites and some Balto-Scandian species. *Norsk Geol. Tidsskr.*, **53**, 283–314.

Bergström, J. (1981). Lower Cambrian shelly faunas and biostratigraphy in Scandinavia. *Open File Rep. US geol. Surv.*, **81–743**, 22–5.

Bland, B. H. (1984). *Arumberia* Glaessner & Walter, a review of its potential for correlation in the region of the Precambrian–Cambrian boundary. *Geol. Mag.*, **121**, 625–33.

Boulton, W. S. (1924). On a recently discovered breccia-bed underlying Nechells (Birmingham), and its relation to the red rocks of the district. *Q. J. geol. Soc. Lond.*, **80**, 343–73.

Boynton, H. E. (1978). Fossils from the Precambrian of Charnwood Forest, Leicestershire. *Mercian Geol.*, **6**, 291–6.

Boynton, H. E. & Ford, T. D. (1979). *Pseudovendia charnwoodensis*—a new Precambrian arthropod from Charnwood Forest, Leicestershire. *Mercian Geol.*, **7**, 175–7.

Brasier, M. D. (1980). The Lower Cambrian transgression and glauconite–phosphate facies in western Europe. *J. geol. Soc. Lond.*, **137**, 695–703.

Brasier, M. D. (1984). Microfossils and small shelly fossils from the Lower Cambrian *Hyolithes* Limestone at Nuneaton, English Midlands. *Geol. Mag.*, **121**, 229–53.

Brasier, M. D. (1985). Evolutionary and geological events across the Precambrian–Cambrian boundary. *Geol. Today*, **1**, 141–6.

Brasier, M. D. (1986). The succession of small shelly fossils (especially conoidal microfossils) from English Precambrian–Cambrian boundary beds. *Geol. Mag.*, **123**, 327–56.

Brasier, M. D. & Hewitt, R. A. (1979). Environmental setting of fossiliferous rocks from the uppermost Proterozoic–Lower Cambrian of central England. *Palaeogeogr., Palaeoclimatol., Palaeoecol.*, **27**, 35–57.

Brasier, M. D. & Hewitt, R. A. (1981). Faunal sequence within the Lower Cambrian 'Non-Trilobite Zone' (s.l.) of central England and correlated regions. *Open File Rep. US geol. Surv.*, **81–743**, 26–8.

Brasier, M. D. & Hewitt, R. A., & Brasier, C. J. (1978). On the Late Precambrian–Early Cambrian Hartshill Formation of Warwickshire. *Geol. Mag.*, **115**, 21–36.

Callaway, C. (1878). On the quartzites of Shropshire. *Q. J. geol. Soc. Lond.*, **34**, 754–63.

Cobbold, E. S. (1919). Cambrian Hyolithidae, etc. from Hartshill in the Nuneaton District, Warwickshire. *Geol. Mag.*, **56**, 149–58.

Cobbold, E. S. (1921). The Cambrian horizons of Comley (Shropshire) and their Brachiopoda, Pteropoda and Gasteropoda etc. *Q. J. geol. Soc. Lond.*, **76**, 325–86.

Cobbold, E. S. (1931). Additional fossils from the Cambrian rocks of Comley, Shropshire. *Q. J. geol. Soc. Lond.*, **87**, 459–512.

Cobbold, E. S. & Pocock, R. W. (1934). The Cambrian area of Rushton (Shropshire). *Phil. Trans. R. Soc. Lond.*, **B223**, 305–409.

Cook, P. J. & Shergold, J. H. (1984). Phosphorus, phosphorites and skeletal evolution at the Precambrian–Cambrian boundary. *Nature*, **308**, 231–6.

Cope, J. C. W. & Gibbons, W. (1987). New evidence for the relative age of the Ercall Granophyre and its bearing on the Precambrian–Cambrian boundary in southern Britain. *Geol. J.*, **22**, 53–60.

Cowie, J. W. & Johnson, M. R. W. (1985). Late Precambrian and Cambrian geological time scale. *Mem. geol. Soc. Lond.*, **10**, 65–72.

Cowie, J. W., Rushton, A. W. A. & Stubblefield, C. J. (1972). A correlation of Cambrian rocks in the British Isles. *Spec. Rep. geol. Soc. Lond.*, **2.**, 42 pp.

Cribb, S. J. (1975). Rubidium–strontium ages and strontium isotope ratios from the igneous rocks of Leicestershire. *J. geol. Soc. Lond.*, **131**, 203–12.

Crimes, T. P. & Anderson, M. (1985). Trace fossils from late Precambrian–early Cambrian strata of southeastern Newfoundland (Canada): temporal and environmental implications. *J. Paleont.*, **59**, 310–43.

Daily, B. (1972). The base of the Cambrian and the first Cambrian faunas. *Centre for Precambrian Res., Univ. Adelaide Spec. Paper*, **1**, 13–37.

Fedonkin, M. A. (1985). Paleoichnology of Vendian Metazoa. In *Vendian System* (ed. B. S. Sokolov), pp. 112–17. Akad. Nauk SSSR, Moscow. [In Russian.]

Fedonkin, M. A. (1988). Paleoichnology of the Precambrian–Cambrian transition in the Russian Platform and Siberia. In *Trace fossils, small shelly fossils and the Precambrian–Cambrian boundary—Proceedings.* (eds E. Landing & G. M. Narbonne). Bull. New York State Mus., **463**. (in press.)

Ford, T. D. (1958). Precambrian fossils from Charnwood Forest. *Proc. Yorks. Geol. Soc.*, **31**, 211–17.

Føyn, S. & Glaessner, M. F. (1979). *Platysolenites*, other animal fossils, and the Precambrian–Cambrian transition in Norway. *Norsk Geol. Tidsskr.*, **59**, 25–46.

Glaessner, M. F. (1984). *The dawn of animal life.* Cambridge University Press.

Greig, D. C., Wright, J. E., Hains, B. A., & Mitchell, G. H. (1968). Geology of the country around Church Stretton, Craven Arms, Wenlock Edge and Brown Clee. *Mem. geol. Surv. GB*, **166**, 8.

Hinz, I. (1983). Zur unterkambrischen Mikrofauna von Comley und Umgebung in Shropshire. Unpubl. Diploma Thesis, Bonn Univ., pp. 1–55.

Hinz, I. (1987). The Lower Cambrian microfauna of Comley and Rushton, Shropshire/England. *Palaeontographica*, **A198**, 41–100.

Hiscott, R. N. (1982). Tidal deposits of the Lower Cambrian Random Formation, eastern Newfoundland: facies and palaeoenvironments. *Can. J. Earth Sci.*, **19**, 2028–46.

Hofmann, H. J., Hill, J., & King, A. F. (1979). Late Precambrian microfossils, southeastern Newfoundland. *Geol. Surv. Can. Pap.*, **79–1B**, 83–98.

Hupé, P. (1952). Contribution à l'étude du Cambrian inférieur et du Pré-Cambrian III de l'Anti-Atlas Marocain. *Notes Mem. Serv. Mines Carte geol. Maroc*, **103**, 402 pp. (Issued 1953).

Hutchinson, R. D. (1962). Cambrian stratigraphy and trilobite faunas of southeastern Newfoundland. *Bull. geol. Surv. Can.*, **88**, 1–156.

James, J. H. (1956). The structure and stratigraphy of part of the Pre-Cambrian outcrop between Church Stretton and Linley, Shropshire. *Q. J. geol. Soc. Lond.*, **112**, 315–35.

Jenkins, R. J. F. (1984). Ediacaran events: boundary relationships and correlation of key sections, especially in 'Armorica'. *Geol. Mag.*, **121**, 635–43.

Krogh, T. E., Strong, D. F., & Papezik, V. (1983). Precise U–Pb ages of zircons from volcanic and plutonic units in the Avalon peninsula. *Geol. Soc. Am. Abstr. Progm.* **15**, 135.

Landing, E. & Benus, A. P. (1984). Lithofacies belts of the Smith Point Limestone (Lower Cambrian, eastern Newfoundland) and the lowest occurrence of trilobites. *Geol. Soc. Am. Abstr. Prog.*, **16**, p. 45.

Landing, E. & Benus, A. P. (1988). Stratigraphy of the Bonavista Group, southeastern Newfoundland; growth faults and the distribution of sub-trilobitic Lower Cambrian. In *Trace Fossils, small shelly fossils and the Precambrian–Cambrian boundary*, (eds E. Landing & G. M. Narbonne). Bull. New York State Mus. **463**. (In press).

Landing, E. & Brett, C. E. (1982). Lower Cambrian of eastern Massachusetts: microfaunal sequence and the oldest known borings. *Geol. Soc. Am. Abstr. Prog.* **14**, p. 35.

Landing, E., Narbonne, G. M., Myrow, P., Benus, P., & Anderson, M. M. (1988). Field trip. Faunas and depositional environments of the Upper Precambrian through Lower Cambrian, southeastern Newfoundland. In *Trace fossils, small shelly fossils and the Precambrian–Cambrian boundary*, (eds E. Landing & G. M. Narbonne). Bull. N.Y. State Mus., **463**. (In press).

Lapworth, C. (1882). On the discovery of Cambrian rocks in the neighbourhood of Birmingham. *Geol. Mag.*, **9**, 563–5.

Martinsson, A. (1974). The Cambrian of Norden. In *Cambrian of the British Isles, Norden and Spitsbergen*, (ed. C. H. Holland), pp. 185–283. John Whiley, London.

Matthews, S. C. (1973). Lapworthellids from the Lower Cambrian *Strenuella* Limestone at Comley, Shropshire. *Palaeontology*, **16**, 139–48.

Matthews, S. C. & Missarzhevsky, V. V. (1975). Small shelly fossils of late Precambrian and early Cambrian age: a review of recent work. *J. geol. Soc. Lond.*, **131**, 289–304.

Meneisy, M. Y. & Miller, J. A. (1963). A geochemical study of the crystalline rocks of Charnwood Forest, England. *Geol. Mag.*, **100**, 507–23.

Narbonne, G. M., Myrow, P. M., & Landing, E. (1987). A candidate stratotype for the Precambrian–Cambrian boundary, Fortune Head, Burin Peninsula, southeastern Newfoundland. *Can. J. Earth Sci.*, **24**, 1277–93.

Notholt, A. J. G. & Brasier, M. D. (1986). Proterozoic and Cambrian phosphorites—regional review. In *Phosphate deposits of the World. Volume 1. Proterozoic and Cambrian phosphorites*, eds P. J. Cook & J. H. Shergold), pp. 91–100. Cambridge University Press.

Odin, G. S., Gale, N. H., Auvray, B., Bielski, M., Doré, F., Lancelot, J.-R., & Pasteels, P. (1983). Numerical dating of Precambrian–Cambrian boundary. *Nature*, **301**, 21–3.

Odin, G. S., Gale, N. H. & Doré, F. (1985). Radiometric dating of Late Precambrian times. *Mem. geol. Soc. Lond.*, **10**, 65–72.

Palmer, A. R. (1981). Subdivision of the Sauk sequence. *Open-File Rep. US geol. Surv.* **81–743**, 160–2.

Peat, C. (1984). Precambrian microfossils from the Longmyndian of Shropshire. *Proc. geol. Assoc.*, **95**, 17–22.

Qian Yi & Zhang Shi-Ben (1983). Small shelly fossils from the Xihaoping Member of the Tongying Formation in Fangxian county of Hubei Province and their stratigraphical significance. *Acta Palaeont. Sinica*, **22**, 82–94. [In Chinese with English summary.]

Rozanov, A. Yu. (1975). The problem of the lower boundary of the Cambrian. *Earth Science Revs.*, **11**, 221–33.

Rozanov, A. Yu. & Sokolov, B. S. (1984). *Lower Cambrian Stage Subdivision. Stratigraphy.* Akademii Nauk SSSR: Izdatelstvo 'Nauka', Moscow. [In Russian.]

Rushton, A. W. A. (1966). The Cambrian trilobites from the Purley Shales of Warwickshire. *Palaeont. Soc. Monogr.*, **511**.

Rushton, A. W. A. (1972). In *Ann. Rep. Inst. geol. Sci. for 1971*, 93.

Rushton, A. W. A. (1974). The Cambrian of Wales and England. In *Cambrian of the British Isles, Norden & Spitsbergen* (ed. C. H. Holland), pp. 43–122. John Wiley, London.

Rushton, A. W. A. (1978). Appendix 3. Description of the macrofossils from the Dolwen Formation. *Bull. geol. Surv. GB*, **61**, 46–8.

Sedgwick, A. & Murchison, R. I. (1835). On the Silurian and Cambrian Systems, exhibiting the order in which the older sedimentary strata succeeded each other in England and Wales. *Lond. Edinburgh Phil. Mag.*, **7**, 483–5.

Shergold, J. H. & Brasier, M. D. (1986). Biochronology of Proterozoic and Cambrian phosphorites. In *Phosphate deposits of the World. Volume 1. Proterozoic and Cambrian phosphorites*, (eds P. J. Cook & J. H. Shergold), pp. 295–326. Cambridge University Press.

Smith, J. D. D. & White, D. E. (1963). Cambrian trilobites from the Purley Shales. *Palaeontology*, **6**, 397–407.

Sokolov, B. S. (1972). The Vendian stage in Earth history. *24th Int. Geol. Congr. Montreal 1972*, **1**, 78–84.

Theokritoff, G. (1968). Cambrian biogeography and biostratigraphy in New England. In *Studies of Appalachian geology*, (eds E-an Zen, W. S. White, J. B. Hadley, & J. B. Thompson Jr), pp. 9–22. Wiley-Interscience, New York.

Thorpe, R. S., Beckinsale, R. D., Patchett, P. J., Piper, J. D. A., Davies, G. R., & Evans, J. A. (1984). Crustal growth and late Precambrian–early Palaeozoic plate tectonic evolution of England and Wales. *J. geol. Soc. Lond.*, **141**, 521–36.

Urbanek, A. & Rozanov, A. Yu. (1983). *Upper Cambrian and Cambrian palaeontology of the East-European Platform.* Publishing House Wydawnictwa Geologiczne, Warszawa.

Vidal, G. (1981). Micropalaeontology and biostratigraphy of the Lower Cambrian sequence in Scandinavia. *Open-File Rep. US geol. Surv.*, **81–743**, 232–5.

Walcott, C. D. (1900). Lower Cambrian terrane in the Atlantic Province. *Proc. Washington Acad. Sci.*, **1,** 301–39.

Walliser, H. Von O. (1958). *Rhombocorniculum comleyense* n. gen., n. sp. (*Incertae sedis*, Unterkambrium, Shropshire). *Paläont. Z.*, **32,** 176–80.

Whitcombe, D. N. & McGuire, P. K. H. (1981). A seismic refraction investigation of the Charnian basement with granitic intrusions flanking Charnwood Forest. *J. geol. Soc. Lond.*, **138,** 643–51.

Whittaker, A. & Chadwick, R. A. (1984). The large-scale structure of the Earth's crust beneath southern Britain. *Geol. Mag.*, **121,** 621–4.

Wood, D. S. (1969). The base and correlation of the Cambrian rocks of North Wales. In *The Precambrian and Lower Palaeozoic rocks of Wales*, (ed. A. Wood), pp. 47–66. University of Wales Press, Cardiff.

Xing Yusheng, Ding Qixiu, Luo Huilin, He Tinggui, & Wang Yangeng (1983). The Sinian–Cambrian boundary of China. *Bull. Inst. Geol., Chinese Acad. geol. Sci.*, **10.** [In Chinese.]

6

Other areas: North-west Canada; California, Nevada, and Mexico; Morocco, Spain, and France

M. D. Brasier & J. W. Cowie

6.1. NORTH-WEST CANADA

This region contains a clastic wedge within the eastern Canadian Cordillera, bordering the North American craton, where fossiliferous successions spanning the late Precambrian and earliest Cambrian are preserved. Important sections occur in the Cassiar Mountains of north-central British Colombia, the Wernecke Mountains of Yukon Territory, and the Mackenzie Mountains of Northwest Territories, Canada (Fritz 1980). Four clastic to carbonate Grand Cycles are seen along the strike, between the late Precambrian tillite and the *Nevadella* Zone, allowing event correlation of cycle tops (e.g. Aitken 1981; Fritz & Crimes 1985; Aitken 1988). These Cordilleran sequences are notable for their combination of Ediacarian fauna, trace fossils, early skeletal fossils, and North American trilobites. Although this might seem to be ideal, many of the sections are relatively inaccessible (requiring a helicopter) and skeletal fossils are sparse. Lateral variations are rapid and correlations of facies and hiatuses are unclear, so the following brief résumé is necessarily simplified and selective.

Map unit 10b or the Blueflower Formation (> 700 m) is a clastic unit containing soft-bodied impressions of *Inkrylovia* sp., *Suzmites*? sp., and *Sekwia excentrica* Hofmann, plus trace fossils *Torrowangea* and *Gordia* in the Mackenzie Mountains (Hofmann 1981). Forms compared with *Sekwia excentrica*, *Beltanelliformis brunsae* Menner, and *Cyclomedusa* sp. also occur at about this level, 350 m below the *Protohertzina–Anabarites* fauna of the Wernecke Mountains (Hofmann *et al.* 1983; Nowlan *et al.* 1985; Narbonne & Hofmann 1987). Arguably Precambrian trace fossils from the lower 400 m of the Stelkuz Formation of the Cassiar Mountains (> 609 m) include *Chondrites*, *Didymaulichnus*, *Gordia*, *Helminthopsis*, *Neonereites*, *Planolites*, *Skolithos*, and *Taphrhelminthopsis*, below a disconformity surface (Fritz & Crimes 1985).

Map unit 11 or the Risky Formation (locally 150–200 m with *Palaeophycus*) represents the upper, carbonate half-cycle, preserved in the northern part of the Cordillera. The early skeletal fossil *Protohertzina* aff. *anabarica* and several other problematical forms were reported from this unit in the Mackenzie Mountains by Conway Morris & Fritz (1980), though Aitken (1984, 1988) suggests that this fauna actually came from the top of the succeeding Ingta Formation cyclothem, above traces *Harlaniella*, *Phycodes*, and others.

The Backbone Ranges or Vampire Formation (Map units 12 to 13) varies in thickness, being up to 1300 m in the Mackenzie Mountains but reducing to 300 m in the Wernecke Mountains. Here it overlies a significant disconformity which may cut out lower parts of the formation and upper parts of Map unit 11. At section 8E, it bears a phosphatic limestone conglomerate at its base, containing an association of early skeletal fossils: *Anabarites trisulcatus*, *Hyolithellus* cf. *isiticus*, *Protohertzina anabarica* group, annulated phosphatic tubes and other problematica including ?*Maikhanella* sp. (G. M. Narbonne, pers. comm. 1987), plus arthropod scratch-marks, *Cruziana* sp., *Rusophycus* sp., and *Palaeophycus* sp. (Nowlan *et al.* 1985). It is thought that the skeletal fossils in this assemblage were not derived from Map unit 11 beneath. If the 'late Tommotian' age of *Cruziana* traces can be substantiated (Crimes, Chapter 8), the Nemakit–Daldyn to Chinese zone I aspect of the skeletal fossils would indicate condensation, hiatuses, and reworking at this level. This postulated hiatus may correlate with the strong erosional event seen widely across the southern part of the eastern Canadian Cordillera, between Precambrian and Cambrian trace-fossil assemblages (e.g. Fritz & Crimes 1985).

The trace fossil *Phycodes* aff. *pedum* occurs about 15 m higher in dark shales of the Vampire Formation at section 8C. The correlative upper Stelkuz and Boya Formations, Cassiar Mts, also contain traces *Arenico-*

lites?, *Cruziana, Diplocraterion, Monomorphichnus, Plagiogmus, Planolites, Skolithos, Treptichnus,* and *Teichichnus,* taken to be of Cambrian age (Fritz & Crimes 1985).

The Sekwi Formation (*c.*800 m) represents the carbonate half-cycle, and follows above a mild disconformity in the Wernecke Mts, where the first trilobites are of late *Fallotaspis* or early *Nevadella* Zone age and pass up into *Nevadella* and *Bonnia–Olenellus* Zones (Fritz 1972; Fritz 1973; Nowlan *et al.* 1985). The low *Nevadella* Zone here contains *Microdictyon* aff. *rhomboidale* Bengtson *et al.* 1986 and roughly correlates with the Rosella Formation of the Cassiar Mountains, where basal strata in the *Nevadella* Zone contain *Lapworthella filigrana* (Conway Morris & Fritz, 1984). The presence of *Serrodiscus mackenziensis* Fritz from the lower part of the Sekwi Formation (Fritz 1973) is either a very early record for this group or indicates significant diachronism for the *Nevadella* Zone (see Brasier, this volume, pp. 150–1).

This sequence may be correlated with Meishucun and the Palaeotethyan scheme (see Brasier, this volume, Chapter 3), tentatively as shown in Table 6.1.

6.2. CALIFORNIA, NEVADA, AND MEXICO

The southern part of the Cordillera exposes a broad embayment preserving Precambrian–Cambrian boundary strata. Here are seen richly fossiliferous Lower Cambrian rocks with *Fallotaspis, Nevadella* and *Bonnia–Olenellus* Zones, together with archaeocyathans, early echinoderms, and trace fossils. There is also a modest sub-trilobitic small shelly fauna. Grand cycles from clastic to carbonate are seen along the strike and the tops of the cycles may have correlation potential (e.g. Aitken 1981; Palmer 1981). Accessibility is fairly good and the biostratigraphical succession from *Fallotaspis* upwards serves as a classic standard for the North American craton. More needs to be known about the

nature and age of the earliest skeletal faunas, however, though occurrences have been reviewed recently (Signor & Mount 1986).

The Wyman Formation (2900 m) of the White-Inyo Mts, California, represents a clastic half-cycle, with trace fossil *Planolites*(?) near the top (Alpert 1977; Nelson 1978).

The Reed Dolomite Formation (650 m) is the upper, carbonate half-cycle, with *Planolites*(?) and the small, hyolith-like calcareous tube of *Wyattia reedensis* Taylor (1966) at the top and in basal limestones of the overlying Deep Spring Formation (500 m). A new fauna from the lower part of the Deep Spring Formation of California and Mount Dunfee, Nevada, includes *Coleoloides* sp., and problematical ?calcareous tubes referred to *Sinotubulites* and other taxa (Signor *et al.* 1987). The problematical shelly fauna from over the border near Carborca, Mexico, may be of similar age (e.g. McMenamin *et al.* 1983; McMenamin 1985).

Clastic–carbonate intercalations occur throughout the Deep Spring Formation, with the burrow of *Plagiogmus*(?) followed by arthropod tracks of *Rusophycus, Diplichnites,* and burrows *Skolithos, Monocraterion,* and *Scolicia* (Alpert 1977; Nelson 1978).

The Deep Spring Formation marks relative emergence, followed by the initiation of 'Grand Cycle A' and the mainly clastic Campito Formation (Mount & Rowland 1981) some 1150 m thick. The agglutinated tube of *Platysolenites* appears in the Andrews Mountain Member, close to the first trilobite *Fallotaspis* cf. *tazemmourtensis* Hupé in the middle, and a rich trace-fossil assemblage of *Rusophycus, Cruziana, Bergaueria, Zoophycos, Arthrophycus, Teichichnus, Phycodes, Diplichnites,* and *Skolithos* occurs near the top (Wigget 1977; Nelson 1978). *Volborthella* also occurs in the formation (Yochelson 1977).

The upper or Montenegro Member of the Campito Formation has *Fallotaspis* cf. *longa,* joined by *Daguinaspis, Judomia*(?) and trace fossils; brachiopod *Nisusia* appears about here, as does *Microdictyon* n. sp. 1. *Nevadia, Holmia, Holmiella,* and brachiopod *Obolella*

Table 6.1. Tentative correlation of sections in north-west Canada with the Meishucun section and the Palaeotethyan scheme.

North-west Canada	China
Bonnia–Olenellus Zone	Zone V upwards
Nevadella Zone	Zones III, IV, V
Fallotaspis Zone	Zone III
upper Vampire Formation	Badaowan Member
—disconformity—	—disconformity—
Map Unit 11—lower Vampire Formation	Zones 0, I, II
Map Unit 10b	Zone 0

appear at the top of the member (Rowell 1977; Nelson 1978; Bengtson *et al.* 1986) comparable with assemblages from the Puerto Blanco Formation, unit 3 of Mexico (McMenamin 1987). Archaeocyathans of Botomian aspect appear above this level (Rozanov & Debrenne 1974).

A return to largely carbonate deposition in the Poleta Formation (350 m), compares with that seen in the Sekwi Formation of north-west Canada. The lower member has yielded the first *Lingulella* (Rowell 1977). The upper member has more clastics and preserves *Nevadella* and echinoderm *Helicoplacus* near the base, followed by mollusc *Stenothecoides*, and then by *Olenellus, Fremontia, Laudonia,* and ptychopariids of the *Bonnia–Olenellus* Zone (Wiggett 1977; Nelson 1978). A succession of trilobite faunules is seen from here upwards into the Harkless, Saline Valley, and Mule Spring Formations (650 m, 275 m, and 300 m respectively; see Nelson 1978; Brasier this volume, p. 148). Also useful for correlation is *Salterella* in the middle of the Harkless Formation (Yochelson 1977).

This sequence may be correlated with Meishucun and with the Palaeotethyan scheme (see Brasier, this volume, pp. 40–74) tentatively as shown in Table 6.2.

6.3. MOROCCO, SPAIN, AND FRANCE

The sections in Morocco are of considerable importance because of the apparent continuity of the succession without disclosed significant breaks in sedimentation, but they suffer because the only fossils found below the earliest trilobites are stromatolites which may range between Precambrian and Cambrian in age assignment. Morocco has been studied by Moroccan and European geologists for some time and so have the Precambrian–Cambrian successions in France (Normandy and Brittany), but outcrops of similar age in Spain have only

received detailed study in the last two decades. In the 1970s it was claimed by some that the tectonically undisturbed sections in the Anti-Atlas could provide a Global Stratotype Section and Point (GSSP) for the Precambrian–Cambrian Boundary. Field studies in 1976 and subsequent research have shown that the Precambrian–Cambrian Boundary cannot be closely bracketed by biostratigraphical or other evidence and consequently is not globally correlatable at the present stage of research. No GSSP candiate is currently being considered by the Working Group on the Precambrian–Cambrian Boundary in Morocco, Spain, or France.

6.3.1. Morocco

The area of main Precambrian–Cambrian boundary interest is in the mountains of the Anti-Atlas and the High Atlas, lying south and north respectively of the great valley plain of Souss, east of Agadir. Pioneer workers in this region were Choubert (1953) on stratigraphy and regional geology and Hupé (1953) on trilobites, but many have followed. In this brief review it is essential to mention the preliminary report by Sdzuy (1978).

The schematic succession of the shallow-water shelf sedimentation of the Precambrian–Cambrian cover overlying the older, Proterozoic, mountain belts of the Anti-Atlas of Morocco, can be summarized as:

5. *Schisto-calcaire* Series (shale and limestone) with archaeocyathans and trilobites: Atdabanian (Cambrian).

4. *Calcaire supérieur* Series with archaeocyathans and trilobites in the upper part only. Atdabanian and ?Tommotian (Cambrian), and ?Vendian (Precambrian).

Table 6.2. Tentative correlation of sections in California with the Meishucun section and the Palaeotethyan scheme.

California	China
Bonnia–Olenellus Zone	Zone V, upwards
Nevadella Zone	Zones III, IV, V
Fallotaspis Zone	Zone III
Andrews Mountain Member	Badaowan Member
—disconformity—	—disconformity—
upper Deep Spring Formation	Zone II
lower Deep Spring Formation	Zone 0 to Zone I
Reed Dolomite Formation	Zone 0 of Dengying Formation

3. *Lie de vin* Series (Talwinian)—mainly continental but developed into marine stromatolitic formations in the west. ?Tommotian and/or ?Vendian.

2. *Calcaire inférieur* Series (*Dolomie inférieur*). Adoudounian/Vendian.

1. Precambrian III—volcanic and molassic formations (Ouarzazate volcano-sedimentary group). Vendian.

On chronometric grounds, the age of the above succession is uncertain (Cowie & Johnson 1985, pp. 56–7). The Bou Ourhioul rhyolite of the Ouarzazate Group in the High Atlas gives an age of 578 ± 15 Ma (Jéury *et al.* 1974) on the basis of U–Pb dating of zircons. This could indicate either a Precambrian or Cambrian correlation in the present state of research, but the consensus of geologists working in Morocco would favour a Precambrian (Vendian) age for subdivisions 1 and 3. Magnetostratigraphical correlation is not available in any detail.

The *Lie de vin* Series of the Adoudounian Stage exhibits stromatolites which may range from Precambrian in the lower beds to Cambrian in the uppermost beds. Schmitt (1978, 1979*a,b*) favoured a Vendian age for the *Lie de Vin* series, but Bertrand–Sarfati (1981) suggested a Tommotian age based on the presence of thrombolitic stromatolites. Many biostratigraphers do not favour the use of stromatolites to differentiate late Precambrian from Lower Cambrian strata, but they can probably give a rough correlation of Proterozoic strata.

During the IGCP Project 29 excursion to the Anti-Atlas in 1976, A. Yu. Rozanov (Moscow, USSR) discovered towards the top of the *Lie de vin* series at Tiout (Anti-Atlas) a horizon with oncolites (assigned to *Azagia*), which could correlate the beds with the Nemakit–Daldyn Series of northern Siberia. The Nemakit–Daldyn (see Chapter 4) Series is assigned by different authors to latest Precambrian or earliest Cambrian. This oncolitic horizon has not been rediscovered by later field investigators, but the level, about 200 m below the base of the *Calcaire supérieur* Series, within the 600–1000 m-thick *Lie de vin* Series, gives further credence, if substantiated, to speculation that the uppermost part of the *Lie de vin* Series could be close to the Precambrian–Cambrian boundary (Choubert 1983, pers. comm.) A Vendian age for the *Lie de vin* Series has been derived from acritarchs (Choubert *et al.* 1979).

In the absence of other biostratigraphical evidence for the age of the *Lie de vin* Series, it can be ?Vendian and/or ?Tommotian by coarse correlation with the USSR stratigraphical classification.

The *Calcaires supérieur* Series contains no recorded 'small shelly fossils' but displays stromatolites such as *Acaciella angepina* which are found in Lower Cambrian beds in Australia (Tommotian or Atdabanian). Unfortunately the association of stromatolites with diagnostic Tommotian–Meishucun–Etcheminian small shelly fossil assemblages has not yet been recorded anywhere in the world, possibly due to palaeoenvironmental and/or palaeoecological factors.

The earliest archaeocyathans occurring in Morocco are of early Atdabanian age but do not indicate the earliest part of the early Atdabanian: these and the earliest trilobites occur in the upper beds of the *Calcaire supérier* Series. The trilobites have a first arrival datum (FAD) earlier in the succession than the archaeocyathan FAD. Thus the lowest part of the Atdabanian and the whole of the Tommotian Stage have to be placed within the *Calcaires supérieur* Series, plus possibly the upper part of the *Lie de vin* Series. There seems to be no reason to consider the earliest Moroccan trilobites to be Tommotian in age. No trilobites are known in the type Tommotian beds in Siberia, and apparently this may be true in equivalent beds elsewhere, but new results are awaited from Australia. The inexactitude at present of early Cambrian correlations leaves some doubt. The Precambrian–Cambrian boundary can therefore in Morocco only be located approximately between the *Calcaire inférieur* Series/*Lie de vin* Series and the *Calcaire supérieur* Series.

The absence of the earliest Cambrian shelly faunas in Morocco (on the basis of reports available to-date) gives great uncertainty in any placing there of the Precambrian–Cambrian boundary. This is disappointing because the Moroccan developments include one of the best sections (at Tiout, 85 km east of Agadir) which spans, in apparently continuous sedimentation, the latest Precambrian and earliest Cambrian times.

Carbon isotope excursions in Precambrian–Cambrian boundary beds are discussed in Chapter 11 of this book: there is an interesting example from Morocco. Tucker (1986) postulated a massive increase in the biomass of the photic zone through phytoplankton blooms as a postulate in late Riphean/early Vendian and early Cambrian times. A graphical plot of carbon isotope ($\delta^{13}C$) values against sediment thickness in Morocco suggests two distinctive positive $\delta^{13}C$ excursions, which Tucker interpreted as records of increased organic productivity and evidence of a Precambrian–Cambrian 'explosion' of life. Although sediment thickness is not directly related to time duration, these carbon isotope excursions could be useful for broad time correlation between boundary sections. The data were obtained from the 'classic' section of Tiout, 25 km east of

Taroudannt and 85 km east of Agadir. A positive peak of $+7‰\delta^{13}C$ is shown near the top of the *Dolomie inférieur* series and in middle beds of the *Calcaire supérieur* series a smaller positive peak of about $+3‰$ of $\delta^{13}C$ was obtained. Tucker considered these to reflect changes in the carbon isotopic composition of sea-water during the critical later Precambrian–Cambrian time.

6.3.2. Spain

Three regions in Spain are known to have important Precambrian–Cambrian boundary sections: the Asturian coast of northern Spain; the Montes de Toledo of central Spain; and the Sierra de Cordoba of southern Spain. The significance of these sections will be briefly reviewed here.

The Asturian section is notable for its thick sequence of clastic strata (Candana Quartzite), bearing trace fossils of increasing diversity, interpreted to reflect evolutionary events (Crimes *et al.* 1977; reviewed more fully in Crimes (Chapter 8). The section has neither a clear basal contact with Precambrian strata nor small shelly faunas. The first trilobites are said to be of low early Cambrian age (Sdzuy *in* Crimes *et al.* 1977).

The Montes de Toledo provides a fuller picture of late Precambrian events, especially in the Rio Uso section (Brasier *et al.* 1979; Liñan *et al.* 1984). The thick Alcudian rocks comprise shales, sandstones, and conglomerates, bearing late Lower Vendian acritarchs *Bavlinella faveolata* and *Octoedryxium truncatum* in the Lower Detrital Beds (>200 m), no recorded fauna in the Conglomerate Beds (100–150 m), and *Bavlinella* plus medusoids, trace fossils, and cyanobacteria in the Upper Detrital Beds (>300 m; Liñan *et al.* 1984). Perconig *et al.* (1986) report other Vendian acritarchs, while de San José (1984) mentions vendotaenid algae.

Cadomian movements preceded and accompanied the accumulation of the Fuentes Olistostrome (30–50 m) mass-flow deposit (Brasier *et al.* 1979; de San José 1984). It bears trace-fossil *Planolites* in the interbeds (Brasier *et al.* 1979), while acritarchs *Bavlinella faveolata* and *Trachysphaeridium laufeldi* occur in

the correlated Cases del Membrillar Olistostrome, and are taken to suggest an early–middle Vendian age (Palacios *in* Perconig *et al.* 1986).

The lower Pusa Shale (over 800 m thick) contains the first arthropodan traces of *Monomorphichnus*. About 100 m higher in the section there are bedding planes with large, ovate traces comparable with the alga *Shouhsenia* from zone 0 of the Yangtze Platform (i.e. *Beltanelloides* of Brasier *et al.* 1979), discs cf. *Chuaria circularis* Walcott, ribbon-like ?vendotaeniids, and seven kinds of trace fossil that include minute (1–2 mm diameter) feeding burrows related to *Phycodes*, *Treptichnus*, and *Diplocraterion* (op. cit.). Acritarchs from the Pusa Shales include *Bavlinella faveolate*, *Polytorama*, *Protosphaeridium*, *Pterospermopsimorpha*, *Kildinella sinca*, and *Leiosphaeridia*, pointing to a late Vendian age according to Perconig *et al.* (1986).

A phosphatic oncolite conglomerate (*c.*130 m), infilling channels at Fontanarejo, is located in the middle of the Pusa Shales by de San José (1984). This deposit clearly shows *Protospongia* spicules in thin-section (e.g. Perconig *et al.* 1986, their fig. 18.8) comparable with boundary spongiolites of Kazakhstan and India.

The Azorejo Sandstone (250–550 m) is a coastal sand body with abundant trace fossils, including *Scolicia*, *Didymaulichnus*, *Astropolichnus*, *Monomorphichnus*, *Psammichnites*, and *Diplocraterion* (Brasier *et al.* 1979). *Rusophycus* gr. *radwanskii* Alpert and *Gordia* sp. also occur (Liñan *et al.* 1984).

The lower Cambrian ends with the La Estrella or Navalucillos Limestone (140–350 m), containing Toyonian archaeocyathans, plus protolenid trilobites *Termierella* sp., *Realaspis strenoides* Sdzuy and *Pseudolenus weggeni* Sdzuy (Gil Cid *et al.* 1976). A tentative correlation with the Chinese sections (see Brasier, this volume, pp. 40–74) may be as shown in Table 6.3.

Another sequence of international interest occurs in the Sierra de Cordoba of Andalusia. This is important for its similarity to the Anti-Atlas sections of Morocco and has the potential for correlation with both China and Siberia. The lower part of the section comprises volcaniclastics of the San Jeronimo Formation

Table 6.3. Tentative correlation of central Spanish sections with the Meishucun section and the Palaeotethyan scheme.

Toledo Mountains	China
Estrella Limestone Formation	*c.* Canglangpuian–Lonwangmiaoian
Azorejo Sandstone Formation	Zones II to V
Fontanarejo Member	Zone I
lower Pusa Shale Formation	Zones 0, I
—Cadomian movements—	Zone 0
Alcudian	Zone 0

(*c.*1200 m), bearing the trace fossil *Neonereites* and acritarchs *Bavlinella faveolata*, *Protosphaeridium flexosum* Timofeev, *Trachysphaeridium* sp., *Phycomicetes?* sp., *Ooidium* sp., and forms close to *Octoedryxium truncatum* Rudavskaya, and is suggested to be of Vendian age (Liñan & Palacios 1983).

The sandstones, conglomerates, and shales of the peritidal Torrearboles Formation (0–450 m) follow above a 'Cadomian' unconformity. They preserve numerous trace fossils, including *Cochlichnus*, *Treptichnus*, *Teichichnus*, *Bergaueria*, *Scolicia*, *Skolithos*, *Monocraterion*, *Monomorphichnus*, possible *Rusophycus*, *Phycodes* aff. *palmatum* Seilacher and *P.* aff. *pedum* Seilacher, which are arguably of Cambrian age (Fedonkin *et al.* 1983).

The conformably succeeding Pedroche Formation (400–500 m) contains four members of limestones and shales, yielding a rich fauna of archaeocyathans, algae, trilobites, and small shelly fossils. Most important is the fauna from the base of the first member, overstepping the Precambrian at Las Ermitas. Here occur small shelly fossils figured by Liñan (1978), that may be compared with chancelloriids *Allonnia*, *Onychia*, and *Chancelloria*; a monoplacophoran resembling *Ceratoconus*; and a pelagiellid. This association needs to be studied further, but helps to characterize the widespread *Allonnia* association of zone III seen throughout the Palaeotethyan region (see Brasier & Singh 1987; Brasier, this volume, pp. 40–74). The brachiopod *Paterina* also occurs in this member (Liñan & Mergl 1982). Archaeocyathan in reefs associated with this fauna are arguably of early Atdabanian age (Rozanov & Debrenne 1974) or belong to its second part (?middle–upper; Zamarreño & Debrenne 1977). The third member contains trilobites *Lemdadella linarese* Liñan & Sdzuy and *Bigotina* sp., which arguably indicate correlation with the *Fallotaspis tazzemourtensis* Zone (Liñan & Sdzuy 1978; Liñan *et al.* 1984).

The Santo Domingo Formation (200 m) consists of sandstones, shales, and peritidal carbonates containing brachiopods, chancelloriids and *Renalcis*. The Los Villares Formation (>450 m) comprises Lower to Middle Cambrian shallow-water sandstones and shales (Liñan 1978). A correlation with Meishucun and the Palaeotethyan scheme (see Brasier, this volume, pp. 40–74) is tentatively give in Table 6.4.

6.3.3. France

In northern France, several important boundary successions are found in the Massif Armoricain. The upper part of the Brioverian flysch of Brittany and Normandy contains acritarchs and simple trace fossils (e.g. *Planolites*) suggesting a Vendian age; these are intruded by granodiorites of the late Cadomian tectonic phase, the Mancellia intrusion giving U/Pb monazite dates of 540 ± 10 Ma (Doré 1985a,b) or the 560 Ma Rb/Sr revised age (Peucat in Doré 1985b).

Cambrian sediments follow above a major discordance. They are well seen in the region of Carteret, Normandy, although the basal contact is concealed. Conglomerates and arkoses in the lower 40–400 m of section are followed by *c.*190 m of sandstones and slates of the de Carteret formation; a fauna with *Allonnia* and orthid brachiopods has been discovered at the base of this, as have traces of *Rusophycus* (Doré 1985a) and *Cruziana*, *Phycodes*, *Taphrehelminthopsis*, and *Skolithos* (Doré 1985b). A similar assemblage, but bearing *Diplocraterion*, occurs at Rozel (Doré 1985b). Higher in the succession, there are sandstones with calcareous nodules (230 m) bearing a fauna of *Allonnia tripodophora* Doré & Reid, *Chancelloria* sp., orthothecids, and hyoliths (Doré & Reid 1965; Doré 1972).

In the correlated Bocaine Zone, *Scolicia* and *Cochlichnus* occur above a basal assemblage with *Monomorphichnus*, *Taphrhelminthopsis*, *Helminthopsis*, *Planolites*, *Phycodes*, and *?Coleoloides* (Doré 1984).

Table 6.4. Tentative correlation of sections in Andalusia with the Meishucun section and the Palaeotethyan scheme.

Cordoba	China
Los Villares Formation	*c.* Maozhuangian
Santo Domingo Formation	*c.* Canglangpuian–Lonwangmiaoian
Pedroche Formation, Member IV	Zones III to V
Pedroche Formation, Member I	Zone III
Torrearboles Formation	Zones I to II
—unconformity—	?
San Jeronimo Formation	Zone 0

About 200 m further up in the Carteret section, above some purple and green slates, are found oolites and algal–archaeocyathan limestones of the St Jean de la Rivière formation; this yields ?Middle Atdabanian archaeocyathan, algae, and the trilobite *Bigotina* (Doré 1972; Rozanov & Debrenne 1974) as well as the gastropod *Aldanella*, the bradoriid *Indianites*, and algae (Doré 1984). Strata with *Circotheca* and *Fordilla* occur at a similar level in the Bocaine Zone (Doré 1984).

The faunas with *Allonnia* have been taken to be late Tommotian in age (e.g. Rozanov & Debrenne 1974; Odin *et al.* 1983, 1985; Doré, 1985*a*) with the implication that the isotopic dates of *c*.540 Ma give, approximately, a minimum isochron for the Tommotian. It is perhaps more accurate to compare the *Allonnia* assemblage with Zone III of Meishucun (?Lower and ?Middle Atdabanian) and the *Bigotina* fauna with the Middle Atdabanian *Pagetiellus anabarus* Zone of Siberia (Brasier, this volume, pp. 40–74). A tentative correlation with the Meishucun sequence is shown in Table 6.5.

Another interesting succession can be seen in the Montagne Noire of southern France. Cambrian strata are here disposed to the north and south of a metamorphic axis and are considerably disrupted by faulting. The basal conglomerates and slates pass up into the *grès de Marcory*, containing 'spicules and Hyolitha' (Boyer *in* Rozanov & Debrenne 1974). Overlying sandstones of the *Pardailhan formation* bear the trilobites *Thoralaspis*, *Ferralsia*, *Blayacina*, and

Alanisia, taken to indicate a Tasousektian age (i.e. Hupé Zone VI) in terms of the Moroccan sequence (Hupé 1960; Doré 1977). Carbonates of the 'Alternances' may be lateral equivalents, and yield Botomian archaeocyathans, *Micmacca*(?), and protolenid trilobites (Cobbold 1935; Rozanov & Debrenne 1974).

Limestones with *Heraultia* appear to occupy a superjacent position, and contain a fauna of small shelly fossils currently being studied by Mr M. Kerber. Originally described by Cobbold (1935), the following seem to be present: *Heraultipegma varensalense* (Cobbold), described by Müller (1975); *Latouchella angusta* (Cobbold), and its possible synonym *L. lata* (Cobbold) (both of the *Latouchella korobkovi* group); *Stenothecopsis heraultensis* Cobbold (cf. *Obtusoconus* sp.); *Pseudorthotheca* spp., and *Torellella gallica* Cobbold (cf. *T. laevigata* Linnarsson). If this fauna is correctly placed in the succession, it indicates a relatively late '*Latouchella korobkovi* assemblage'. A similar situation is indicated in the Elburz Mountains of Iran (Hamdi *in* Brasier, this volume, p. 55).

This ?late *Latouchella korobkovi* assemblage reminds us that biostratigraphical correlations of Marker B (zone II) of Meishucun on the basis of the first appearance of 'diverse' molluscs, or the *L. korobkovi* group, is too simplistic. But one should not be too pessimistic. Early and late latouchellids and pelagiellids may yet be distinguished by morphological and evolutionary analysis (e.g. Brasier, this volume, pp. 136–8).

Table 6.5. Tentative correlation of sections in Northern France with the Meishucun section and the Palaeotethyan scheme.

Armorica	China
Bigotina fauna	Zone IV, V
Allonnia fauna	Zone III
—Cadomian events—	?
Late Brioverian	Zone 0

REFERENCES

Aitken, J. D. (1981). Generalizations about Grand Cycles. *US geol. Surv. Open-File Rep.*, **81–743**, 8–14.

Aitken, J. D. (1984). Strata and trace fossils near the Precambrian–Cambrian boundary, Mackenzie, Selwyn and Wernecke Mountains, Yukon and Northwest Territories: Discussion. *Current Research, Part B*; *Geol. Surv. Can. Paper* **84–1B**, 401–7.

Aitken, J. D. (1988). First appearance of trace fossils in Mackenzie Mountains, Northwest Canada, in relation to the highest glacial deposits and the lowest small shelly fossils. *Bull. N.Y. State Mus*, in press.

Alpert, S. P. (1977). Trace fossils and the basal Cambrian boundary. In *Trace fossils* 2, (eds T. P. Crimes & J. C. Harper), pp. 1–8. Seel House Press, Liverpool.

Bengtson, S., Matthews, S. C., & Missarzhevsky, V. V. (1986). The Cambrian net-like fossil *Microdictyon*. In *Problematic fossils*, (eds H. Hoffman & M. Nitecki). Oxford Monogr. Geol. & Geophys., pp. 97–115, Oxford University Press.

Bertrand-Sarfati, J. (1981). Problème de la limité Precambrien–Cambrien dans la section de Tiout (Maroc). *Newsl. Stratigr.* **10**, 20–26.

Brasier, M. D., Perejón, A., & de San José, M. A. (1979). Discovery of an important fossiliferous Precambrian–Cambrian sequence in Spain. *Estud. geol. Inst. Invest. geol. 'Lucas Mallada'*, **35**, 379–83.

Brasier, M. D., & Singh, P. (1987). Microfossils and Precambrian–Cambrian boundary stratigraphy at Maledota, Lesser Himalaya. *Geol. Mag.*, **124**, 323–45.

Choubert, G. (1953). Le Précambrien III et le Géorgien de l'Anti-Atlas. *Notes Mém. Serv. Géol.* **103**, 7–39.

Cobbold, E. S. (1935). Lower Cambrian faunas from Herault, France. *Ann. Mag. natur. Hist., Lond., Ser.*, **10**, (16), 25–48.

Conway Morris, S. & Fritz, W. H. (1980). Shelly microfossils near the Precambrian–Cambrian boundary, Mackenzie Mountains, northwestern Canada. *Nature*, **286**, 381–4.

Conway Morris, S. & Fritz, W. H. (1984). *Lapworthella filigrana* n. sp. (*incertae sedis*) from the Lower Cambrian of the Cassiar Mountains, northern British Columbia, Canada, with comments on the possible levels of competition in the early Cambrian. *Paläont. Z.*, **58**, 197–209.

Cowie, J. W. & Johnson, M. R. W. (1985). Late Precambrian and Cambrian geological time scale. *Mem. geol. Soc. Lond.*, **10**, 65–72.

Crimes, T. P., Legg, I., Marcos, A., & Arboleya, M. (1977). ?Late Precambrian–low Lower Cambrian trace fossils from Spain. In *Trace fossils 2*, (eds T. P. Crimes & J. C. Harper) pp. 91–138. Seel House Press, Liverpool

Doré, F. (1972). La transgression majeure du Paleozoïque inférieur dans le Nord-Est du Massif Armoricain. *Bull. Soc. géol. France*, 7th series, **14**, 79–93.

Doré, F. (1977). L'Europe moyenne Cambriennes. Les modèles sédimentaires, leur zonalité, leur contrôle. *Colloq. Int. Centre Nat. Recherche Scient.* **243**, 143–55.

Doré, F. (1984). The problem of the Precambrian–Cambrian boundary in the Armorican Massif. *Bull. Liais. Inf. IGCP Proj.* 196, **2**, 39–43.

Doré, F. (1985*a*). Premières meduses et premières faunes à squelette dans le Massif Armoricain; problème de la limite Précambrian/Cambrien. *Terra Cognita*, **5**, p. 237.

Doré, F. (1985*b*). Recherches biostratigraphiques dans le Massif Armoricain près de la limite Précambrien–Cambrien. *Bull. Liais. Inf. IGCP Proj.* 196, **5**, 7–10.

Doré, F. & Reid, R. E. (1965). *Allonnia tripodophora* nov. gen., nov. sp., nouvelle éponge du Cambrien inférieur de Carteret (Manche). *C. R. somm. Séanc. Soc. géol. Fr.*, 1965, 20–1.

Fedonkin, M., Liñan, E., & Perejón, A. (1983). Icnófosiles de las rocas precambrico–cambricas de la Sierra de Cordoba, España. *Bol. R. Soc. Española Hist. nat. (geol.)*, **81**, 125–38.

Fritz, W. H. (1972). Lower Cambrian trilobites from the Sekwi Formation type section, Mackenzie Mountains, northwestern Canada. *Bull. geol. Surv. Canada*, **212**.

Fritz, W. H. (1973). Medial Lower Cambrian trilobites from the Mackenzie Mountains, northwestern Canada. *Geol. Surv. Can. Pap.*, **73–24**, 42 pp.

Fritz, W. H. (1980). International Precambrian–Cambrian boundary working group's 1979 field study to Mackenzie Mountains, Northwest Territories, Canada. *Geol. Surv. Can. Pap.*, **80–1A**, 41–5.

Fritz, W. H. & Crimes, T. P. (1985). Lithology, trace fossils, and correlation of Precambrian–Cambrian boundary beds, Cassiar Mountains, north-central British Columbia. *Geol. Surv. Can. Pap.*, **83–13**.

Gil Cid, D., Perejón, A., & de San José, M. A. (1976). Estratigrafia y paleontologia de las calizas Cambricas de Los Navalucillos (Toledo). *Tecniterrae*, **13**, 1–19.

Hofmann, H. J. (1981). First record of a late Proterozoic faunal assemblage in the North American Cordillera. *Lethaia*, **14**, 303–10.

Hofmann, H. J., Fritz, W. H., & Narbonne, G. M. (1983). Ediacaran (Precambrian) fossils from the Wernecke Mountains, northwestern Canada. *Science*, **221**, 455–7.

Hupé, P. (1953). Contributions à l'étude du Cambrien inférieur et du Précambrien III de l'Anti-Atlas marocain. *Protectorat République franç. Maroc. Serv. geol., Notes Mém.* **103**, 1–402.

Hupé, P. (1960). Sur le Cambrien inférieur du Maroc. *21st Int. geol. Congr.*, Norden, at Copenhagen. VIII, 75–85.

Jéury, A., Lancelot, J. R., Hamet, J., Proust, F., & Allègre, C. J. (1974). L'âge des rhyolites du Précambrian II du Haut-Atlas et le probleme de la limite Précambrian–Cambrian. *2ᵉ Reun. annu. Sci. Terre, Nancy*, p. 230.

Liñan, E. (1978). Bioestratigrafia de la Sierra de Cordoba. Published Ph.D. Thesis, Univ. Granada.

Liñan, E. & Mergl. M. (1982). Lower Cambrian brachiopods of Sierra Morena, SW Spain. *Bol. R. Soc. Española Hist. nat. (geol.)*, **80**, 207–20.

Liñan, E & Palacios, T. (1983). Aportaciones micropaleontologicas para el conocimiento del limite Precambrico-–Cambrico en la Sierra de Cordoba, España. *Comm. Serv. geol. Portugal*, **69**, 227–34.

Liñan, E. & Sdzuy, K. (1978). A trilobite from the Lower Cambrian of Córdoba (Spain) and its stratigraphical significance. *Senckenbergiana Lethaea*, **59**, 387–99.

Liñan, E., Palacios, T., & Peréjon, A. (1984). Precambrian–Cambrian boundary and correlation from southwestern and central part of Spain. *Geol. Mag.*, **121**, 221–8.

McMenamin, M. A. S. (1985). Basal Cambrian small shelly fossils from the La Ciénega Formation, northwestern Sonora, Mexico. *J. Paleont.*, **59**, 1414–25.

McMenamin, M. A. S. (1987). Lower Cambrian trilobites, zonation and correlation of the Puerto Blanco Formation, Sonora, Mexico. *J. Paleont.*, **61**, 738–49.

McMenamin, M. A. S., Awramik, S. M., & Stewart, J. H. (1983). Precambrian–Cambrian transition problem in western North America: Part II. Early Cambrian skeletonized fauna and associated fossils from Sonora, Mexico. *Geology*, **11**, 227–30.

Mount, J. F. & Rowland, S. M. (1981). Grand Cycle A (Lower Cambrian) of the southern Great Basin: a product of differential rates of relative sea-level rise. *US geol. Surv. Open-File Rep.*, **81–743**, 143–6.

Müller, K. J. (1975). ‘*Heraultia*’ *varensalensis* Cobbold (Crustacea) aus dem Unteren Kambrium, der alteste Fall von Geschlechtsdimorphismus. *Paläont. Z.*, **49**, 168–80.

Narbonne, G. M. & Hofmann, H. J. (1987). Ediacaran biota of the Wernecke Mountains, Yukon, Canada. *Palaeontology*, **30**, 647–76.

Nelson, C. (1978). Late Precambrian–Early Cambrian stratigraphic and faunal succession of eastern California and the Precambrian–Cambrian boundary. *Geol. Mag.*, **115**, 121–6.

Nowlan, G. S., Narbonne, G. M., & Fritz, W. H. (1985). Small shelly fossils and trace fossils near the Precambrian–Cambrian boundary in the Yukon Territory, Canada. *Lethaia*, **18**, 233–56.

Odin, G. S., *et al.* (1983). Numerical dating of Precambrian–Cambrian boundary. *Nature*, **301**, 21–3.

Odin, G. S., Gale, N. H., & Doré, F. (1985). Radiometric dating of Late Precambrian times. In *The chronology of the geological record*, (ed. N. J. Snelling), pp. 65–72. Mem. geol. Soc. Lond., 10.

Palmer, A. R. (1981). On the correlatability of Grand Cycle tops. *US geol. Surv. Open-File Rep.*, **81–743**, 156–9.

Perconig, E., Vasquez, F., Velando, F., & Leyva, F. (1986). Proterozoic and Cambrian phosphorites: Fontanarejo, Spain. In *Proterozoic and Cambrian phosphorites*, (eds P. J. Cook & J. H. Shergold), pp. 220–34. Cambridge University Press.

Rowell, A. J. (1977). Early Cambrian brachiopods from the southwestern Great Basin of California and Nevada. *J. Palaeont.*, **51**, 68–85.

Rozanov, A. Yu. & Debrenne, F. (1974). Age of archaeocyathid assemblages. *Am. J. Sci.*, **274**, 833–48.

de San José, M. A. (1984). Los materiales anteordovicicos del anticlinal de Navalpino (Provincias de Badajoz y Ciudad Real, España central). *Cuadernos Geología Ibérica, Madrid*, **9**, 81–117.

Schmitt, M. (1978). Stromatolites of the Tiout section Precambrian–Cambrian boundary beds, Anti-Atlas, Morocco. *Geol. Mag.*, **115**, (2), 95–100.

Schmitt, M. (1979*a*). The section of Tiout (Precambrian–Cambrian boundary beds), Anti-Atlas, Morocco: stromatolites and their biostratigraphy. *Arb. Paläont. Inst. Wurzburg*, **2**, 1–118.

Schmitt, M. (1979*b*) New stromatolites from the late Precambrian of the Anti-Atlas and from the Lower Cambrian of the High Atlas, Morocco. *Senckenbergiana Lethaea*, **60**, (1–3), 39–49.

Sdzuy, K. (1978). The Precambrian–Cambrian boundary beds in Morocco (Preliminary Report). *Geol. Mag.*, **115**, 83–94.

Signor, P. W. & Mount, J. R. (1986). Lower Cambrian stratigraphic paleontology of the White-Inyo Mountains of eastern California and Esmeralda County, Nevada. In *Natural history of the White-Inyo Range, Eastern California, Western Nevada and high altitude physiology* (eds C. A. Hall Jr & D. J. Young), pp 6–15. Univ. California, White Mountain Res. Station Symp., Vol. 1.

Signor, P. W., Mount, J. M., & Onken, B. R. (1987). A pre-trilobite shelly fauna from the White-Inyo region of eastern California and western Nevada. *J. Paleont.*, **61**, 425–38.

Taylor, M. E. (1966). Precambrian mollusc-like fossils fron Inyo County, California. *Science*, **153**, 198–201.

Tucker, M. E. (1986). Carbon isotope excursions in Precambrian/Cambrian boundary beds, Morocco. *Nature*, **319**, (6948), 48–50.

Wiggett, G. J. (1977). Late Proterozoic–Early Cambrian biostratigraphy, correlation, and paleoenvironments, White-Inyo facies, California–Nevada. *Spec. Rep. California Div. Min. Geol.*, **129**, 87–92.

Yochelson, E. I. (1977). Agmata, a proposed extinct phylum of early Cambrian age. *J. Paleont.*, **51**, 437–54.

Zamarreño, I. & Debrenne, F. (1977). Sédimentologie et biologie des constructions organogènes du Cambrien inférieur du Sud de l'Espagne. *Mém. BRGM.*, **89**, 49–61.

PART II
SPECIALIZED TOPICS

7

Towards a biostratigraphy of the earliest skeletal biotas

M. D. Brasier

7.1. INTRODUCTION

Major evolutionary and geological events across the Precambrian–Cambrian boundary present us with three main puzzles. Firstly, what was the sequence of events? This is a matter of stratigraphical detective work, capable of reasonable solution. Secondly, what was the nature of these events? This is a complex question of special interest to biologists, palaeobiologists, and palaeoceanographers, of which much must remain unsolved. And thirdly, what brought these changes about? This is probably a large-scale geological question, complex and still obscure. The second and third questions have provided much of the focus of Western literature on the Precambrian–Cambrian boundary over the previous decade (e.g. Sepkoski 1979; Brasier 1979), with much of the evidence for the sequence of events being provided by Soviet workers (e.g. Rozanov *et al.* 1969). Progress on biostratigraphy has, however, been substantial in areas outside of the Soviet Union over the last few years. The time is therefore ripe to put aside philosophy and focus upon the first, fundamental question: 'what was the sequence of events'?

7.2. BOUNDARY SECTIONS

The dispositions of continents in late Precambrian and early Cambrian times are poorly known, though various attempts have been made (e.g. Ziegler *et al.* 1979; Piper 1982; Rozanov 1984; Zhuravlev 1986; Parrish *et al.* 1986; Donovan 1987). The main crustal blocks appear to have been Laurentia (including North America, Greenland, and north-west Scotland) Baltica, Siberia, Gondwana (including Africa, South America, southern Europe, Asia Minor, India, southern China, Australia, and Antarctica), Avalonia, and Kazakhstan. These are thought to have lain mainly at low latitudes, in relatively close proximity during a major phase of oceanic growth. Fig. 7.1 shows important boundary sections in their present-day settings.

Cambrian rocks of the North American continent

provide a clear sedimentological model. Thin, incomplete but fossiliferous sequences of clastics and carbonates accumulated over the craton (Lochman-Balk 1971). Thick intercalations of clastics and limestones were deposited in the inner detrital belt, middle carbonate belt, and outer detrital belt, especially in basins around the edges of the craton; these can be relatively fossiliferous (Palmer 1971, 1974). Because late Precambrian–Middle Cambrian sediments barely overstepped on to the cratonic interior of North America, boundary sequences must be sought in marginal basins and allochthonous terranes, notably the Cordilleran geosyncline along the western margin. This provides two main areas with boundary reference sections: the Mackenzie, Selwyn, and Wernecke Mountains, northwest Cordillera, Canada (Fig. 7.1, loc. 40; e.g. Fritz 1972, 1973, 1980; Fritz & Crimes 1985; Nowlan *et al.* 1985); and the White-Inyo Mountains of the Great Basin, California and Nevada, USA; and Sonora in Mexico (Fig. 7.1, locs 39, 38; e.g. Taylor 1966; Nelson 1978; McMenamin *et al.* 1983; Signor *et al.* 1987). Both areas have sparse small shelly fossils, good trace fossils, sporadic and rather unusual archaeocythan assemblages, and North American olenellid trilobite assemblages that are not easily correlated with other continents.

Along the eastern margin of North America lie allochthonous terranes, such as the Avalon and Taconic allochthons, accreted by easterly subduction and obduction. These are displaced slices of continental slopes, arc basins, island-arc volcanics, oceanic volcanics, and sediments. Prior to opening of the Atlantic, this zone continued into England and Wales (Brasier 1980) which was the original type area for the Cambrian System (Sedgwick & Murchison 1835). There are two regions with boundary reference sections. The excellent sections of the Burin, Avalon, and Bonavista Peninsulas of south-east Newfoundland, Canada, provide a stratotype candidate in the Fortune Bay area (Fig. 7.1, loc. 35; e.g. Bengtson & Fletcher 1983; Crimes & Anderson 1985; Narbonne *et al.* 1987). The Charnwood–Nuneaton and Comley–Wrekin sections of the English Mid-

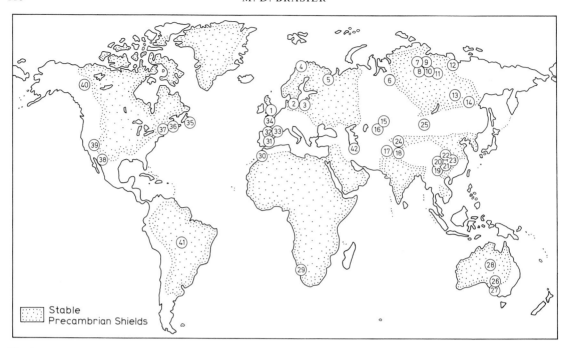

Fig. 7.1. Present-day distribution of some important Precambrian–Cambrian boundary sections (circled) and Precambrian cratons (stippled). 1, North Wales; Shropshire and Nuneaton, England; 2, Bornholm and southern Sweden; 3, northern Poland; 4, Trøms, Norway, and Finnmark; 5, Onega Peninsula; 6, Sukharika River; Igarka region. 7–10, Anabar region: 7, Eriechka River; 8, Kotui River; 9, Fomitch and Rassokha Rivers; 10, Kotuikan River. 11, Olenek Uplift. 12, Chekurovka, lower reaches of Lena River; 13, middle reaches of Lena River; 14, Aldan River. 15, Kuznetask Alatau and north-eastern Sayan; 16, Karatau, southern Kazakhstan; 17, Salt Range and Hazara district, Pakistan; 18, Mussoorie, Lesser Himalaya of India; 19, Meishucun, near Kunming, eastern Yunnan; 20, Maidiping, near Emei, south-western Sichuan; 21, north-western Guizhou; 22, south-western Shaanxi; 23, eastern Yangtze Gorges, western Hubei; 24, western Xinjiang; 25, Salanygol, Mongolian People's Republic; 26, Ediacara, Flinders Range, South Australia; 27, Mount Lofty and Yorke Peninsula, South Australia; 28, Amadeus and Georgina Basins, Northern Territory; 29, Nama Group, Namibia; 30, Anti-Atlas and High Atlas, Morocco; 31, Sierra Morena and Montes de Toledo, Spain; 32, Cantabria and Asturia, northern Spain; 33, Montagne Noire, Herault, France; 34, Brioverian of Normandy and Brittany, France; 35, Fortune Bay, Burin, Bonavista, and Avalon Peninsulas, south-east Newfoundland; 36, St John, New Brunswick; 37, Nahant and North Attleborough, Massachusetts; 38, Carborca, Sonora, Mexico; 39, Mount Dunfee, Nevada, and White-Inyo Mountains, eastern California; 40, Mackenzie, Selwyn, and Wernecke Mountains of Yukon and Northwest Territories, north-western Canada; 41, Corumba Group, State of Matto Grosso, Brazil; 42, Elburz Mountains, northern Iran. Localities 5–16 are in the USSR and 19–24 are in the People's Republic of China.

lands provide comparable reference sections in inland quarries (Fig. 7.1, loc. 1; Brasier *et al.* 1978; Brasier & Hewitt 1979, 1981; Brasier 1984, 1985, 1986; Hinz 1987). Archaeocyathans are absent from the Avalon terrane, but small shelly fossils and trilobites are relatively widespread forms of value for international correlation. Trace fossils are also well developed.

The smaller Baltic craton was largely emergent until mid-Cambrian times. Clastic sediments of cool-water aspect predominate (Theokritoff 1979; Brasier 1980). Cratonic sequences are thin, often condensed, with phosphatic and conglomeratic intervals (Notholt &

Brasier 1986). Marginal basins with thicker clastic sediments occur in Finnmark (Fig. 7.1, loc. 4), the Sparagmite basin of the Mjøsa district of Norway, the Lublin Slope of Poland (e.g. the Radzyn borehole, Fig. 7.1, loc. 3) and the Moscow syneclise, providing important Vendian–Cambrian reference sections. Fossiliferous occurrences are sparse and sporadic, and assemblages tend to be composite (e.g. Bengtson 1977*a*; Føyn & Glaessner 1979; Bergström & Ahlberg 1891; Bergström 1981; Vidal 1981; Volkova *et al.* 1983). Trace fossils and acritarchs are excellently developed here, while sparse small shelly fossils are useful wide-ranging forms; scarce trilobites tend to be rather provincial taxa. Marked sedimentary cycles also provide potential markers for event stratigraphy.

By late Precambrian times the Siberian Platform was already flooded (Rozanov *et al.* 1969) and provides richly fossiliferous boundary sections of stable platform type. Dolomites predominate in the late Precambrian, while limestones, argillites, and evaporites accumulated in the Cambrian. Sections are magnificently exposed in river-cliffs such as the lower and middle reaches of the Lena River, which provides reference sections for much of the Lower Cambrian (Fig. 7.1, loc. 12, 13). Equally important sections spanning the Yudomian–Tommotian boundary occur along the Aldan River (with a stratotype candidate at Ulukhan–Sulugur; Fig. 7.1, loc. 14), the Fomitch, Kotuikan, Kotui, Rassokha, and Eriechka Rivers of the remote Anabar Shield, the Olenek Uplift, and the Sukharika River of the Igarka region (Fig. 7.1, loc. 6–11; e.g. Rozanov *et al.* 1969; Missarzhevsky 1983; Sokolov & Zhuravleva 1983; Rozanov & Sokolov 1984; Sokolov & Fedonkin 1984; Rozanov 1984; Khomentovsky 1986). Assemblages include important admixtures of archaeocyathans, algae, and small shelly fossils; trilobites include many wide-ranging polymerids and eodiscinids; trace fossils and acritarchs are poorly known.

The Cambrian of China is also richly fossiliferous, with two major platform areas: the North China Platform and the Yangtze Platform of southern China, with a stratotype candidate from the latter area at Meishucun, near Kunming, Eastern Yunnan. On these stable platforms light-coloured clastics, limestones, dolomites, phosphorites, and intervals of black shales accumulated. South-east of the Yangtze Platform are dark argillites and limestones of the deeper-water Jiangnan facies, while in the extreme south-east flysch and volcanic rocks of geosynclinal type are found, representing allochthonous terranes of island arcs and arc basins. Important sections spanning the Sinian–Cambrian transition are found in the Yangtze Platform deposits of southern China at Meishucun, Yunnan, and in sections in Sichuan, Shaanxi, Guizhou, and the Yangtze Gorges of Hubei (Fig. 7.1, loc. 19–22; e.g. Xiang *et al.* 1981; Luo *et al.* 1982, 1984; Xing *et al.* 1983, 1984). The Krol-Tal succession of Lesser Himalaya, India, is remarkably similar (Fig. 7.1, loc. 18; Azmi & Pancholi 1983; Brasier & Singh 1987) and this Yangtze type of succession stretches through Pakistan to the Elburz Mts of Iran (Fig. 7.1, loc. 42; see Brasier, Chapter 3 of this volume), providing evidence for a 'Palaeotethyan belt' of Precambrian–Cambrian boundary strata. The thick and deformed boundary section of Salanygol in Mongolia lay along the southern margin of the Siberian Platform. This has lithological and faunal affinities with Chinese and Siberian sections (Fig. 7.1, loc. 25; Voronin *et al.* 1982), while the condensed sections of Maly Karatau, Southern Kazakhstan (Fig. 7.1, loc. 16; Missarzhevsky & Mambetov 1981) compare with Himalayan, Siberian, and Chinese sections. In all 'Palaeotethyan' sections, small shelly fossils and trilobites are well developed but include many forms endemic to Asia; algae and archaeocyathans are found only in certain palaeogeographical settings, but trace fossils are also present.

Although the various blocks of southern Gondwana provide very important late Precambrian sections, such as the Ediacarian of South and Central Australia and the Nama Group of Namibia in southern Africa (Fig. 7.1, loc. 26–29), rich assemblages of early skeletal fossils are poorly known or lacking. The Flinders Range succession of South Australia is nevertheless of great importance for its characterization of the Ediacarian fauna (Glaessner 1984) and for the archaeocyathans of the Ajax Limestone (e.g. Debrenne 1969). Mid-Lower Cambrian assemblages of small shelly fossils occur in South and Central Australia (e.g. Bischoff 1976; Laurie & Shergold 1985; Laurie 1986). Here, the lowest Cambrian beds have good trace-fossil assemblages and poorly known small shelly fossils; higher up are good archaeocyathans, plus trilobites and small shelly fossils that allow comparison with China and Mongolia. The Atlas Mountains of Morocco provide successions of trilobites of mixed-realm affinities (Fig. 7.1, loc. 30; Hupé 1952; Sdzuy 1978) as do those of the Iberian Peninsula (Fig. 7.1, loc. 31, 32; Sdzuy 1971), with many wide-ranging genera partly calibrated against archaeocyathan assemblages (e.g. Debrenne & Debrenne 1978). Small shelly fossils are sparse (e.g. Liñan 1978) but include Chinese forms.

To summarize, no section yet provides a complete view of the succession. The Siberian Platform is undoubtedly important in linking archaeocyathans to small shelly fossils, and these to olenellid and eodiscid trilobites. The Avalon terrane of south-east Newfoundland and England, links small shelly fossils to olenellid trilobites and to trace fossils; they are also demonstrably above early *Charnia* faunas. The Baltic and East European Platform sections link similar suites of small shelly fossils and trace fossils to acritarchs, and again demonstrate the post-*Charnia* and post-*Vendotaenia* age of the assemblages. The Mongolian section at Salanygol is very important in helping to link Siberian with Chinese small shelly fossils, while that in southern Kazakhstan also links Avalonian and Siberian small shelly fossils. The Krol-Tal section in India aids correlation of southern Kazakhstan with China, while the Iranian sections provide a link between Mongolia and other Palaeotethyan sections.

The accessible Chinese sections provide good reference stratigraphy for the Palaeotethyan margin of Gondwana, traceable through southern China, India, and Pakistan to Iran and probably into southern Europe. Most importantly, stratigraphical correlation along this belt seems to be feasible at relatively low biostratigraphical levels, within the interval that contains the earliest skeletal faunas, the major phosphogenic event, and the carbon isotope anomalies.

7.3. THE PROBLEMS

Progress in answering the question about the sequence of events has been gradual for several good reasons: classic boundary sequences are uncommon and often remote; most early skeletal fossils are small in size, very localized, and biologically problematical, and there are relatively few specialists who are able to work on them full time. This lack of specialists, for example, explains the extraordinary confusion over the age of the Tal Formation at Mussoorie in India, where good boundary assemblages previously had been described in terms of Cretaceous worms, Permian foraminifera, or algae and Ordovician conodonts (see Brasier & Singh 1987). The last two decades have nevertheless seen considerable efforts to establish the regional sequence of biostratigraphically useful fossils in boundary beds, notably of archaeocyathans, 'small shelly fossils' (SSFs), trilobites, acritarchs, and trace fossils. Unfortunately, it has become apparent that each of these groups has its own problems of correlation.

Archaeocyathan assemblages have been dated with some confidence but their distribution is patchy in time and space, their first appearance is diachronous, and there is too much reliance on the Siberian succession. Trilobites are very provincial at this time and their biomineralization may have been affected by environmental conditions, so that they appear and disappear from the fossil record. Acritarch successions are poorly preserved except on the East European Platform and the contiguous Baltic Shield, where numerous hiatuses and poor control by other 'yardstick' biostratigraphy renders their wider use uncertain. Trace fossils have potential for palaeoecology, and possibly also for biostratigraphy (see Crimes, this volume), but they await a rigorous application of biostratigraphical techniques.

So-called 'small shelly fossils' (Matthews & Missarzhevsky 1975) have hitherto provided the main hope for a viable international biostratigraphy, with a decision at

Cambridge in 1978 to place the Precambrian–Cambrian boundary as close as practicable to the base of the oldest stratigraphical unit 'to yield Tommotian (*sensu lato*) fossil assemblages' (Cowie 1978). Even small shelly fossils have their problems, however, and disagreements have arisen over correlation between Chinese and Siberian boundary sequences (e.g. Cowie 1985). Some of the problems and misconceptions leading to confusion are as follows:

1. Absence of trilobite remains: 'Tommotian type' assemblages without trilobites can form a biofacies that continues through the Lower Cambrian, especially in phosphatic facies such as the *Lapworthella* Limestone at Comley, and the early Cambrian of Mongolia (e.g. Brasier 1979, fig. 10; Voronin *et al.* 1982). Hence it may be misleading to call associations of simple tubes (e.g. *Hyolithellus*, *Coleoloides*, or *Wyattia*) either 'Tommotian' or even 'Atdabanian' without corroborative evidence. Even the use of 'Tommotian type' or 'non-Trilobite' *sensu* Cowie *et al.* (1972) and Brasier & Hewitt (1981) is liable to be misleading. Conversely, the presence of trilobite-like traces well before their skeletal remains suggests there may have been Tommotian trilobite-like arthropods. Trilobite remains in rocks equivalent to the Siberian Tommotian would not be unexpected, though reports of lower Tommotian trilobites from Siberia have proved to be the small shelly fossil *Tumulduria* (Bengtson *et al.* 1986*b*). Suggestions that *Buccanotheca* from the second fauna of Meishucun may represent genal spines of trilobites (probably incorrect) and are therefore possibly Atdabanian (Bengston *et al.* 1984) have no stratigraphical implication. In short, the appearance of trilobites was a facies-controlled, diachronous biomineralization event, and a simplistic view of 'Non Trilobite Zone' faunas as Tommotian (or vice versa) cannot be supported.

2. 'Long-ranging' Tommotian taxa: recent studies have reported high stratigraphical occurrences of taxa previously thought to be indicative of Tommotian beds in Siberia (e.g. Landing *et al.* 1980; Bengtson & Fletcher 1983; Brasier 1984). This has led to the impression that Siberian forms such as *Sunnaginia imbricata* are restricted to the Tommotian of Siberia but range into Atdabanian or higher beds elsewhere (e.g. Bengtson *et al.* 1984). The ranges of some taxa (e.g. *Coleoloides typicalis*, *Hyolithellus micans*, and *Camenella baltica*) may certainly be of secondary value for stratigraphy. But evidence presented in Brasier (1986), and below, indicates arguable to clear differences between older and younger examples of several SSFs, especially the

conoidal phosphatic microfossils. This means there is now an imperative need to distinguish various successive assemblages or zones of SSFs, based on evolutionary series wherever possible. While hardly unexpected, it does mean that specific rather than generic determinations are needed to distinguish older and younger assemblages.

3. Diachronous first appearance: this seems to be a particular problem with calcareous forms. Thus *Coleoloides typicalis* (s.l.) first appears close to the base of the Tommotian in Siberia (Missarzhevsky 1969) and Himalaya (Azmi & Pancholi 1983); in the *Aldanella* Zone of the Chapel Island Formation of Newfoundland (Narbonne *et al.* 1987, *contra* Bengtson & Fletcher 1983); with *Sunnaginia neoimbricata* and below *Rhombocorniculum isolutum* in the Home Farm Member of Nuneaton (Pl. 7.4, part 10; Brasier 1986); in the *Rhombocorniculum insolutum* fauna of the Green Shales of Bornholm (Poulsen 1967; Brasier 1984, 1986); and with the *Lapworthella cornu* assemblage at the base of the Yu'anshan Member of Yunnan (Luo *et al.* 1982). It is unknown in Mongolia (Voronin *et al.* 1982). The first appearance of trilobites and archaeocyathan taxa is also diachronous, but evidence of evolving lineages makes that more obvious. A valuable test of diachronism is the homotaxial method advanced by Scott (1985) and Brasier (1986).

4. Condensation, reworking, and hiatuses: SSFs are frequently associated with phosphatic sediments (Shergold & Brasier 1986) and are prone to reworking (e.g. Brasier & Hewitt 1979; Azmi, 1983), so taphonomic and sedimentological studies are essential for reliable stratigraphy. Hiatuses are also an aspect of the eustatic sea-level changes at this time (Notholt & Brasier 1986). Failure to recognize and emphasize the inevitable reworking and lacunae in several phosphatic, conglomeratic, and calcareous sequences must have confused the literature with doubtful ranges and correlations. For example, it is now clear there is a major hiatus between Atdabanian and Botomian sediments in the Salanygol section of Mongolia, which has probably introduced reworked phosphatic fossils such as *Torellella lentiformis*, *Sunnaginia acuta*, and *Lapworthella tortuosa* into Botomian beds. This was not appreciated until after the publication of Voronin *et al.* (1982). Acritarch studies on the Lublin Slope succession of Poland (Moczydłowska & Vidal 1986) and the Vendian of the Siberian Platform (Khomentovsky 1986) effectively cast doubt on the restricted range of the Lontova and Lukati Formation microfloras and the value of the 'Baltisphaeridium' datum; a large hiatus may be present between some Lontova and Lukati beds. More integrated studies of biostratigraphy and facies analysis are required.

5. 'Premature' event stratigraphy: while the potential for 'event stratigraphy' e.g. tectonic events, eustatic events, isotope events, and geomagnetic events) is considerable, interpretations that are not based on good biostratigraphical control may actually be damaging to developing biostratigraphy. An example of this was the suggestion that the appearance of a *Coleoloides* fauna in south-east Newfoundland correlates with the post-Lontova palaeogeographical change on the East European Platform, and therefore is early Atdabanian in age (Bengtson & Fletcher 1983). This speculation has now appeared in the literature as a biostratigraphical interpretation (Cowie & Johnson 1985; Crimes & Anderson 1985). Since the post-Lontova palaeogeographical change cannot yet be precisely dated because of hiatuses, and a good SSF assemblage has not yet been described from those *Coleoloides* beds, this Atdabanian interpretation is premature. Correlations between positive or negative $\delta^{13}C$ anomalies in boundary sections (e.g. Hsü *et al.* 1985; Lambert *et al.* 1987) must also remain doubtful in the absence of good biostratigraphical calibration of the curves.

6. Nomenclatural problems: historically different sets of nomenclature have been used in the Atlantic area, Siberia–Mongolia and China, tending to obscure some faunal similarities. This is particularly true for monoplacophorans, hyoliths, and hyolithelminthes, which thus may appear to be more endemic than was probably the case. Many conoidal phosphatic microfossils exhibit considerable variability, including symmetry transition series, but there has been minimal illustration and discussion on the degree of variation admitted. Size range, morphological variation, and skeletal composition should be more often referred to in descriptions. One result has been to obscure, for example, the wide occurrence of the *Latouchella korobkovi* group across Asia. More consideration also needs to be given to the vexed problems of multi-element taxonomy, which in turn requires informed palaeobiological approaches to the material (e.g. Bengtson 1985).

7. Inaccurate range citations: accidents of drafting may be responsible for inconsistencies between the ranges of taxa in Missarzhevsky (1982), for example, and their actually recorded levels. Some of these are quite significant: *Rhombocorniculum insolutum* is cited as Lower Atdabanian (ibid., Missarzhevsky & Mambetov 1981) leading to this interpretation for the English occurrence (Brasier 1984); but its first appearance is

actually in the Upper Tommotian (Sokolov & Zhuravleva 1983; Rozanov & Sokolov 1984). The ranges of *R. cancellatum*, *Kijacus kijanicus*, and possibly of *Fomitchella infundibuliformis* given by Missarzhevsky (1982) also do not tally with the data, with bases being too high. Ranges of taxa in the Meishucun section also vary between published lists and diagrams.

8. Reference sections and sampling: the ranges of SSFs need to be tied to reference sections that ideally have evolutionary series, 'yardstick' stratigraphical control, and taphonomic research. Important yardsticks include archaeocyathan assemblages and the appearance of the *Serrodiscus bellimarginatus* or *Calodiscus lobatus* trilobite fauna *circa* uppermost Atdabanian–lowermost Botomian, e.g. Robison *et al.* 1977). Composite range charts that do not relate to successions, and thereby conceal implicit assumptions of correlation, are also more likely to be misleading and should be regarded with caution. Random spot or channel samples are also misleading because of the extreme condensation of many Tommotian and Atdabanian levels. For example, confusion may be caused by acritarch studies in which associated skeletal fossils are noted from the same formation but not the same sample. Geochronology and geomagnetism (e.g. Odin *et al.* 1983, Kirschvink & Rozonov 1984) and geochemical events (Hsü *et al.* 1985; Magaritz *et al.* 1986; Tucker 1985; Lambert *et al.* 1987) have many problems to overcome before they can be used for correlation, not the least of which is the lack of an agreed biostratigraphical scale.

7.4. BIOSTRATIGRAPHY

Problems of biostratigraphical correlation require abundant, widespread, evolving, and easily used indices (e.g. Fig. 7.2). As at other levels in the geological column, the abundance of microfossils and the potential high quality of modern electron micrography gives small fossils a clear advantage, and there seems little doubt that archaeocyathans, skeletal microfossils, and acritarchs will together eventually provide an important international biostratigraphical scale. Surprisingly few trained micropalaeontologists have turned their hand to microfossils at this level, however, and this achievement is still some time away. Informal groupings of fossils ('small shelly fossils' and 'earliest skeletal fossils'), although convenient, unscientifically lump together microfossil problematica (i.e. protoconodonts, tommotiids, lapworthellids, and halkieriids) and micro- to small molluscs, hyoliths, sponges, and other remains.

TEMPORAL CHANGES	Sunnaginia	Protohertzina	Chancelloria	Torellella	Latouchellids	Mobergella	Lapworthella	Anabaritids	Pelagiellids	Rhombocorniculum
SHAPE	●	●		●	●	●	●	●	●	●
SYMMETRY			●					●		
COILING				●	●				●	●
SCULPTURE						●	●	●	●	●
WALL STRUCTURE										●

Fig. 7.2. A table outlining the pattern of temporal changes in some potentially useful small shelly fossils from Precambrian–Cambrian boundary to Lower Cambrian strata. Those with more of these morphological characteristics, such as *Rhombocorniculum*, may prove to be more reliable as stratigraphical indices.

Although this heterogeneous assemblage has the obvious potential to build up integrated, cross-referenced assemblage zones, SSF elements vary widely in ecological and preservational character and their biostratigraphical potential is therefore equally variable. This analysis will therefore concentrate on those microfossils which the writer believes have the potential to provide a biostratigraphy through a series of morphological forms (e.g. *Protohertzina*, *Rhombocorniculum*, and *Lapworthella*) or first appearances. Such phosphatic microfossils have many of the advantages of conodonts: they are widespread, occur in clastic, carbonate, and phosphatic facies of different provinces, they suffer little from diagenesis, and they evolved into distinctively sculptured morphotypes. A disadvantage is that, in the earliest Cambrian, they are better represented in open-marine and phosphatic facies than in middle or inner carbonate facies. Fortunately, the evolutionary lineage of Siberian archaeocyathans also provides such a backbone for tropical carbonate facies, though the earliest forms are unknown outside of Siberia.

This fundamental 'backbone' is punctuated at intervals by the appearance of useful marker fossils, reasonably well separated in time, but whose evolutionary pattern is not yet clear (e.g. *Fomitchella*, *Platysolenites*, *Mobergella*, *Volborthella*, *Microdictyon*, *Tannuolina*, *Hadimopanella*, and *Salterella*). Calcareous forms are dealt with last, including anabaritids, small and distinctive pelagiellid and latouchellid molluscs, and chancelloriids.

The series of morphological forms provided by

archaeocyathans which give a backbone stratigraphy will be discussed first. Phosphatic, agglutinated, and calcareous small shelly fossils that help to calibrate successions are followed with a review of the trilobite and acritarch succession. A series of SSF intervals is then outlined, to serve as the basis for future biostratigraphical discussion. Finally, attention is turned towards the potential of first-appearance datum points (biohorizons) for finer-scale event stratigraphy.

7.4.1. The archaeocyathan basis for biostratigraphy

Archaeocyatha left behind them the most promising and diverse fossil record of any group in early Cambrian times, and yet were largely restricted to that epoch. Their skeletons were of microgranular calcite—perforate and generally conical with two walls connected by septa, tabulae, or both, though the morphology is highly variable (Hill 1972). Their lack of spicules contrasts with non-calcareous sponges, suggesting their status as an independent phylum (Hill 1972), a superphylum of the kingdom Archaeozoa (Zhuravleva 1970) or Archaeata (Zhuravleva 1974). Palaeobiological studies of Brasier (1976) and Debrenne & Vacelet (1984) however, have tended to confirm their sponge-like organization, and they are best regarded as an extinct class of the Porifera.

Archaeocyathans were the first reef invertebrates, thriving in shallow waters on the flanks of carbonate belts (e.g. Debrenne & James 1981). At first they were subordinate to the algae *Epiphyton* and *Renalcis* as mound-builders, but later bioherms were constructed largely by Irregulares such as *Metaldetes* and *Archaeocyathus*. This group could be compared with Tertiary larger foraminifera in ecology, strongly geometrical architecture, and global utility for biostratigraphy of tropical and subtropical limestones from orientated thin-sections. They also show a tendency for rapid, progressive, and convergent patterns of evolution, so that an experienced worker can readily estimate the age of a deposit from the morphology of the archaeocyathan assemblage, while their widespread nature allows genera to serve for international correlation. A generic range chart and brief palaeogeographical analysis are given by Debrenne & Rozanov (1983). Evolutionary patterns in regular Archaeocyatha have been reviewed by Zhuravleva (1960), Hill (1972), Rozanov (1973), and Rozanov & Debrenne (1974). They include trends from single to double wall, insertion of septa and/or tabulae, trends from simple to tumulate outer wall, from simple to reticulate outer wall, reduction in the number of vertical pore-rows per intersept, addition of a restrictive outer wall envelope, finer porosity of outer wall pores, and enlargement and elaboration of the inner wall pores

leading to ethmophyllid-like pore-tubes or annulae. There is also a trend towards colonial development. The end result seems to have been more efficient entrainment of water currents from the outer wall, through the intervallum and out of the central cavity. These trends provide the basis for the standard zonation of Lower Cambrian rocks on the Siberian Platform and Altai-Sayan Fold Belt (e.g. Fig. 7.11) and for some widely used stage names: Tommotian, Atdabanian, Botomian (formerly Lenian), and Toyonian (formerly Elankian or Lenian; Rozanov *et al.* 1969; Hill 1972; Rozanov & Debrenne 1974; Sokolov & Zhuravleva 1983; Rozanov 1984; Rozanov & Sokolov 1984). Their evolution and biogeography have been reviewed recently in English by Zhuravlev (1986).

Tommotian

The base of the Tommotian Stage is taken at the appearance of the first archaeocyathan assemblage with the index *Aldanocyathus sunnaginicus* (Zhur.). The forms here are simple, with single or double walls and simple pores, and several lineages—especially of *Aldanocyathus*—show a trend to reduce the number of longitudinal rows of outer wall pores during the Tomotian ('oligomery'; Hill 1970; Rozanov 1973; Rozanov & Debrenne 1974). The *Aldanocyathus sunnaginicus* Zone also contains *A. virgatus* (Zhur.), *Archaeolynthus*, *Cryptoporocyathus*, *Nochoroicyathus*, *Dokidocyathus*, and the Irregulares *Okulitchicyathus*, suggesting a preceding period of cryptic evolution.

The mid-Tommotian *Dokidocyathus regularis* Zone marks the appearance of the index, plus more developed, forms of *Aldanocyathus* such as *A. anabarensis* (Vol.) and *A. tkatschenkoi* (Vol.). Several important genera also appear: *Robustocyathus*, *Retecoscinus*, the Irregulares *Dictyocyathus* and the first forms with porous tabulae, represented by *Coscinocyathus*. Tumulate *Kotuyicyathus* and *Tumuliolynthus* made an appearance in the upper part of the *regularis* Zone.

The late Tommotian *Dokidocyathus lenaicus* Zone is marked by evolution of *D. lenaicus* Roz., plus ajacicyathids with fewer outer wall pores per intersept such as *Aldanocyathus turbidus* (Roz.), the tumulate monocyathid *Tumuliolynthus primigenius* Zhur., and tumulate *Tumulocyathus*, plus the irregular *Protopharetra*. Tommotian archaeocyathan assemblages are not confirmed outside of Siberia (Rozanov & Debrenne 1974; Debrenne & Rozanov 1983).

Atdabanian

The base of the Atdabanian Stage is defined by the appearance of new archaeocyathan assemblages, representing a second step in archaeocyathan evolution.

Envelopes were added to the outer walls of many Ajacicyathidae, Robertocyathidae, and Coscinocyathidae; the oligomeric trend in outer wall pores (e.g. *Aldanocyathus arteintervallum*) was joined by oligomery of envelope micropores (Rozanov 1973). Archaeocyathans also became more widespread at this time, appearing in Europe, Morocco, and Australia (Rozanov & Debrenne 1974). Four zones are recognized on the Siberian Platform. As well as the foregoing genera, typical new forms found widely include: *Fransuasaecyathus*, *Loculicyathus*, *Tumuliocyathellus*, *Sibirecyathus*, and *Taylorcyathus*.

Botomian

The Botomian (formerly Lenian) Stage is characterized by the demise of genera that began in the Tommotian, and also by the appearance of novel skeletal elements. Forms with elaborate or net-like outer walls or thick inner walls with pore-tubes are characteristic. Tabulate Regulares of the Coscinocyathidae are common. Taxa typical of this stage include Regulares *Ajacicyathus* and *Ladaecyathus*, and Irregulares *Pycnoidocyathus*, *Flindersicyathus* and *Syringocnema*, but only one zone is recognized in Siberia. The Atdabanian–Botomian boundary also has the status of a Subseries (Aldanian–Lenian) boundary in Siberia. This was the time of widest distribution and greatest diversity in both Regulares and Irregulares, coinciding with a wide extent of suitable facies (Brasier 1981; Debrenne & James 1981).

Toyonian

The end of the Botomian saw the first major extinction of archaeocyathans, and the ensuing Toyanian marked their twilight, with low-diversity Irregulares such as *Archaeocyathus*, *Claruscyathus*, and *Erbocyathus* predominating. *Irinaecyathus grandiperforatus* provides a zonal index in Siberia (Sokolov & Zhuravleva 1983). Toyonian archaeocyathans are associated with *Bonnia–Olenellus* trilobites in Labrador (Palmer & Rozanov 1976; Debrenne & James 1981). The group suffered widespread extinction at the end of the Toyonian, though simple irregular Archaeocythidae survived through middle into late Cambrian times in Antarctica (Debrenne *et al.* 1984).

The general approach to archaeocyathan biostratigraphy has been to assess the developmental condition of an assemblage and then compare it (usually at generic level) with a stage or part of a stage in Siberia (e.g. Rozanov & Debrenne 1974). This method tends to overlook the risks of parallel evolution in separate provinces. It must be admitted that proposed comparisons have seemed to conflict with concepts of trilobite biostratigraphy (e.g. Sdzuy 1978). A major problem is

the lack of experienced workers and the lack of zonal schemes for many regions and continents.

Although archaeocyathans are not found in Tommotian equivalents outside of Siberia, they emphasize a very important point: the simple organizational grade of Tommotian archaeocyathans confirms the real status of this stage, as distinct from Atdabanian and later assemblages. The morphological variety found even in the first assemblages, however, shows that the Tommotian *s.s.* does not preserve the earliest stage of evolution. Thus the appearance of the *sunnaginicus* Zone fauna was a migratory event, as well as an evolutionary event. This major wave of new immigrants is also seen in the small shelly fossils.

7.4.2. Selected phosphatic microfossils

A necessary step in the correlation of Precambrian–Cambrian boundary beds is the development and testing of zonal schemes that employ 'evolutionary' successions of microfossils and small shelly fossils through the boundary beds (e.g. Fig. 7.2). A step in this direction has been taken by Missarzhevsky (1982), noting the vertical ranges of selected hyoliths, molluscs, protoconodonts, tommotiids, and anabaritids in Siberia, showing how each may be arranged into a series of (undefined) partial or total range zones. These supposedly allow the recognition of twelve successive assemblage zones of small shelly fossils, ranging from the earliest assemblages of the 'Manykayan Stage' (roughly equivalent to the Nemakit–Daldyn Series or 'Horizon') to the Lower Botomian. The data of Missarzhevsky (1983), Sokolov & Zhuravleva (1983), Rozanov & Sokolov (1984), and Khomentovsky (1986) allow for further refinement. Fig. 7.11 shows the Siberian ranges of some taxa that are also found in other boundary sections.

A similar approach has been pursued by the writer with respect to conoidal phosphatic microfossils and other problematica in England and Siberia (Brasier 1986). Here is seen a succession of species, notably *Sunnaginia*, *Eccentrotheca*, *Torellella*, and *Rhombocorniculum*. Further work on Baltic, Canadian, and Palaeotethyan material confirms similarities in the order of appearance of species in these different areas, and an attempt is made (Figs 7.12 to 7.15) to illustrate their comparative ranges. Uncertainty about the upper ranges of some of the taxa, however, means that greater emphasis should be given to the point of first appearance, as discussed in a later section.

The phosphatic microfossils selectively discussed and illustrated below are relatively well-known taxa notable for their biostratigraphical potential. In most cases, rigorous taxonomic and morphological work remains to be done, and this treatment does not examine the many local variants and synonyms. Descriptions are informal, except where otherwise stated.

Protohertzina *and related forms*

The small phosphatic spines of *Protohertzina anabarica* Missarzhevsky and related forms are of considerable interest for correlation and delimitation of Precambrian–Cambrian boundary beds, since they are cosmopolitan and appear widely in the earliest skeletal assemblages (Figs 7.11, 7.14, 7.15). *Protohertzina* Missarzhevsky, *Hertzina* Müller, and *Amphigeisina* Bengtson are placed in the 'protoconodonts' by Bengtson (1976, 1983), being slender phosphatic cones with deep basal cavities and only basal–internal growth increments of mainly fibrous structure. Evolutionary modifications to the earliest protoconodont taxon could prove of great importance to biostratigraphy, since they would provide the clearest evidence for a second, developmental stage in the Cambrian fauna. The possibility that the variable *P. anabarica* plexus was ancestral to several stocks, through reduction of the posterior keel (i.e. *Hertzina* or *Amphigeisina*), reduction of the posterolateral keels (i.e. *Protohertzina siciformis* group), and broadening of the base and cusp (i.e. the *Maldeotaia bandalica* group) certainly deserves further research (Fig. 7.3). But it is also possible that *P. siciformis* is an element in the *P. anabarica* group (e.g. the 'siciform' elements of Brasier & Singh 1987). The relationship of *Hertzina/Amphigeisina* to *Protohertzina* is less clear, but the latter occurs with transitional 'hertziniform' elements in India (Brasier & Singh 1987) and elsewhere. The 'siciform' but broad *Maldeotaia–Ganloudina* morphotypes certainly seem to represent a younger element that deserves further study.

Protoconodonts of the *Protohertzina anabarica* Missarzhevsky group are reported from the lowest skeletal assemblages around the world. Considerable ontogenetic variation and possible symmetry transitions (e.g. Nowlan *et al.* 1985; Brasier & Singh 1987) indicate that *P. unguliformis* Miss. is not readily distinguishable from *P. anabarica*, and the two are combined here: symmetrical, proclined, slender, gently curved protoconodont elements with large basal cavity; anterior rounded, posterior median keel prominent to weak but not absent; posterolateral keels strong in 'anabariform' elements (pl. 7.1, part 1), weaker in 'unguliform' elements (pl. 7.1, part 3); lateral sulci weak in 'anabariform' elements, stronger in 'unguliform' elements; cross-section compressed in antero-posterior plane, in lateral plane, or isometric; ultrastructure fibrous in inner layers (Bengtson 1983) but may appear smoothly microgranu-

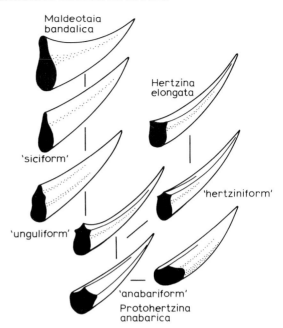

Fig. 7.3. Some morphotypes of *Protohertzina* and related forms. The cross-sectional profile is shown in solid black, though the specimens may contain a large internal cavity.

lar on the exterior. This *Protohertzina anabarica* group is known from the Manykayan/Nemakit–Daldynian assemblages of the Anabar and Olenek regions, Siberia (Missarzhevsky 1973; Sokolov & Fedonkin 1984), the upper Yudoma Formation at Tommot on the River Aldan, Siberia (ibid; Dr S. Bengtson pers. comm. 1986), the Kyrshabakty Formation of southern Kazakhstan (Missarzhevsky & Mambetov 1981), the first skeletal assemblage in Iran (Dr B. Hamdi, pers. comm.), the Chert–Phosphoritie Member of the Lower Tal Formation, India (Azmi 1983), the first, *Tiksitheca licis–Maikhanella multa* assemblage at Salanygol, Mongolia (Missarzhevsky 1982), the Zone I assemblage of the Meishucunian Stage in China (Luo *et al.* 1982; Xing *et al.* 1983), and the basal Vampire Formation of Yukon Territory, Canada (Nowlan *et al.* 1985).

Protohertzina siciformis Missarzhevsky appears to be related to the 'unguliform' type: elongate, 'siciform' *Protohertzina* with rounded anterior, posterior extended into a sharply pointed keel; cross-section teardrop shaped; posterolateral keels lacking. It is known from beds above *P. anabarica* in the Chulaktau Formation of Kazakhstan (Missarzhevsky & Mambetov 1981) and also as rare 'siciform' elements from the Lower Tal formation of India (pl. 7.1, part 5), associated with *P. anabarica* (Azmi 1983; Brasier & Singh 1987). The same form may be recorded as *P. ?cultrata* Miss. from the upper Zhongyicun Member of Meishucun, China (Jiang 1980; Xing *et al.* 1983). Although its range is given as Lower and perhaps Middle Tommotian (Missarzhevsky 1982), it ranges into the *Bercutia cristata* Zone in southern Kazakhstan (Missarzhevsky & Mambetov 1981) and many arguably, therefore, range through the Tommotian into the Lower Atdabanian. *Protohertzina* sp. from the trilobite-bearing Parara Formation of South Australia are close to *P. unguliformis* and *P. siciformis* in shape, though the base appears to be more expanded and they may be associated with broad elements (material of Dr S. Bengtson).

'Hertziniform' elements transitional between *P. anabarica* and typical *Hertzina/Amphigeisina* also appear to be present in the Lower Tal assemblage (pl. 7.1, part 4; Brasier & Singh 1987). These have a posterior keel that is weak and passes apically into a flattened posterior face. Similar forms may be present in the Vampire Formation assemblage (Nowlan *et al.* 1985, figs 8, 9).

Forms variously attributed to *Hertzina* and *Amphigeisina* also occur in early assemblages. *Hertzina elongata* Müller ranges from the Bed 1ii–iii assemblage at Nuneaton (pl. 7.1, part 7; Brasier 1984) into Beds Ac4 and Ac5 at Comley (Hinz 1987) and up into the Upper Cambrian (Müller 1959). Further subdivision of this group may be possible, but Lower Cambrian forms are broadly as follows: nearly symmetrical, slender, gently curved protoconodont; posterior face flat to gently concave, bounded by two distinct posterolateral carinae; lateral sides flat to weakly concave near the carinae; anterior surface convex. As well as the above material, this form is also known from Chapel Island Member 4 and the lower Bonavista Formation to just below *Serrodiscus bellimarginatus* in the Redlands Limestone of south-east Newfoundland (unpublished data of E. Landing; and Brasier) and from the Ac3 to Ad Limestones at Comley (Hinz 1983). Forms referred either to *Hertzina* or *Amphigeisina* sp. also occur in the first assemblage in northern Iran (material of Dr B. Hamdi), and in zone I assemblages of Sichuan (material of Dr He Ting-gui).

The broad elements of *Maldeotaia bandalica* Singh & Shukla (1981), *Ganloudina symmetrica* He (in Xing *et al.* 1983), and *Mongolodus rostriformis* Missarzhevsky (1977) are similar: they are symmetrical, proclined, gently curved elements with an expanded base and deep central cavity; the anterior side is rounded, the posterior side is more angular; the cross-section is rounded, elliptical, or ovate at the apical end and elliptical, teardrop-, dumb-bell-, or slightly U-shaped at the basal end; the surface has fine longitudinal fibres. This type is found in the Lower Tal Formation (pl. 7.1, part 6) and first assemblage of Maidiping, Sichuan (op. cit.) at levels that are probably in the lower to upper part of Chinese zone I. Other species of *Ganloudina* described from northern Sichuan by Yang *et al.* (1983) may be synonyms of similar age.

The internal–basal secretion of these protoconodonts, and the evidence for symmetry transitions, suggests they were borne externally, as clusters of tooth-like spines: much as in modern chaetognaths (e.g. Bengtson 1976, 1983).

Rhombocorniculum

The distinctive elements of *Rhombocorniculum* provide one of the most promising indices for global biostratigraphy of Lower Cambrian strata (Fig. 7.4; pl. 7.2). Three successive species (*insolutum*, *cancellatum*, and *spinosus*) can now be defined in the boundary beds of Nuneaton and Shropshire, while the first two are found in Siberia and Bornholm–Scania; *R. cancellatum* has a nearly global distribution. Allowances must be made for natural variation and symmetry transition series, but there are distinct evolutionary changes in rhomboidal sculpture, ultrastructure, profile, cross-section, and curvature.

Rhombocorniculum insolutum Missarzhevsky appears in Bed 10iii at Nuneaton (pl. 2, part 2; Brasier 1986) and ranges through the upper *lenaicus* Zone and the Lower Atdabanian *pinus* Zone in Siberia (Sokolov & Zhuravleva 1983; Rozanov & Sokolov 1984): assemblages comprise gently curved to straight elements, generally with subcircular to ovate cross-section and a wide subcentral basal cavity; broad flat elements lacking; sculpture flat, a cut-glass like rhomboid pattern, formed by diagonally intersecting grooves; ultrastructure essentially of very elongate coaxial fibres with microgranular outer sculptured layer. This form also occurs in the Green Shales of Bornholm (pl. 7.2, part 1, 3,), with probable *Mobergella* sp., trilobite fragments (unpublished data), plus a skeletal assemblage close to that of bed 10iii, Nuneaton (Poulsen 1967). It is also known with upper *Camenella baltica* assemblages in the upper 50 m of the Bonavista Formation, south-east Newfoundland (Landing in press). This fossil provides a useful index for early Atdabania assemblages.

Rhombocorniculum cancellatum (Cobbold) occurs in Ac2 to Ad beds in Shropshire (pl. 7.2, part 7–9; Cobbold 1921; Brasier 1986) and in the Atdabanian *pinus* Zone to Botomian of Siberia (Sokolov & Zhuravleva 1983; Rozanov & Sokolov 1984); *Rhombocorniculum* ex. gr. *cancellatum* occurs in the Botomian of the Siberian Platform according to Missarzhevsky (1982): assemblages contain more strongly curved elements, generally

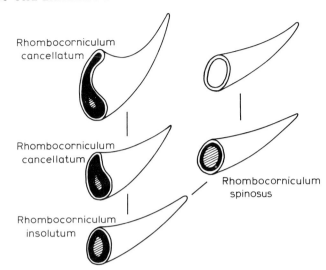

Rhombocorniculum
cancellatum

Rhombocorniculum
cancellatum

Rhombocorniculum
spinosus

Rhombocorniculum
insolutum

Fig. 7.4. Some typical morphotypes of *Rhombo-corniculum*. The fibrous inner layer is shown in black (in cross-section) and the microgranular outer layer is shown in white. External sculpture is omitted.

with compressed cross-section, narrow subcentral to peripheral basal cavity in slender elements; broad flat elements with a central basal cavity are present; sculpture varies from flat to raised rhomboid. The broad triangular elements referred to *R. walliseri* by Mambetov (1977) are also regarded as part of the *R. cancellatum* apparatus (see Landing *et al.* 1980). As well as the Siberian and English occurrences, the taxon occurs in the Brigus Formation of south-east Newfoundland (material of Dr S. Bengtson) and supposedly in the upper Bonavista Formation (Landing in press); those found 6 m above the base of the Brigus Formation have a rather conservative shape but true *cancallatum* sculpture. The species also occurs in the Gislöv Formation of Scania (material of Dr S. Bengtson), Botomian Salanygol Formation of Mongolia (Voronin *et al.* 1982), with *Cambroclavus* in the lower Shabakty Formation of southern Kazakhstan (Missarzhevsky & Mambetov 1981), with *Cambroclavus* in the correlated Xihaoping Member of the Tongying Formation, Hubei (Qian & Zhang 1983), below the *Serrodiscus* Slate of Gorlitz, GDR (Rozanov 1973), with a *Callavia* fauna in the Hoppin Slate of Massachusetts and with a *Serrodiscus bellimarginatus* fauna in Nova Scotia (Landing *et al.* 1980). The taxon therefore provides a useful index to middle Atdabanian to Botomian assemblages.

Rhombocorniculum spinosus (Hinz) occurs with *R. cancellatum* in the *Serrodiscus bellimarginatus* Limestone Bed Ac3 (pl. 7.2, parts 4–6) and *Lapworthella* Limestone Bed Ad at Comley (Brasier, Chapter 5 of this volume). This is described as follows: nearly straight to curved element with rounded cross-section, base flared and trumpet-like in one specimen; wide basal cavity; sculpture relatively flat, of scale-like to flute-mark like or

linguoid cusps which are asymmetrical, not strictly rhomboid. The wall is thin and lamellar (from one to three lamellae seen), the non-lamellar fibrous layer being weak or absent. This is the form referred to *R.* aff. *insolutum* by Brasier (1986). The holotype is from an undifferentiated level in the Comley Limestone of Rushton, Shropshire (Hinz 1987). The latter author placed these specimens in *Rushtonites* n. gen.

Ultrastructure and morphology are not inconsistent with a relationship between *Rhombocorniculum* and protoconodonts such as *Protohertzina* (e.g. Bengtson 1983).

Torellella

Further work on the abundant Nuneaton material of the phosphatic ?worm tube *Torellella* Holm indicates the presence of two successive morphotypes in the Home Farm Member; older forms are found in Siberia and Nuneaton, and younger forms in Siberia, Nuneaton, Shropshire, the Balto-Scandinavian area and south-east Newfoundland. Apparent trends include an increase in the length and regularity of the tube, a reduction in the curvature and angle of divergence, and a change from a sharply lenticular to a rounded ovate cross-section (Fig. 7.5). The sequence of *Torellella curva–T. lentiformis–Torellella biconvexa* is found both in Siberia (Missarzhevsky 1969) and England (Brasier 1986), but not in eastern North America (Landing in press). Differentiation of these simple tubes is problematical, and more evidence is needed to confirm the trends.

Torellella curva Missarzhevsky is abundant in the basal Tommotian but probably ranges higher (e.g. Zhuravleva 1975): it has small, smooth tubes with a

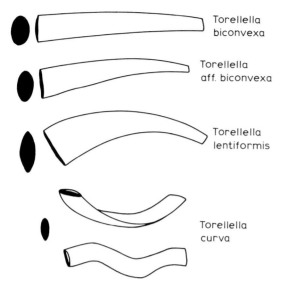

Torellella
biconvexa

Torellella
aff. biconvexa

Torellella
lentiformis

Torellella
curva

Fig. 7.5. Some typical morphotypes of *Torellella*. The cross-sectional profile is shown in solid black, though the specimens may contain a large internal cavity.

compressed lenticular cross-section, a low angle of divergence, or parallel sides, with slight and random lateral and dorsoventral curvature in some tubes. Larger forms, but with a similar shape, occur in Bed 1ii–iii at Nuneaton (pl. 7.1, part 14), though they are referred by Brasier (1986) to *T. lentiformis*.

Torellella lentiformis (Syssoiev) ranges from Bed 1ii to Bed 10ii of the *Hyolithes* Limestone at Nuneaton and from the basal Tommotian *sunnaginicus* Zone to the Upper Atdabanian *kokoulini* Zone in Siberia (pl. 7.1, part 13; Sokolov & Zhuravleva 1983; Rozanov & Sokolov 1984): it has a lenticular cross-section, weak lateral keels, curvature in the lateral and dorsoventral planes, a conspicuous angle of divergence, and weak sculpture of growth striae and corrugations. The length of the tubes does not exceed 10 mm at Nuneaton. This form also occurs in the *Aldanella* fauna from the Hoppin Slate, Massachusetts (material of Dr E. Landing) and questionably in the Ab1 Beds of the Lower Comley Sandstone the Ercall, Shropshire (Brasier, unpublished).

Torellella aff. *biconvexa* Missarzhevsky ranges from Bed 8 to 10ii of the Hyolithes Limestone at Nuneaton; this shows features intermediate between *T. lentiformis* and *T. biconvexa*. The latter ranges from Bed 10iii to the ?*Callavia*-bearing nodules at the base of the Purley Shale in Nuneaton: it has a rounded to ovate cross-section without keels, a gentle lateral curvature, and a low angle of divergence. The length does not exceed

10 mm at Nuneaton. It also occurs in the Comley Limestone, Shropshire, and in the Redland Limestone Member of the Brigus Formation of south-east Newfoundland, just below the appearance of *Serrodiscus bellimarginatus*. Forms close to *T. biconvexa* occur lower, however, in the *Aldanella* fauna of Massachusetts (material of Dr E Landing). In Siberia, the species ranges commonly from the Upper Tommotian *lenaicus* Zone (pl. 7.1, part 12; Missarzhevsky 1969) to the Lower Atdabanian *zegebarti* Zone (Sokolov & Zhuravleva 1983; Rozanov & Sokolov 1984). *T. biconvexa* is also a typical element of the *Bercutia cristata* Zone at the top of the Chulaktau Formation in southern Kazakhstan (Missarzhevsky & Mambetov 1981), regarded as late Tommotian to early Atdabanian (Missarzhevsky 1982; Rozanov & Sokolov 1984).

Torellella laevigata (Linnarsson) is similar to *T. biconvexa* and may also be a senior synonym of the other species (e.g. Landing in press). It is longer and relatively narrow, however, reaching up to 20 mm in the type area of Vastergötland in Sweden (Lindström 1977). It usually appears above a *Mobergella* fauna, associated with a *Mickwitzia monilifera–Volborthella tenuis*–olenellid fauna in Scandinavia (e.g. Martinsson 1974). A fragment that may be this form is known from the *Rhombocorniculum insolutum* assemblage in the Green Shales of Bornholm (Brasier 1986).

Palaeoecological studies of Nuneaton material suggest a possible reclining or attached, worm-like habit for *Torellella*.

Sunnaginia

In the earlier opinion of the writer (Brasier 1986), the various phosphatic sclerites attributed to the problematical tommotiid *Sunnaginia* Missarzhevsky may not all belong to *S. imbricata* Missarzhevsky. Allowing for symmetry transition series, at least two species were suggested at Nuneaton with another present at higher levels in the Comley Limestone and in Nova Scotia. Postulated trends included a reduction in size, an increase in relative breadth, a stronger sulcus, increased development of anterior and posterior arches, accentuation of the rostrate first lobe (L1 following the terminology of Landing *et al.* 1980) and attenuation of the umbo-like L3 lobe from rounded obtuse to right-angled in early forms to sharply obtuse to acute in later forms (Fig. 7.6). *Sunnaginia* is a highly variable form, however, and larger assemblages are needed to quantify changes in this cline of closely related sclerites. Neither Hinz (1987) nor Landing (in press) have yet favoured the 'splitting' of *Sunnaginia imbricata*.

Sunnaginia imbricata Missarzhevsky occurs from the *sunnaginicus* to *regularis* Zones in Siberia (pl. 7.1, part

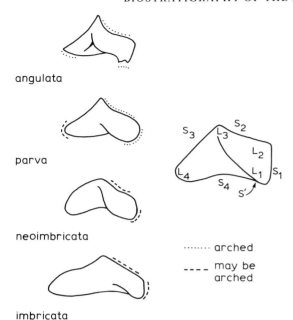

angulata

parva

neoimbricata

········ arched

- - - - may be arched

imbricata

Fig. 7.6. Some typical morphotypes of the *Sunnaginia imbricata* group, showing the nomenclature followed here. Concepts after Brasier (1986).

11; Sokolov & Zhuravleva 1983; Rozanov & Sokolov 1984): assemblage of large, elongate forms with rounded lobes, coarse concentric ribs and tubercles, obtuse to right-angled L3 lobe; anterior arch weak but may be present(?); posterior arches weak(?); sulcus weak(?). This form is still poorly known, and attribution of other forms to this taxon (e.g. Landing *et al.* 1980, Hinz 1987; Landing in press) is questionable.

Sunnaginia neoimbricata Brasier occurs in the Quartzose Conglomerate, is common in Bed 1ii–iii, and may include questionable large fragments up to Bed 10iii at Nuneaton (pl. 7.1, part 10): it is a large form with rounded lobes, an obtuse to right-angled L3 lobe, rarely acute; the anterior and posterior arches are usually poorly developed; a fold-like sulcus and a ridge cross the sclerite from S1 to L3. Landing (in press) retains this form in *S. imbricata*.

Sunnaginia parva Brasier ranges from Bed 10iii to 12 at Nuneaton (pl. 7.1, part 9): assemblage with smaller elongate forms, with obtuse to acute L3 lobe but other lobes relatively rounded; anterior arch better developed than posterior arch; L1 may be rostrate, overhanging a sulcus. This form also occurs with a *Coleoloides* fauna in south-east Newfoundland, below the *callavia* Zone (Bengtson & Fletcher 1983), and specimens with an *Aldanella* fauna in Massachusetts may also be close to

this species (material of Dr E. Landing). Landing (in press) places this morphotype in *S. imbricata*.

Sunnaginia angulata Brasier has generally more equant proportions, more accentuated L3 and rostrate L1 lobes, and strong posterior and anterior arches, sulcus distinct. It occurs with a *Serrodiscus bellimarginatus–Rhombocorniculum cancellatum* fauna in the Ac3 Beds at Comley, Shropshire (pl. 7.1, part 8), and in Nova Scotia (e.g. Landing *et al.* 1980). Hinz (1987) and Landing (in press) place this form in *S. imbricata*.

Sunnaginia acuta Grigorieva is closest to *S. parva*, but differs in having a straight, truncated anterior margin. It is described from a mixed Tommotian–Atdabanian–Botomian assemblage in Mongolia (Grigorieva *in* Voronin *et al.* 1982) and may have been reworked.

These problematical sclerites have been placed in the Sunnaginiidae (Landing 1984) and may be related to other multi-element, lamellar phosphatic sclerites of the Tommotiidae, Lapworthellidae, Tannuolinidae, and Kelanellidae in an order Mitrosagophora (Bengtson 1977, 1986).

Lapworthella

The phosphatic cornute sclerites of *Lapworthella* vary greatly in their cross-section, angle of divergence, and curvature (Matthews 1973; Bengtson 1980) and sculpture (Conway Morris & Fritz 1984; Hinz 1987) but species have been broadly distinguished by biometry of surface sculpture and aperture length to width. *Lapworthella* species have also proved useful for stratigraphy in Siberia (Missarzhevsky 1966; Rozanov *et al.* 1969; Missarzhevsky 1982). The series is best demonstrated from the Siberian Platform. Trends seem to include development of pronounced transverse ribs and the development of granular to denticulate sculpture, with some retrogression towards smooth ridges in *L. cornu nigra*. Only a selection of forms is discussed here, pending further information on the many species reported from China and elsewhere.

Lapworthella ludvigseni Landing is a form described from an *Aldanella attleborensis* assemblage in eastern Massachusetts (Landing 1984): it has rounded transverse ribs of varying strength crossed by irregularly spaced, weak longitudinal ridges; the cross-section is circular to compressed slit-like; and internal septa may be present (Landing 1984). *Lapworthella* n. sp. reported from Chapel Island Member 4, south-east Newfoundland (Bengtson & Fletcher 1983) is ascribed to this form by Landing (1984; the taxon ranges into the upper Bonavista Formation, Landing pers. comm. 1986), as is that figured from the Zone II fauna in eastern Yunnan (Luo *et al.* 1982, p. 186, pl. 18, figs 1, 2). Although Hinz

(1987) questions the validity of this taxon, it is accepted herein, pending further studies.

Lapworthella tortuosa Missarzhevsky ranges from the ?lower, and certainly the upper *sunnaginicus* Zone into the *lenaicus* Zone of Siberia (Rozanov *et al.* 1969; Sokolov & Zhuravleva 1983; Rozanov & Sokolov 1984; Khomentovsky 1986): it has broad transverse granular bands of low relief; the cross-section is circular, oval to subtriangular in the adult. The taxon is also reported from the upper part of the second faunal zone at Salanygol, Mongolia (Voronin *et al.* 1982).

Lapworthella bella Missarzhevsky ranges from the *bella* Subzone (*regularis* Zone) to the basal Atdabanian (Missarzhevsky 1982; Sokolov & Zhuravleva 1983; Rozanov & Sokolov 1984): it has low-relief transverse ribs; elongate, longitudinally disposed nodes associated with the ribs; and the cross-section is circular to oval in the adult. A *Lapworthella* sp. close to this form is associated with a *Serrodiscus bellimarginatus–Rhombocorniculum cancellatum* fauna in Nova Scotia (Landing *et al.* 1980).

Lapworthella filigrana Conway Morris & Fritz is known from the Rosella Formation, low in the *Nevadella* Zone of British Columbia, of presumed late Atdabanian age (Conway Morris & Fritz 1984) and correlated levels in California and Mexico (McMenamin 1987): denticulate ribs and inter-rib areas with a scattered ornament of larger and smaller denticles, the latter tending to be arranged in subcircular arrays towards the apex; the cross-section is variable, being provided with spur-like extensions near the aperture; a longitudinal furrow is sometimes present. This form is close to *L. lucida* Meshkova from the Anabar and Khara–Ulakh region (upper *Pagetiellus anabarus* to upper *Judomia* Zones), though the latter form lacks a subcircular pattern of inter-rib smaller denticles (Meshkova 1969). *L. rete* Yue is a similar form from Zone III of Sichuan, China (Yue 1987).

Lapworthella cornu (Wiman) was first described from the 'Olenellid Sandstone' of the south Bothnian area, associated with *Torellella laevigata* and *Proampyx? balticus*. Reports of *Lapworthella dentata* from the Ac4 to Ad beds at Comley (Matthews 1973) were later referred to *L. cornu* (Bengtson 1980) and are here referred to *L. cornu cornu* (pl. 7.3, part 6): denticulate ribs, over 14 ribs per mm, *c.*10–20 denticles per mm, Bengtson 1980). This form also occurs in the Gislöv Formation with *Holmia sulcata* (Bengtson 1980). Forms placed in non-Linnaean 'Lapworthella types A–D' from the *Callavia* to *Lapworthella* Limestones of Shropshire (Hinz 1987) may belong here.

Lapworthella dentata Missarzhevsky appears to be a geographical subspecies, here placed in *L. cornu dentata*. It ranges from the upper Atdabanian *lermontovae* Zone

to the Lower Botomian in Siberia (Missarzhevsky 1982; Sokolov & Zhuravleva 1983; Rozanov & Sokolov 1984): denticulate ribs, *c.*8 ribs per mm, *c.*25–45 denticles per mm; cross-section ovate-rectangular to circular.

A closely related form described as *Lapworthella schodackensis* (Lochman-Balk 1956) comes from the *Elliptocephala asaphoides* assemblage of eastern New York, and was recently redescribed by Landing (1984). This is here placed in *L. cornu shodackensis*: denticulate ribs, *c.*14–50 ribs per mm; *c.*35–144 denticles per mm; the cross-section is a circular to strongly compressed rectangular and oval shape.

Lapworthella nigra Cobbold was first described from the Ad beds at Comley (Cobbold 1921); its range was later extended down to the Ac4 beds in the Rushton area (Cobbold & Pocock 1934). It is tentatively regarded here as *Lapworthella cornu nigra* (pl. 7.3, part 7): smooth transverse ribs, 4–15 per mm; cross-section circular to subrectangular. This form is also reported from Lower Botomian rocks in Siberia (Missarzhevsky 1982) and in the lower part of the Salanygol Formation of Mongolia, of supposed early Botomian age (*L. nigra* of Voronin *et al.* 1982). *Lapworthella* cf. *nigra* occurs with a *Calodiscus lobatus* fauna in Jamtland, central Swedish Caledonides (Ahlberg 1984). The forms referred to *L. cornu*, *L. dentata*, and *L. bella* from the base of the Yu'anshan Member, Qiongzhusi Formation in Eastern Yunnan (Luo *et al.* 1982) may also be closely related.

Lapworthella sclerites, like those of *Sunnaginia*, formed part of a multi-element skeleton with continuous symmetry transitions (unimembrate), placed in the Lapworthellidae (e.g. Landing 1984) and the order Mitrosagophora (e.g. Bengtson 1986).

Fomitchella *and trumpet-shaped elements*

Fomitchella Missarzhevsky is a taxon of phosphatic conoidal microfossils with a rather acicular apex, round to ovate cross-section, and smooth curvature. The walls are lamellar, of microgranular apatite with a smooth or finely striated surface (Missarzhevsky 1969; Bengtson 1983). Progenitors of *Fomitchella* from the earlier part of the *Protohertzina anabarica* range are not yet known, but *Fomitchella*-like forms of trumpet-shaped elements range at least as high as beds with *Rhombocorniculum cancellatum*.

Fomitchella infundibuliformis Missarzhevsky is a euconodont-like conoidal phosphatic microfossil (Bengtson 1983): it has straight to curved proclined, symmetrical elements of lamellar apatite with narrow elongate apex and trumpet-like basal extension; its cross-section is circular to ovate; and its external surface is smooth or with faint longitudinal striations. On the Anabar Platform, the species is known from the *sunnaginicus*

Zone (Rozanov *et al.* 1969; Missarzhevsky & Mambetov 1981), ranging up into the *regularis* Zone and down into the upper Manykayan/Nemakit–Daldynian (Missarzhevsky 1983; Khomentovsky 1986). This taxon is arguably present from lower *Aldanella* to *Camenella* assemblages of Newfoundland and Massachusetts (Landing in press). Similar *F.* aff. *infundibuliformis* occurs in the phosphorite unit of the Chulaktau Formation in southern Kazakhstan (Missarzhevsky & Mambetov 1981). Superficially similar trumpet-shaped elements also occur in the Chert–Phosphorite Member of the Lower Tal Formation (Brasier & Singh 1987), although they are thicker walled. Trumpet-shaped elements also occur in the second assemblage of Huitze County, Yunnan, and the third assemblage at Meishucun (material of Dr Jiang Zhiwen).

Fomitchella acinaciformis Missarzhevsky is probably restricted to the *sunnaginicus* Zone of Siberia (Missarzhevsky 1982), but differs in having a lower rate of expansion and greater lateral compression. *F.* cf. *acinaciformis* occurs in the *Aldanella* Zone of south-east Newfoundland (Bengtson & Fletcher 1983), and similar forms (though they may be *Kijacus* sp. or *Torellella* sp.) range up into the Smith Point Limestone (Dr E. Landing, pers. comm. 1986; Narbonne *et al.* 1987); unfortunately, their ultrastructure is poorly preserved.

Various forms referred to *Fomitchella* from the Meishucunian of China are questioned by Bengtson (1983) and require further study.

Camenella

Camenella Missarzhevsky is a distinctive multi-element, bimembrate assemblage of tommotiid sclerites, widespread from the upper *sunnaginicus* Zone to about mid-Atdabanian times. In this genus are placed those phosphatic sclerites that are four-sided, with one side (the 'duplicature' of Bengtson 1970) deeply infolded. Cornute 'mitral' sclerites and saddle-shaped 'sellate' sclerites are present, as are right- and left-handed variants; symmetry transition series are also present within these bimembrate assemblages.

Camenella garbowskae Missarzhevsky was originally described for the sellate form, which ranges from the upper part of the *sunnaginicus* Zone to the *zegebarti* Zone of Siberia (Sokolov & Zhuravleva 1983) and possibly into the *pinus* Zone (Missarzhevsky 1982). The mitral form is here taken to be *Tommotia kozlowskii* (Missarzhevsky) which ranges from the upper *sunnaginicus* to *zegebarti* Zones, in association with *C. garbowskae* (Sokolov & Zhuravleva 1983; Rozanov & Sokolov 1984): it has a sculpture of radial ribs and concentric growth scarps; usually a zone of multiple radial folds on the larger lobe of sellate forms, or obplicate side of mitral

forms; and also a zone of weak to no radial sculpture on the duplicature (all forms), on the smaller lobe (sellate forms), or plicate and accrescent sides (mitral forms).

Camenella baltica (Bengtson) is an Acado-Baltic species with a slightly longer time-span. In south-east Newfoundland, it appears above the *Aldanella attleborensis* assemblage in the Bonavista Formation, and ranges into beds with *Serrodiscus bellimarginatus* (Dr E. Landing, pers. comm. 1986; 1987). In England, it appears in Bed 1ii–iii of the *Hyolithes* Limestone, Nuneaton (pl. 7.3, Fig 7.3; Brasier 1984); it also occurs in the *Mobergella* Zone of Småland, Sweden (Bengtson 1970, 1977a) and equivalent beds in Shropshire and Nuneaton, England (Brasier 1984): *Camenella baltica* has a plaid-like reticulate sculpture of prominent growth ridges with finer striations between, and finely spaced radial growth costae; the radial sculpture is absent over the duplicature, weak on the decrescent side, and otherwise rather uniformly developed; multiple folds are absent from the larger lobe of sellate forms; the mitral form is often a rectangular-sided pyramid with a large aperture and a less-infolded duplicature. The species shows little tendency towards helical growth. Forms similar to the mitral sclerite, from the lowest, *Tiksitheca licis–Maikhanella multa* assemblage of Salanygol, Mongolia, have been referred to *T.* cf. *baltica* (Missarzhevsky 1982; Voronin *et al.* 1982). This may be related to *Camenella parilobata* Bengtson, a form ?reworked into the Botomian Chairchanskaya Formation, and having a V-shaped trough between the radial ridges (Bengtson 1986). Its precise age is unknown.

Camenella korolevi Missarzhevsky is a strongly plicate form, known only in the *Bercutia cristata* Zone of the upper Chulaktau Formation of southern Kazakhstan, associated with *Torellella biconvexa* (Missarzhevsky & Mambetov 1981).

The bimembrate, multi-element apparatus of *Camenella* (including *Tommotia*), with symmetry transitions within both mitral and sellate sclerites, is a distinctive feature of mitrosagophorans in the Tommotiidae (Landing 1984; Bengtson 1986); their wider affinities are still unknown.

Mobergella

Mobergella is a problematical discoidal, operculum-like sclerite of lamellar phosphatic composition. It is widespread in strata of late Tommotian and younger age whose occurrence has allowed the recognition of a *Mobergella* Zone in Scandinavia and the East European Platform (e.g. Bengtson 1977a; Lendzion 1983a), more-or-less coincident with the appearance of olenellid trilobites. Higher forms confirm a trend towards a reduction in the number of muscle scars from fourteen to ten (Fig. 7.7). Unfortunately, no single succession shows this trend.

Mobergella radiolata Bengtson appears near the base of the *lenaicus* Zone in Siberia, and is either restricted to that zone (Bengtson 1970) or ranges up into the *polyseptatus* (cf. *zegebarti*) Zone (Sokolov & Zhuravleva 1983). It also occurs at Aspelund, Öland, and Gotland in Sweden (Bengtson 1968) and more doubtfully from the upper part of the Klimontovian Stage, *Mobergella* Zone, of northern and Central Poland (Urbanek & Rozanov 1983): the thin-shelled *Mobergella* is usually flat but sometimes moderately arched; it has an apex with a central concavity; the muscle scars are narrow ridges radiating from the apex and are fourteen in number (i.e. seven pairs).

Mobergella radiolata is associated with *M. holsti* (Moberg) in Sweden (Bengtson 1970). The latter occurs throughout the West Baltic area and also in beds just above *Schmidtiellus mickwitzi* and *Platysolenites antiquissimus* in the Mjøsa district of Norway (Åhman & Martinsson 1965; Bengtson 1968; Bergström 1981). It is

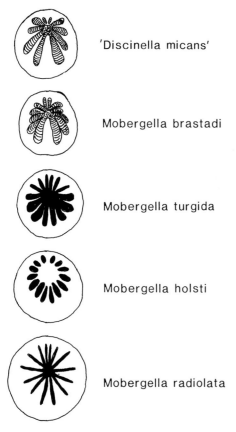

'Discinella micans'

Mobergella brastadi

Mobergella turgida

Mobergella holsti

Mobergella radiolata

Fig. 7.7. Changes in muscle-scar pattern in *Mobergella*.

also recorded from the upper part of the Klimontovian Stage, *Mobergella* Zone, of the Prabuty borehole, northern Poland (Urbanek & Rozanov 1983): *Mobergella holsti* has a strongly convex apex; the shell flattens towards the margin; and the muscle marks commonly form relatively broad impressions, fourteen in number.

Mobergella turgida Bengtson is known from *Mobergella* Zone glacial erratics in the West Baltic area (Bengtson 1968), while *M.* cf. *turgida* occurs also in bore-holes at this level in northern Poland (Volkova *et al.* 1983), at the base of the Lower Comley Sandstone, Shropshire, England (Bengtson 1977, and confirmed herein): this small, relatively thick-shelled *Mobergella* has an apex with a central concavity; faint growth lines on the exterior; and elevated 'swollen' muscle attachments, usually fourteen in number.

Mobergella sp. seen by Bengtson (1970) from the Atdabanian of Siberia are said to have only six pairs of muscle prints. *Mobergella brastadi* (Poulsen) occurs in the Bastion and Ella Island Formations of East Greenland, associated with *Bonnia–Olenellus* Zone assemblage with *Calodiscus lobatus* (Poulsen 1932; Cowie 1971a): the *Mobergella* here has broad, petal-like muscle scars, twelve in number (i.e. six pairs). At least one of the figured specimens has narrow raised muscle scars, resembling those of *M. radiolata* (Poulsen 1932, pl. 7, fig. 9).

'*Discinella micans*' (Billings) has been used to describe *Mobergella* sclerites that are similar to *M. brastadi* but show only ten petaloid muscle scars, marked by concentric striations. These occur in the Bastion and Ella Island Formations of East Greenland (Poulsen 1932; Cowie 1971a) and in the *Elliptocephala asaphoides* fauna of the Taconic allochthon, eastern USA (Theokritoff 1964), the latter again containing *Calodiscus lobatus* (Hall) and *Serrodiscus speciosus* (Ford). This suggests a latest Atdabanian to Toyonian age for *M. brastadi* and '*D. micans*'.

Suggestions that *Mobergella* formed the operculum of *Hyolithellus* tubes are not confirmed by size or distribution (Bengtson 1968), and affinities remain unknown.

Hadimopanella

This button-like phosphatic microfossil ('*Lenargyrion*' of Bengtson 1977*b*; Sokolov & Zhuravleva 1983) is so far unknown from strata with *Rhombocorniculum insolutum* or earlier assemblages, but has been widely found in strata of late Atdabanian to mid-Cambrian age. The several described species may have some stratigraphical value.

Hadimopanella knappologica (Bengtson (1977*b*) ranges from the *lermontovae* to *B. micmacciformis–Erbiella* Zones in Siberia (Bengtson 1977*b*; Sokolov & Zhuravleva 1983): it has a small (50–460 μm) circular to oblong disc with one smooth slightly convex side ('smooth face') and an opposite side comprising a nearly conical 'girdle' culminating in a flat-topped 'nodular face' bearing small nodes; the outer nodes are arranged in a ring but the inner nodes are less regular; nodes 10–15 μm diameter and c.50 per 0.01 mm^2; internal structure with porous core and dense outer layer.

Hadimopanella apicata Wrona occurs in the Lower Cambrian of Spitsbergen and North Greenland, in both cases associated with *Serrodiscus bellimarginatus*, *Calodiscus*, and olenellids of the *Bonnia–Olenellus* Zone (Wrona 1982; Peel & Larsen 1984). The small *Hadimopanella apicata* ($<140\ \mu$m) have a broad, strongly conical 'girdle' with a well-developed marginal 'brim' with slight radial corrugation; the 'nodular face' is not flat-topped but forms an apex to the girdle, with up to four (possibly five) small nodes; more than half may show no nodes. A similar element occurs in the Ac4 beds of the Comley Limestone in England (Hinz 1987). *H. antarctica* Wrona are smaller (55–93 μm) forms from the mid Lower Cambrian of Antarctica, with nodes around the girdle (Wrona 1987).

Hadimopanella oezgueli Gedik is the type species of the genus; it occurs in the Middle (and ?Lower) Cambrian beds of Turkey (Gedik 1977, 1981) and the Middle Cambrian Lancara Formation of Spain (van den Boorgard 1983) and similar strata in Sardinia (Cherchi & Schroeder 1985): similar to *H. knappologica* but with coarser nodes (c.20 μm in diameter) and more widely spaced (c.15 per 0.01 mm^2; Bengtson 1977b).

Hadimopanella sp. from the Middle Cambrian Exsulans Limestone Formation of Bornholm (Berg-Madsen 1985) need further description; these resemble *H. knappologica* with small (c.1 μm?) densely packed nodes, lacking an organized outer ring of nodes.

The possibilities of vertebrate affinities were briefly discussed by Bengtson (1977b), though the evidence is inconclusive.

Microdictyon

The net-like phosphatic structure of *Microdictyon* Bengtson, Matthews, & Missarzhevsky (*in* Missarzhevsky & Mambetov 1981; Bengtson 1986a) appears widely at about the same level as *Lapworthella cornu*, *Hadimopanella*, and *Serrodiscus bellimarginatus*. Its stratigraphical range, however, is from Tommotian to mid-Cambrian.

Microdictyon effusum Bengtson, Matthews, & Missarzhevsky occurs in the lower Shabakty Formation of southern Kazakhstan (Missarzhevsky & Mambetov 1981; Bengtson *et al.* 1986a, op. cit.): structure is oblong, domed, 0.5 to 2.5 mm in diameter; the surface is perforated by round holes 10 to 130 μm (decreasing towards the periphery) in diameter, with a regular diagonal arrangement, separated from each other by apatite walls with a smooth outer and porous inner layer; the floor of each pore may have a hemispherical basal plate; the outer wall of the pore usually has six hexagonally arranged nodes with mushroom-shaped tops; these may abrade to form false pores; the edge of the plate has an imperforate rim. *Microdictyon* cf. *effusum* occurs at uncertain levels in the South Bothnian area of Sweden (Bengtson *et al.* 1986, op. cit.), in the Redlands Limestone at the top of the Brigus Formation, south-east Newfoundland (pl. 7.3, part 1, unpublished data), and the *Strenuella* Limestone Ac4 Beds of Comley, Shropshire (cf. Matthews & Missarzhevsky 1975, pl. 4, figs 2, 5, 8; *M. sphaeroides* Hinz (*pars*) in Hinz (1987)). Some of those from the Redlands Limestone have the centre raised into a broad conical spine.

Other related forms of *Microdictyon* are reviewed in Bengtson *et al.* (1986, op. cit.) and a preview of their text and specimens were kindly shown to the writer. These include *Microdictyon*? *tenuiporatum* Bengtson *et al.*: Tommotian locality at Isit, Lena River: this species has a round, low domed shape with small, rounded to tetragonal holes of variable size; four or five nodes around each hole are short pillars with flat tops. *Microdictyon* n. sp. 1 occurs in the lower *Nevadella* Zone, Montenegro Member of Nevada; similar to *M. effusum* but highly convex with very thick and robust walls. *M. rhomboidale* Bengtson *et al.* comes from the Upper Atdabanian of Tamdytau, Kazakhstan: it is weakly convex, with a rhombic outline and thin walls; the holes are round, not closed; and the nodes are mushroom shaped. Comparable forms occur in the lower part of the *Nevadella* Zone, Mackenzie Mountains of Canada. *M. robisoni* Bengtson *et al.* comes from the *Ptychagnostus gibbus* Zone or just below, Swasey Limestone, Utah: it is strongly convex with spike-shaped nodes that have chiselled ends.

Microdictyon spp. also occur in the following: the *Callavia* Zone of the Hoppin Formation of eastern Massachusetts (Bengtson *et al.* 1986, op. cit.); Middle Cambrian lag deposits of Bornholm and Scania, possibly reworked from the underlying Lower Cambrian (op. cit.); the *Elliptocephala asaphoides* fauna of New York (Dr E. Landing, pers. comm. 1986); the lower *Nevadella* Zone, Rosella Formation of the Cassiar Mountains, British Columbia (Bengtson *et al.* 1986, op. cit.); the Puerto Blanco Formation (cf. Poleta Formation) of the Carborca region, Mexico (op. cit.); the Upper Atdabanian and Lower Botomian of the Siberian platform (op. cit.; Sokolov & Zhuravleva 1983); the Wilkawillina and Parara Limestones of S Australia (op. cit.); the Comley Limestone of England (Hinz 1987; her holotype of *M. sphaeroides* from Nuneaton is herein questionably regarded as a contaminant).

The palaeobiology remains problematical (Bengtson
et al. 1986, op. cit.) though one might note an intriguing
comparison between *M. effusum* and pagetinid 'abathochroal' eye structures (cf. Jell 1975), although the
analogy does not clearly extend to all other species.

Tannuolina

The phosphatic sclerites of *Tannuolina* Fonin & Smirnova have a pseudoporous lamellar ultrastructure (pl.
7.3, part 5) and assemblages can be separated into
distinct convex and flat valves. Tannuolinidae appear at
about the level of *Rhombocorniculum cancellatum* and
their occurrence in Australian, Chinese, and Siberian
sections gives the group some potential for international
correlation. *Tannuolina* may be described as: phosphatic
sclerites with lamellar ultrastructure bearing pores on
and/or pseudopores over parts of the shell; a bimembrate apparatus of 'convex' and 'flat' sclerites: the
convex sclerites are asymmetrical, bearing growth lines,
with a smooth undersurface bisected by a carina; flat
sclerites are compressed and triangular. *Tannuolina
multiforata* Fonin & Smirnova was first described from
the Altai–Sayan Fold Belt of Siberia (Fonin & Smirnova
1967) in strata now held to be of Botomian age (Dr A.
Yu Rozanov, pers. comm. 1985). This form is also
reported from the base of the Yuhucun Member of the
Qiongzhusi Formation at Meishucun, China, just below
the first trilobite *Parabadiella* (Luo *et al.* 1982; Jiang
1985). Specimens from similar levels in Sichuan (pl. 7.3,
part 4) include more laterally elongate forms (Yue 1987);
further work is needed on the morphology of Chinese
Tannuolina. Assemblages with *Micrina etheridgei* (Tate)
in South Australia (Bischoff 1976) and Central Australia
(Laurie & Shergold 1985; Laurie 1986) differ in having a
bilaterally symmetrical conical sclerite and convex flat
sclerite; these occur at a similar stratigraphical level,
associated with the tommotiid *Dailyatia*.

Tannuolina and *Micrina* are mitrosagophoran problematica, placed in the Tannuolinidae because of their
distinctive wall structure (e.g. Landing 1984; Bengtson
1986; Laurie 1986).

7.4.3. Selected agglutinated microfossils

Platysolenites

The flattened, siliceous-agglutinated tubes of *Platysolenites antiquissimus* Eichwald (pl. 7.4, part 5) are
abundant in some early skeletal successions. Irregular,
transversely ridged *P. lontova* Opik, spiral *Spirosole-
nites spiralis* Føyn & Glaessner and small *Yanishev-
skyites petropolitanus* (Yanischevsky) are merely morphological variants of *P. antiquissimus* according to
Rozanov (*in* Urbanek & Rozanov 1983). Thus the broad

description is briefly as follows: cylindrical to narrowly
tapering siliceous agglutinated tubes, tending to break
along transverse wrinkles; up to 30 mm long and
0.9 mm thick, usually flattened by compaction, obscur-
ing the internal cavity. The wall structure of Estonian
specimens is curiously two layered: an inner layer of
minute spherulites or panidiomorphic crystals of cristo-
balite, and an outer layer of hypidiomorphic tridymite
(?) or quartz (op cit.). Material from Leningrad,
sectioned by Glaessner (1978), and in the Dolwen Grits
of North Wales (Rushton 1978), shows transverse
layering and the presence of clay or heavy-mineral
grains. Norwegian specimens have detrital grains in
silica cement (Føyn & Glaessner 1979). Silica crystals
and spherulites also occur in a specimen from Nuneaton. Rozanov suggested that the spherulites were
secreted by the organism and later became recrystallized
in the outer layers, but this seems doubtful. Volcanic
crystals, spherulites, and even gypsum occur commonly
in the sediment at Nuneaton and an agglutinated origin
(cf. Føyn & Glaessner 1979) with grain selection is
preferred. Their suggestion of foraminiferid affinities
now seems doutful.

The taxon shows little temporal change in morphology, but its appearance and disappearance are datum
points of stratigraphical interest. The lowest record
seems to be in the upper part of the Włodawa (cf. Rovno)
Formation of Poland, at a similar level to *Aldanella
polonica* Lendzion and *Sabellidites* (Urbanek & Rozanov 1983; Lendzion 1983*a*). A similar association is
found in northern Scandinavia (Føyn & Glaessner
1979) and Rozanov has tentatively identified the taxon
with *Aldanella* in Chapel Island Member 4, south-east
Newfoundland (Dr T. P. Fletcher, pers. comm. 1986;
now confirmed in the field). It occurs widely through the
Lontova Formation and supposed equivalents of the
East European Platform, but is restricted to a few
individual specimens in the Talsy and Vergale Formations (Rozanov *in* Urbanek & Rozanov 1983). This is
consistent with sparse records from the *Schmidtiellus
mickwitzi* Zone of Mjøsa, Norway (e.g. Bergström,
1981), the Home Farm Member, the basal Lower
Comley Sandstone (new data) and the base of the Purley
Shales, Nuneaton (Brasier 1986). Specimens reported
from the Comley Limestone (Hinz 1987) are not this
taxon and are 'dubiofossils'. The occurrence in California (Rozanov op. cit., p. 100) appears to be from beds
with *Fallotaspis* cf. *tazemmourtensis* and above the first
Diplocraterion and *Skolithos* (Wiggett 1977; Signor &
Mount 1986).

Platysolenites therefore appears at a similar horizon
to *Aldanella* and ranges into assemblages contemporaneous with *Rhombocorniculum insolutum*.

Plates

Plate 7.1.   1. *Protohertzina anabarica*, 'anabariform' element; cross-section and posterior views. Chert–Phosphorite Member, Maldeota, Uttar Pradesh, India. Length 0.8 mm.

2. *Protohertzina anabarica*: specimen with strong lateral and posterior keels; cross-section and lateral views. Chert–Phosphorite Member, Maldeota, Uttar Pradesh, India. Length 0.6 mm.

3. *Protohertzina anabarica*, 'unguliform' element; cross-section and lateral views. Chert–Phosphorite Member, Maldeota, Uttar Pradesh, India. Length 0.58 mm.

4. *Protohertzina anabarica*, 'hertziniform' element; cross-section and posterior views. Chert–Phosphorite Member, Maldeota, Uttar Pradesh, India. Length 0.7 mm.

5. *Protohertzina* cf. *siciformis*, or 'siciform' *P. anabarica*; cross-section and lateral views. Chert–Phosphorite Member, Maldeota, Uttar Pradesh, India. Length 0.55 mm.

6. *Maldeotaia bandalica*, lateral and postero-basal views. Chert–Phosphorite Member, Maldeota, Uttar Pradesh, India (type locality). Length 1.5 mm.

7. *Hertzina elongata*, basal and lateral views. Bed 1, Home Farm Member limestone, Nuneaton, England. Length 1.9 mm.

8. *Sunnaginia angulata*, holotype, dorsal and outer lateral views. Bed Ac3, Comley Limestone, Shropshire, England. Length 0.9 mm.

9. *Sunnaginia parva*, holotype, dorsal and inner lateral views. Bed 10iii, Home Farm Member limestone, Nuneaton, England. Length 1.1 mm.

10. *Sunnaginia neoimbricata*, holotype, dorsal view. Bed 1, Home Farm Member limestone, Nuneaton, England. Length 2.26 mm.

11. *Sunnaginia imbricata*, dorsal view of broken specimen. Basal *sunnaginicus* Zone, Ulukhan Sulugur, Aldan, Siberia, USSR. Length 1.75 mm.

12. *Torellella biconvexa*, lateral view. From *lenaicus* Zone, unknown locality, Siberia, USSR. Length 1.02 mm.

13. *Torellella lentiformis*, lateral view. From *regularis* Zone, *tortuosa* Subzone, Tiktirikteekh, Lena River, Siberia, USSR. Length 1.7 mm.

14. *Torellella* cf. *curva*, lateral view. Bed 1, Home Farm Member limestone, Nuneaton, England. Length 1.9 mm.

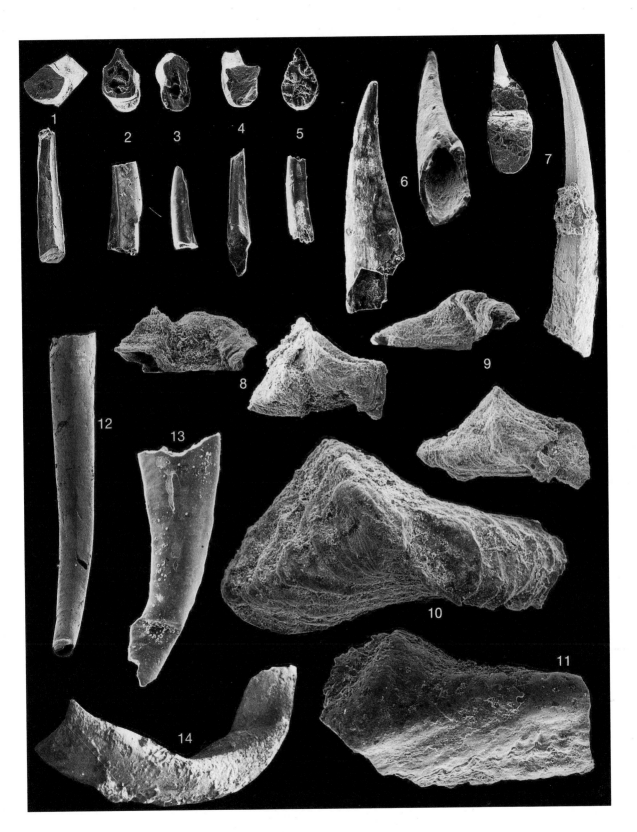

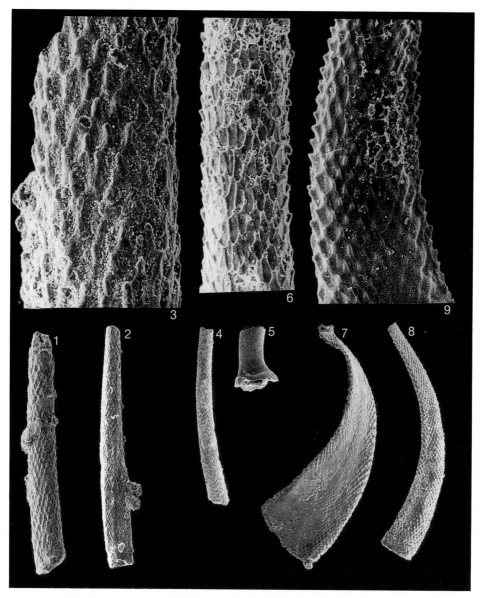

Plate 7.2. 1. *Rhombocorniculum insolutum*, lateral view. Green Shales, Bornholm, Denmark. Length 1.95 mm.

2. *Rhombocorniculum insolutum*, lateral view. Bed 10iii, Home Farm Member limestone, Nuneaton, England. Length 1.81 mm.

3. *Rhombocorniculum insolutum*, detail of part 1. Length of view 0.5 mm.

4. *Rhombocorniculum spinosus*, lateral view of HUP 79/1/274. Bed Ac3, Comley Limestone, Comley Quarry, Shropshire, England. Length 1.42 mm; width of base 0.16 mm.

5. *Rhombocorniculum spinosus*, lateral view of trumpet-based element. Bed Ac3, Comley Limestone, Comley Quarry, Shropshire, England. Length 0.5 mm.

6. *Rhombocorniculum spinosus*, detail of part 4. Length of view 0.47 mm.

7. *Rhombocorniculum cancellatum*, lateral view. Bed Ac3, Comley Limestone, Comley Quarry Shropshire, England (type locality). Length 2.1 mm.

8. *Rhombocorniculum cancellatum*, lateral view. Bed Ac3, Comley Limestone, Comley Quarry, Shropshire, England. Length 2.0 mm.

9. *Rhombocorniculum cancellatum*, detail of specimen in part 8. Length of view 0.49 mm.

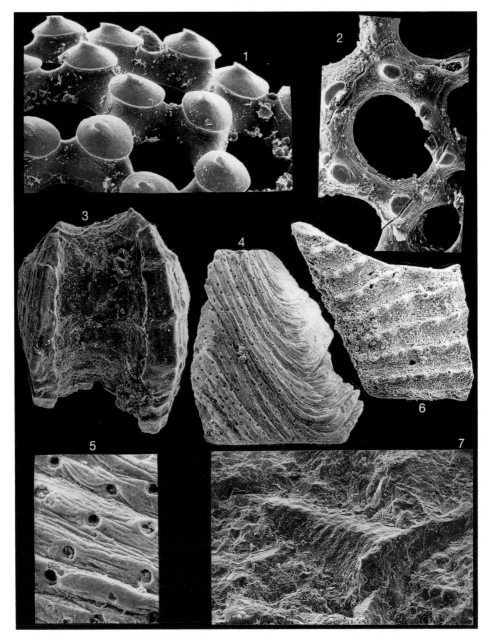

Plate 7.3.   1. *Microdictyon* cf. *effusum*, oblique view of outer side. Redlands Limestone, Brigus Formation, south-east Newfoundland, Canada. Width of view 0.3 mm.

2. *Microdictyon* sp., view from above of outer side. Redlands Limestone, Brigus Formation, south-east Newfoundland, Canada. Length of view 0.16 mm.

3. *Camenella baltica*, dorsal view of sellate L-form sclerite. Bed 1, Home Farm Member limestone, Nuneaton, England. Length 1.2 mm.

4. *Tannuolina* sp., outer surface of fragment from Bed 49, Maidiping, Sichuan, China. Length 1.46 mm.

5. *Tannuolina* sp., detail of part 4. Width of view 0.18 mm.

6. *Lapworthella cornu cornu*, lateral view of fragment. Bed Ad, Comley Limestone, Comley Quarry, Shropshire, England (type locality of genus). Length of fragment 0.8 mm.

7. *Lapworthella cornu nigra*, oblique view of specimen on rock. Bed Ad, Comley Limestone, Comley Quarry. Shropshire, England (type locality). Width of view 1.9 mm.

Plate 7.4. 1. *Bemella jacutica*, lateral view of worn, phosphatic steinkern, from the *regularis* Zone, *tortuosa* Subzone, Tiktirikteekh, Lena River, Siberia, USSR. Length 1.1 mm.

2. *Latouchella korobkovi*, lateral view of worn, phosphatic steinkern. Bed 37, Maidiping, Sichuan, China. Length 1.46 mm.

3. *Obtusoconus honorabilis*, lateral view of phosphatic steinkern. Soltanieh Formation, Elburz Mountains, Iran. Length 1.2 mm.

4. *Pelagiella emeishanensis*, dorsal and side views of phosphatic steinkern. Bed 68, Maidiping, Sichuan, China. Length 1.0 mm.

5. *Platysolenites antiquissimus*, end view of broken tube. Lontova Formation, near Talyn, Estonia. Width of tube 0.95 mm.

6. *Volborthella tenuis*, lateral view. Lukati Formation, near Talyn, Estonia. Length 1.6 mm.

7. *Chancelloria*, ex. gr. *lenaica*, oblique view of specimen with six radial spines (central spine removed). Basal *sunnaginicus* Zone, Ulukhan Sulugur, and Aldan, Siberia, USSR. Width 1.15 mm.

8. *Allonnia erromenosa*, basal view of fragmentary rosette. Bed 49, Maidiping, Sichuan, China. Length 1.0 mm.

9. *Anabarites trisulcatus*, lateral view. From *trisulcatus* Zone, unknown locality, Siberia, USSR. Length 2.17 mm.

10. *Coleoloides typicalis*, lateral view. Bed 1, Home Farm Member limestone, Nuneaton, England. Length 1.1 mm.

Volborthella *and* Salterella

Volborthella tenuis Schmidt (pl. 7.4, part 6) tends to succeed *Platysolenites* as an agglutinated microfossil in clastic strata from the Balto-Scandinavian and North American areas. Morphology and stratigraphical distribution have been reviewed in a series of papers by Yochelson (1977, 1981, 1983); Yochelson *et al.* (1977); Fritz & Yochelson (1988). The latter author considers them as evidence for an extinct phylum Agmata, while Lipps & Sylvester (1968) and Glaessner (1976) suggested that they may have been constructed by annelids.

Volborthella tenuis from Mjøsa are described by Yochelson *et al.* (1977): they are subcylindrical to conical, siliceous agglutinated fossils with thick walls of layered quartzose and mineral silt in a partly calcareous matrix; the layers are inclined from the outside towards a narrow central canal; the wider end of the test forms a shallow v-shaped apertural concavity; the narrow end is rounded; a thin outer calcium carbonate shell is suspected but not preserved. In the Balto-Scandinavian area, *Volborthella tenuis* is first known from the Talsy Formation and equivalent strata that bear a *Schmidtiellus mickwitzi* fauna (Bergström 1981; Urbanek & Rozanov 1983) and ranges up into the *Holmia kjerulfi* group Zone in Scandinavia (Bergström 1981). It is unconfirmed from the *Proampyx linnarssoni* Zone (e.g. Yochelson *et al.* 1977) and the Middle Cambrian (Yochelson 1977), although rare specimens are said to occur into the *Protolenus* Zone of the East European Platform (Rozanov, cited *in* Yochelson 1977). On the North American platform it occurs with *Fallotaspis* cf. *tazemmourtensis* from the Campito Formation (*Campitius* of Firby & Durham 1974 according to Yochelson *et al.* 1977) to Harkless and Saline Valley Formations (Lipps & Sylvester 1968; Signor & Mount 1986) associated with the *Ogygopsis* Faunule of the mid-*Bonnia–Olenellus* Zone (Nelson 1978). These are poorly preserved however, and may be *Salterella* (Yochelson 1977, p. 450; Signor & Mount 1986). In north-west Scotland, a similar form occurs in the *Salterella* Grit (Yochelson 1983) above a low *Bonnia–Olenellus* Zone assemblage (Cowie 1974). At this level the range of *Volborthella* overlaps with *Salterella*.

Salterella Billings is a calcareous agglutinated relative of *Volborthella*, generally found in calcareous substrates (Yochelson 1983), though it also occurs in siliciclastic rocks (Yochelson 1977). The rough wall of the type species *S. rugosa* Billings is due to differential weathering (Yochelson *et al.* 1977), and this is therefore a junior synonym of *S. maccullochii* (Murchison), as are other described species (Yochelson 1983). The latter author came to the conclusion that *S. maccullochii* varied its wall composition according to substrate; siliceous forms are indistinguishable from *V. tenuis*, which was therefore suggested to be the junior synonym. Even so, the calcareous forms are all typical of the middle part of the *Bonnia–Olenellus* Zone (Fritz & Yochelson 1988) and it seems better to retain the distinction at present.

7.4.4. Calcareous microfossils

Anabaritids

These problematical tubes with trimeral symmetry appear widely in the first skeletal assemblages. Their palaeobiology is little understood; Missarzhevsky (1969) tentatively suggested polychaete affinities, while Val'kov (1982) placed them in an extinct scyphozoan subclass. Their walls appear to be of lamellar calcium carbonate, in some forms with external sculpture, suggesting to Abaimova (1978) that the tube was an inner skeleton. The several genera and species appear to have biostratigraphical value in Siberia (Missarzhevsky 1982), and there are some apparent trends that deserve investigation. Firstly, there seems to be a series from shallow to deeper longitudinal grooves, and from three to six grooves in Siberian strata (Fig. 7.8). Secondly, the appearance of non-grooved *Tiksitheca licis* suggests a 'degenerate' trend, if it belongs here at all.

The *Anabarites trisulcatus* Missarzhevsky group (pl. 7.4, part 9) ranges through the early Manykayan (Missarzhevsky 1982) into the *sunnaginicus* Zone in Siberia (Sokolov & Zhuravleva 1983): it is a tube with three rounded lobes and intervening v-shaped depressions forming longitudinal grooves; co-marginal growth-lines curve gently towards the apex on convex sides. Most records of *A. trisulcatus* (including *A. signatus* Miss. and *A. primitivus* Qian & Jiang) are from the earliest skeletal fossil assemblages, with *Protohertzina anabarica*: e.g. Siberia (Missarzhevsky 1969), Mongolia (Voronin *et al.* 1982), southern Kazakhstan (Missarzhevsky & Mambetov 1981), Lesser Himalaya (Azmi & Pancholi 1983; Kumar *et al.* 1987), Yangtze Platform (Xing *et al.* 1983), first assemblage in northern Iran (Hamdi *et al.* in prep.) and north-western Canada (Nowlan *et al.* 1985). Related forms also appear with the *Aldanella* fauna in the lower Hoppin Slate of Massachusetts (but with from three to twelve ribs?), and the form ranges from the Chapel Island Member 4 to the top Smith Point Limestone in south-east Newfoundland (Dr E. Landing, pers. comm. 1986; Landing in press). *Anabarites compositus* Missarzhevsky from the Lower Comley Limestone (Hinz 1987) seems to be a very questionable identification.

Cambrotubulus decurvatus Missarzhevsky is a tube superficially resembling *Anabarites trisulcatus* in rate of

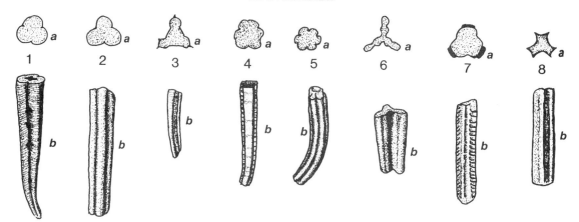

Fig. 7.8. Anabaritid morphotypes from Siberian Precambrian–Cambrian boundary strata, showing the cross-section as typically seen in infilled steinkerns or replicas (a), and lateral or oblique-lateral view (b). 1, *Anabarites trisulcatus* (Missarzhevsky), *trisulcatus* to *sunnaginicus* Zones; 2, *Anabarites tripartitus* Missarzhevsky, *sunnaginicus* Zone; 3, *Anabarites tricarinatus* Missarzhevsky, *sunnaginicus* to *regularis* Zones; 4, *Jakutiochrea tristicha* (Missarzhevsky), *sunnaginicus* to *regularis* Zones; 5, *Anabaritellus hexasulcatus* Missarzhevsky, *sunnaginicus* Zone; 6, *Anabarites ternarius* Missarzhevsky, *sunnaginicus* to *tortuosa* Zones; 7, *Aculeochrea composita* (Missarzhevsky), *regularis* Zone; 8. *Anabaritellus isiticus* Missarzhevsky, *regularis* to *lenaicus* Zones.

taper and irregular curvature, and commonly occurs with it in the earliest skeletal assemblages. Longitudinal grooves are lacking, however, and growth-lines are straight. This form ranges from low Manykayan to the *regularis* Zone in Siberia (Missarzhevsky 1982), and appears in the first skeletal assemblage in Iran (data of Dr Hamdi) and Shaanxi (Xing *et al.* 1983).

Tikistheca licis Missarzhevsky has an intermediate morphology, with a rounded triangular cross-section and straight growth-lines, and ranges from Upper Manykayan to the *Aldanocyathus sunnaginicus–Tiksitheca licis* Zone. This index appears in the second assemblage in northern Iran (Dr B. Hamdi, unpublished) and Kazakhstan (Missarzhevsky & Mambetov 1981), in Zone II assemblages of Shaanxi and Hubei (Xing *et al.* 1983), in the first assemblage in Mongolia (Voronin *et al.* 1982) and lower *Aldanella* to *Camenella* assemblages in Newfoundland (Landing in press). The type species *T. korobovi* (Miss.) is more typical of *regularis* Zone strata in Siberia (Missarzhevsky 1982) and is supposedly differentiated on the basis of more-regular curvature (Sokovov & Zhuravleva 1983). Incorporation of *A. trisulcatus* into *T. korobovi* (Landing in press) is not followed here, pending further studies.

From the *sunnaginicus* Zone onwards, forms appear which have modified lobes, such as *Anabaritellus hexasulcatus* Missarzhevsky with additional grooves down each of the main lobes. Hexasulcate forms, but with strong concentric ribs on the exterior, occur in the

Parara Formation in South Australia, (Drs S. Bengtson, P. Jell, S. Conway Morris pers. comm. 1986) probably to be correlated with China Zones IV–V.

Pelagiellids

Aldanella Missarzhevsky (Fig. 7.9) is a small, dextrally coiled, helical shell resembling a gastropod (Missarzhevsky 1969; Pojeta & Runnegar 1976) of which the first species typically appear in the first Tommotian assemblages. The type species is *Aldanella attleborensis* (Shaler & Foerste), which occurs with *Fomitchella* in the earliest skeletal assemblages in south-east Newfoundland (Bengtson & Fletcher 1983; Narbonne *et al.* 1987), in the lower Hoppin Slate of Massachusetts (Grabau 1900; Landing in press) and in the *sunnaginicus* Zone of Siberia (Missarzhevsky 1982): it has a small, low trochospiral, dextrally coiled shell with two to three whorls and an elongate-oval aperture; the surface (especially the upper) has recurved transverse folds, and growth-lines with an antispiral sinus. In northern Siberia it first occurs with *Tiksitheca licis* and other forms that suggest comparison with the Sunnagin Horizon of the Aldan and Lena Rivers (Missarzhevsky 1969, 1982), where the closely related *A. rozanovi* Miss. first appears (Sokolov & Zhuravleva 1983). Variation from 'snail-like' to bizarre heteromorph shapes is seen in '*Aldanella*' from Nahant, Massachusetts and suggested to Landing & Brett (1982) that the pelagiellids are neither snails nor molluscs, or suffered post-mortem

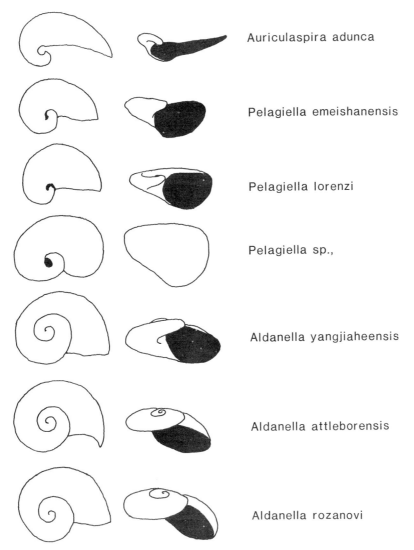

Auriculaspira adunca

Pelagiella emeishanensis

Pelagiella lorenzi

Pelagiella sp.,

Aldanella yangjiaheensis

Aldanella attleborensis

Aldanella rozanovi

Fig. 7.9. Sketches of dorsal and apertural views of roughly successive forms of pelagiellid (from published illustrations). Surface sculpture is omitted and the cross-sectional view of the aperture is shown in solid black. Source of the morphotypes is as follows: *Aldanella rozanovi, sunnaginicus* Zone, Siberia; *A. attleborensis*, from *A. attleborensis* assemblage in south-east Newfoundland; *A. yangjiaheensis* from Zone II in Hubei, China; *Pelagiella* sp. from the basal Pedroche Formation, Cordoba, Spain (*c.* Zone III; after Liñan 1978); *Pelagiella lorenzi* from upper Atdabanian or Lower Botomian of Siberia; *Pelagiella emeishanensis* from Zone V, Maidiping, Sichuan; *Auriculaspira adunca* from above Zone V, Anhui.

deformation (Landing in press). This variation also throws some doubt on the status of other Siberian species, and it may be preferable to refer them all to the *Aldanella attleborensis* group (see Landing in press). This includes *Aldanella yanjiaheensis* Chen Ping, which first appears in Zone II of Hubei, while *Aldanella* sp. also occurs in Zone II of Shaanxi (Xing *et al.* 1983). The genus ranges into higher levels with *Rhombocorniculum* sp. in Newfoundland (Bengtson & Fletcher 1983; Landing in press) and into Upper Tommotian strata in Siberia (Sokolov & Zhuravleva 1983).

Yunnanospira regularis Jiang is a related form from

Zone II in the Meishucun section of Yunnan (e.g. Luo *et al.* 1984). This has a rather flat dorsal surface, wider expansion rate, and closely spaced growth-marks.

Pelagiella differs from *Aldanella* in its much wider expansion rate and fewer whorls. It seems to occur widely with a pandemic assemblage close to the appearance of *Serrodiscus bellimarginatus* or other eodiscids (i.e. a pelagic or epiplanktonic life habit?). *Pelagiella* spp. are reported from the Upper Atdabanian to Lower Botomian (Missarzhevsky 1982) or Lower Botomian (Sokolov & Zhuravleva 1983) of Siberia; the Strenuella Limestone Ac4 Beds of Shropshire (Matthews & Missarzhevsky 1975) to the Lapworthella Limestone Ad Beds (Hinz 1987); the *Serrodiscus bellimarginatus* beds of the Purley Shale, Nuneaton (Rushton 1966); the *S. bellimarginatus* assemblage from Nova Scotia (Landing *et al.* 1980); ranging from close to the appearance of *S. bellimarginatus* in the Buen Formation of Greenland (Peel, pers. comm. 1986); ranging from just below the level of *Serrodiscus bellimarginatus* in Morocco (Geyer, pers. comm. 1986); in the upper assemblage of northern Iran (Hamdi, unpublished data); in the Calcareous Member of the Lesser Himalaya with brachiopod *Diangongia pista* Rong and chancelloriid *Dimidia* sp. (Kumar *et al.* 1987); and with the *Eoredlichia–Wutingaspis* Zone assemblage in south-west Sichuan (pl. 7.4, part 4), also with *Diandongia pista* (Xing *et al.* 1983). The latter indicate correlation with Zone V at Meishucun.

A culmination of evolution in this dextrally coiled lineage appears to have been *Auriculaspira* Zhou & Xiao from the Qiongzhusian to Canglangpuian of the Yangtze Platform (He *et al.* 1984; Zhou & Xiao 1984). This pustulose genus has a flat to concave dorsal surface with one to two rapidly expanding whorls, the early whorl semi-evolute and curved like a hook.

Barskovia Golubev resembles *Aldanella* but is sinistrally coiled. It occurs with the second and third assemblages at Salanygol, Mongolia (Voronin *et al.* 1982) and in the *sunnaginicus* Zone in Siberia (Sokolov & Zhuravleva 1983). More rapidly expanding *Pelagiella*-like sinistral forms of *Nomgoliella* range from the second to fifth assemblages in the same area (ibid.).

Biometric analysis will be necessary before these pelagiellids can be used with confidence for biostratigraphy.

Latouchellids

Pojeta & Runnegar (1976) suggested bellerophontacean monoplacophoran affinities for the small planispiral shells of *Latouchella*, of which the *Latouchella korobkovi* (Vostokova) group has considerable biostratigraphical potential (pl. 7.1, part 2). They appear widely in the *sunnaginicus* Zone of the Anabar region, and the lower

regularis Zone of the Aldan–Lena region (e.g. Missarzhevsky 1969; Sokolov & Zhuravleva 1983): they have a small planispiral shell of one to two evolute whorls, are openly coiled; the folds are prominent, transverse, and evenly spaced folds which may be raised into crests; their apertures are flat with an oval cross-section. Wide variability in the strength of sculpture, spacing of sculpture, number of whorls, and tightness of the whorls indicates that a broad taxonomic concept should apply. The group merits further special studies, but the following are suggested to belong to the group: *Latouchella* sp. from the *sunnaginicus* Zone of the Maya River (Rozanov *et al.* 1969); *Latouchella memorabilis* Miss. from the *sunnaginicus* Zone of the Eriechka River, Anabar (ibid.); *Latouchella korobkovi*, *L. minuta* Zhegallo, *L. gobiica* Zhegallo, and *L. sibirica* (Vost.) from the second zone in Mongolia (Voronin *et al.* 1982); *Shabaktiella shabaktiensis* Miss. from the phosphorite 'horizon' of Maly Karatau, Kazakhstan (Missarzhevsky & Mambetov 1981); *Latouchella* sp. from the Upper Shale Member of northern Iran (Hamdi, unpublished data); *Latouchella korobkovi* from Zone II of Yunnan (Luo *et al.* 1982) and Sichuan (Yin *et al.* 1980); *Yangtzespira regularis* Jiang and *Y. exima* Yu from Zone II of Yunnan (ibid.); *Igorella planumbonia* Jiang (ibid.; Luo *et al.* 1984); *Latouchella* cf. *costata* Cobbold from Zone II in Shaanxi (e.g. Xing *et al.* 1983); *Latouchella sonlinpoensis* Chen & Zhang from Zone II of Hubei (ibid.); and *Maidipingoconus maidipingensis* (Yu) from Zone II in Hubei, Yunnan, Sichuan, Hubei, and Shaanxi (ibid.). Further work is needed to confirm whether the less-enrolled *Bemella* spp. (pl. 7.4, Fig. 7.1) and *Igorella hamata* Yu belong to the same plexus; they have a similar stratigraphical distribution. Almost entirely uncoiled forms of *Obstusoconus honorabilis* (Qian) may be younger variants (pl. 7.4, Fig. 7.3).

The type species of *Latouchella costata* Cobbold, from the Ac3 and Ac5 beds of the Comley Limestone, is closely related. In this form, transverse folds are developed on the sides but not over the convex margin of the shell (Cobbold 1921).

From this it seems that the group was long ranging and numerous. Its early appearance may have been associated with a rapid migration, making a useful datum for correlation. Biometric studies may enhance our understanding of their evolution and biostratigraphical potential.

Chancelloriids

Another group with potential for correlation are sclerites of chancelloriids, perhaps related to the *Wiwaxia* from the Burgess Shale (Bengtson & Missarzhevsky 1981; Bengtson & Conway Morris 1985). Their calcareous, articulated, rosette-like sclerites occur widely in warm- and cool-water facies of most regions and there are some forms with distinctive symmetry (Fig. 7.10), best seen in the redlichiid faunal province from France to China.

Chancelloria ex. gr. *lenaica* Zhuravleva & Korde have rosettes with up to nine radial spines plus a central spine, and first appear in the Sunnagin 'horizon' of Siberia (pl. 7.4, part 7), ranging up to the top of the Tommotian (Sokolov & Zhuravleva 1983). A more important section for chancelloriids is at Meishucun, Yunnan. Here, *Chancelloria altaica* Romanenko illustrated with about eight radial spines and a central spine, appears in Bed 8 in the upper part of Zone II (Luo *et al.* 1982; Xing *et al.* 1984). In the overlying Badaowan Member of the Qiongzhusi Formation, Bed 11 contains *Chancelloria* sp. plus several more-developed forms (Qian *et al.* 1985). *Archiasterella* cf. *pentactina* Sdzuy (1969) has a bilaterally symmetrical rosette, with the central spine apparently displaced into the radial rosette. *Adversella mountanoides* Jiang has reduced radial spines. *Onychia*

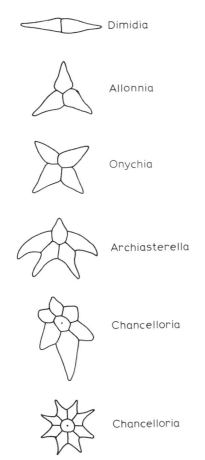

Fig. 7.10. Some distinctive morphotypes of chancelloriid sclerite rosettes.

tetrithallis Jiang has four upraised spines. *Allonnia erromenosa* Jiang (= *tripodophora* Doré & Reid 1965) has three radial spines and no central spine (pl. 7.4, part 8). This assemblage also contains *Tannuolina* sp. and *Sinosachites* sp. (Qian *et al.* 1985). A similar assemblage occurs at this level in Maidiping, Sichuan, while an assemblage from the Arenaceous Member of the Lesser Himalaya, at a similar stratigraphical horizon, has *Allonnia erromenosa* and *Dimidia simpleca* Jiang (Kumar *et al.* 1987).

At the base of the overlying Yu'anshan Member (Bed 13), the Badaowan assemblage is joined by *Dimidia simpleca* with two upraised spines. This assemblage is associated with *Lapworthella cornu* and *Tannuolina multiforata* (cf. late Atdabanian–Botomian in age).

It is clear that considerable variation occurs, with perhaps four, three, and two-rayed sclerites occurring within the scleritome. Three-rayed *Allonnia*, however, appears in Chinese Zones III to V and widely correlated strata (see Brasier, this volume, Chapter 3), and may provide a useful datum.

7.4.5. Range charts of small shelly fossils

The composite ranges of a variety of 'small shelly fossils' from successions around the world are given in Figs 7.11–7.15. Also included are wiwaxiids, bradoriids, brachiopods, hyoliths, sponges, and other taxa which have not been reviewed here. In these figures the Siberian successions are used for comparison and for tentative correlation (see Section 7.4.9).

7.4.6. Trilobites

Major adaptive radiations of arthropods occurred during the Cambrian, including mineralized polymerid trilobites, miomerid trilobites and bradoriid crustaceans, plus non-mineralized forms such as those found in the Burgess Shale (e.g. Whittington 1979). Abundant and diverse trilobite remains provide the basis of Cambrian stratigraphy around the world, greatly assisted by pandemic miomerid, eodiscid, and agnostid 'trilobites'; the latter appear to have been enrolled, blind, pelagic arthropods (e.g. Robison 1972). Agnostids are not found below upper Lower Cambrian rocks, but related eodiscids appeared earlier and probably had a similar life habit. Polymerid trilobites with more restricted distributions were probably benthic or nektobenthic. Cowie (1971*a*) has drawn attention to Lower Cambrian faunal provinces, comprising an Olenellid Realm (initially dominated by Olenellina lacking facial sutures: northern Europe and the Americas) and a Redlichiid Realm (initially dominated by Redlichiina with facial sutures (Palaeotethyan Europe, Asia, Aus-

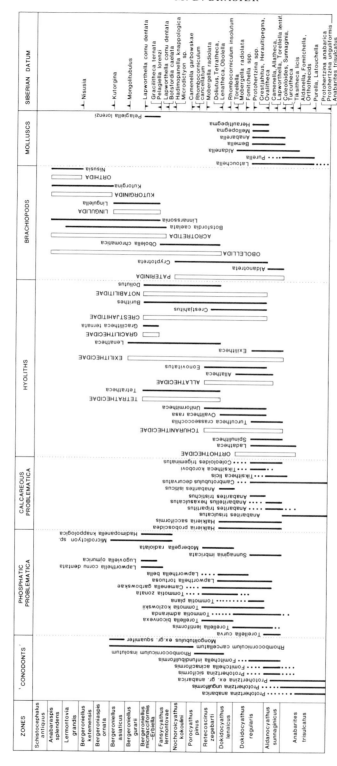

Fig. 7.11. Composite ranges of some widely distributed 'conodonts' (plus some conoidal phosphatic microproblematica), phosphatic and calcareous problematica, hyoliths, brachiopods, and molluscs from the Siberian Platform. These are based on data in Missarzhevsky (1982), Sokolov & Zhuravleva (1983), and Rozanov & Sokolov (1984). Much of the evidence for the first zone hinges on the Anabar Massif and Missarzhevsky's interpretation of the various sequences there, but the pattern is broadly like that since published for this region by Khomentovsky (1986). A selection of first- and last-appearance datum points are shown, that permit correlation with strata beyond the Soviet Union. The candidate Precambrian–Cambrian boundary point at the base of Bed 8, Ulukhan Sulugur, lies at the base of the *sunnaginicus* Zone. As can be seen, a large number of taxa first appear at about this level, close to the *Tiksitheca licis* to *Camenella* datum points.

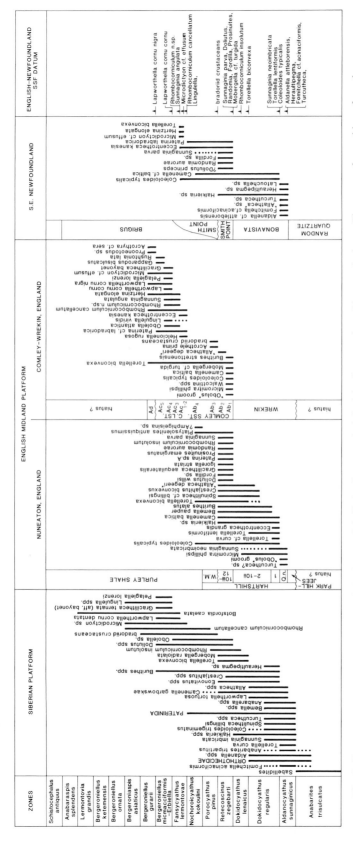

Fig. 7.12. Composite ranges of a selection of Siberian small shelly fossil taxa on the left of the diagram, alongside a zonal scheme. The ranges of assemblages from the Avalon terrane (Nuneaton, Comley–Wrekin, and the Bonavista and Avalon Peninsulas of south-east Newfoundland) are then calibrated against the Siberian chart, and against each other. *Rhombocorniculum* n. sp. = *R. spinosus* Hinz. First appearances with potential for correlation (on the right) are assumed to have been slightly diachronous across the traverse. This correlation requires to be tested against further information from Newfoundland (since listed by Narbonne *et al.* 1987) and other Avalonian sections (Landing in press). A variety of events can be correlated with greater certainty from England to Newfoundland (see Brasier, this volume, Chapter 5).

According to this correlation, the basal *sunnaginicus* Zone time-plane (and Marker A) may lie close to the base of the Hartshill formation and correlated strata in Newfoundland, requiring correlation with trace fossils (see Crimes, Chapter 8 of this volume). Marker B of Meishucun may correlate with the *Aldanella attleborensis* or *Torellella lentiformis* datum points. A more distinct horizon for correlation is that with *Rhombocorniculum insolutum* and *Mobergella* cf. *turgida*, here taken at the Tommotian–Atdabanian boundary. Biohorizons are reasonably comparable from this level upwards, aided by trilobites. Compiled from data in the text.

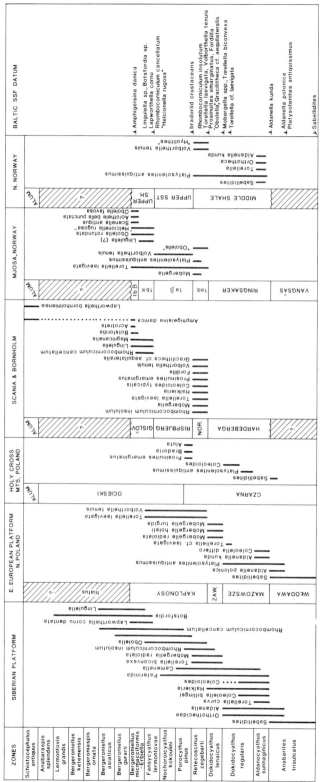

Fig. 7.13. Composite ranges of a selection of Siberian small shelly fossil taxa on the left of the diagram, compared with the ranges of small shelly fossils from north Poland, Holy Cross Mountains (southern Poland), Scania and Bornholm, Mjøsa in southern Norway, and northern Norway. First-appearance datum points on the right-hand side are assumed to be roughly isochronous. They show, at least, that a measure of correlation is possible, not only with Siberia but also with Avalonia (cf. Fig. 7.12).

A major problem here is the scarcity of skeletal fossil remains (especially in the lower parts of the succession), suspected large hiatuses, largely provincial trilobites, and uncertainty about acritarch ranges. Correlation with the *sunnaginicus* Zone or Marker A of China will depend upon the potential of trace-fossil and event stratigraphy. Correlation with Siberia and England becomes a little firmer from the level with *Mobergella* spp. and *Rhombocorniculum insolutum* upwards. Compiled from data in the text.

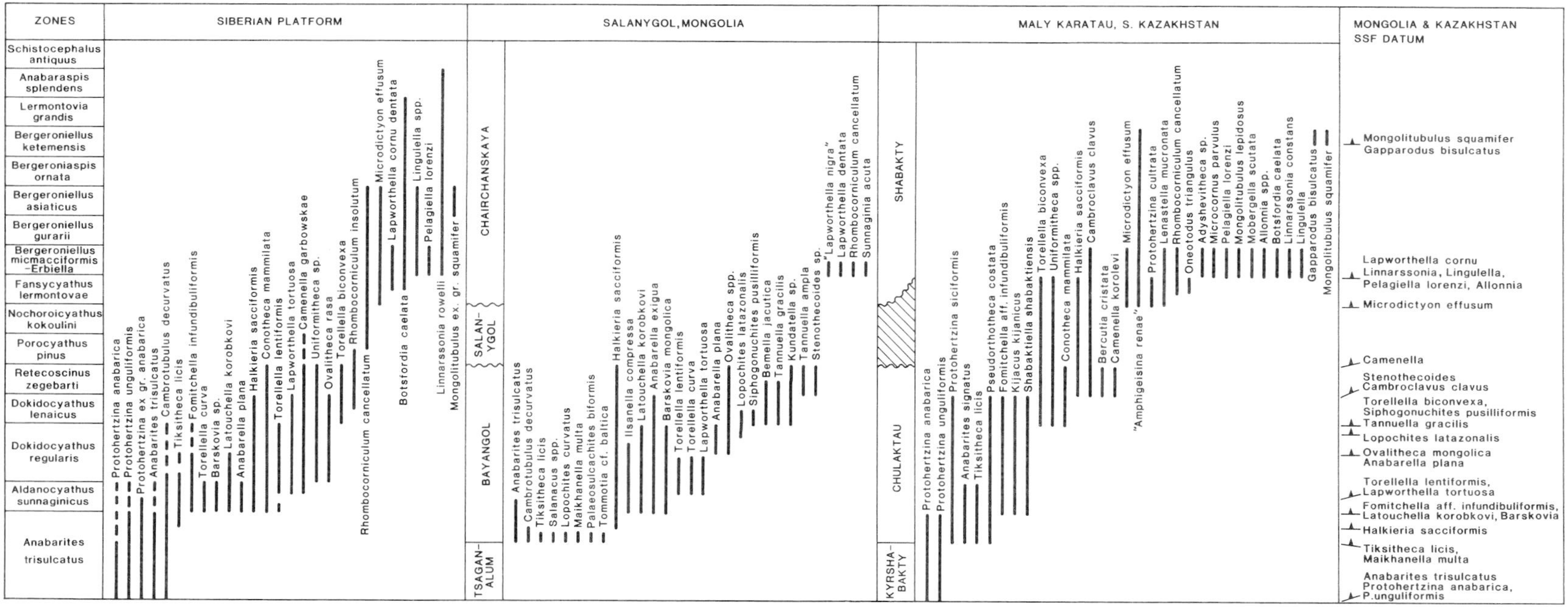

Fig. 7.14. Composite ranges of a selection of small shelly fossils from the Siberian Platform compared with those from Salanygol, Mongolia (on the southern margins of the platform) and Maly Karatau in southern Kazakhstan (from the Palaeotethyan Ocean). First-appearance datum points on the right-hand side are assumed to be roughly isochronous. Correlation of the basal *sunnaginicus* Zone is clearly feasible, forming an important link between Siberian and Chinese successions. Compiled from data in the text.

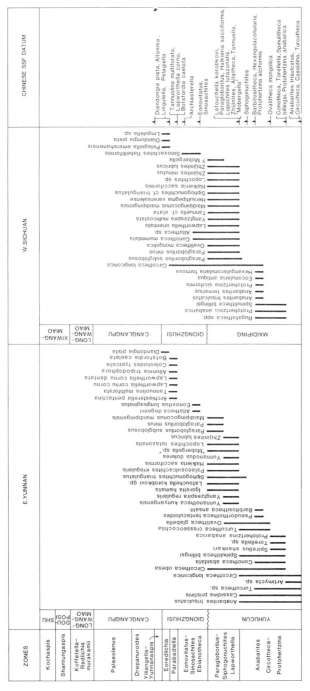

Fig. 7.15. Composite ranges of a selection of small shelly fossils from Meishucun, East Yunnan, correlated against those from Maidiping, western Sichuan in China. At right are shown first-appearance data points with potential for widespread correlation.

Marker A lies at the base of the first zone; the presence of *Cassidina* (aff. *Maikhanella*) and hyoliths attributed to '*Circotheca*' and *Turcutheca* arguably suggest a position close to the base of the first assemblage in Mongolia, which can be traced in Iran (see Brasier, this volume, Chapter 3). The phosphogenic event (with negative $\delta^{13}C$ anomalies) coincides roughly with the level with *Barbitositheca*, traceable into India and probably well beyond. Marker B is the extant candidate level and lies at the base of the second zone. As can be seen, a rich variety of skeletal fossils appear here (mostly molluscs, including the *Latouchella korobkovi* group) through their appearance is erratic and ?diachronous on a global scale. These datum points can be traced along the Palaeotethyan Belt from the Yangtze Platform, India, Pakistan, Iran, Kazakhstan, and Mongolia (see Brasier & Singh 1987; Brasier, this volume, pp. 66–7) and may extend into Oman and Turkey. Compiled from data in the text.

tralia, and Antarctica). Provincial differences within these realms are due to geographical and facies controls (e.g. Lochman-Balk & Wilson 1958; Palmer 1969). Thus the faunas dominated by endemic forms (e.g. *Olenellus*) are typical of inner cratonic regions, perhaps isolated by outer- shelf carbonate barriers; eodiscids are scarce in such facies. Outer shelf, slope, and allochthonous terrane polymerids include *Elliptocephala*, *Callavia*, and *Kootenia*, plus more pandemic eodiscinids.

Strong provincialism has deprived the Lower Cambrian of an international trilobite zonation, though local schemes (e.g. Figs 7.16–19) are used. This problem is serious because the oldest known trilobitic sequences are preserved in inshore facies where endemic polymerids offer little chance of precise inter-continental correlation. Where possible, such correlation is usually made at generic level with polymerids. More pandemic eodiscid species appear in mid and late Lower Cambrian rocks, providing valuable chronostratigraphical markers (Robison *et al.* 1977), though higher than levels likely to be chosen for the Precambrian–Cambrian boundary (e.g. Cowie 1978, 1984).

In this short space an attempt will be made to outline the sequence of faunas, calibrated where possible against small shelly fossil indices discussed above. Eastern Siberia has well-documented trilobites of mixed-realm affinities and provides an important reference zonation. Morocco, south-east Newfoundland, and England have relatively pandemic faunas that allow correlation to be tested, particularly within the Olenellid Realm. The Pacific faunal province of North America provides a test of correlation with a cratonic area. Further work is still needed to develop correlation with the Redlichiid Realm (e.g. Fig. 7.19) and little will be said about that here.

Profallotaspis *age trilobites*

Elongate olenellids of the Fallotaspinae are the first trilobites to appear in eastern Siberia, the Anti-Atlas of Morocco, in North America, and possibly in England, though their range was probably not synchronous. The earliest forms are probably those in Siberia and Morocco, with *Profallotaspis* and *Eofallotaspis*.

Siberia *Profallotaspis* sp. and *P. jacutensis* Repina appear in the *P. jacutensis* Zone, shortly above the first Atdabanian archaeocyathan assemblages and below the appearance of *Fallotaspis* (Repina 1981; Sokolov & Zhuravleva 1983). Supposed ?*Bigotinops* sp. from the *sunnaginicus* Zone of the Tommotian is dismissed by Rozanov as the problematical sclerite *Tumulduria* (Sokolov & Zhuravleva 1983, p. 121). This early assemblage appears somewhat above the start of

Rhombocorniculum insolutum and *Mobergella radiolata* in Siberia (Rozanov & Sokolov 1984).

Morocco In the Anti-Atlas, *Fallotaspis* ranges from trilobite Zones I through V of Hupé (1952, 1960). Zone I contains *Fallotaspis tazemourtensis* Hupé, *Pararedlichia*, *Bigotinops*, *and Tazzemourtia* (Hupé 1952; Boudda *et al.* in Cowie & Glaessner 1975). Although archaeocyathans found below Zone I were compared with middle Atdabanian assemblages of Siberia by Rozanov & Debrenne (1974), an earlier age was preferred by Sdzuy (1978). He has found *Eofallotaspis tioutensis* Sdzuy, *E. prima* Sdzuy, and opistaparian *Hupetina antiqua* Sdzuy below *Fallotaspis* of Zone I, ranging below the first archaeocyathan of the region. Debrenne & Debrenne (1978) suggest a lowest Atdabanian age for the base of this trilobite assemblage.

Fallotaspis *to* Pagetiellus anabarus *age trilobites*

The more widespread *Fallotaspis* faunas are later than *Profallotaspis*, but trilobite assemblages with *Fallotaspis* may not all be of the same age. The latter disappeared first in Siberia but ranged into *Judomia* Zone equivalents or higher in North America, Morocco, and Scandinavia. It is still tempting to recognize a *Fallotaspis* fauna at the lower end of its range, characterized by low-diversity assemblages of *Fallotaspis* and related daguinaspids. Its validity, however, depends on corroborative evidence from archaeocyathans and small shelly fossils.

In several areas, this first *Fallotaspis* assemblage is diachronously(?) joined or replaced by one with subovate olenellids (e.g. *Schmidtiellus*, *Holmia*, *Holmiella*, *Archaeaspis*, and *Nevadella*), plus Pagetidae with facial sutures (e.g. *Pagetiellus*, *Pagetides*, *Triangulaspis*, and *Dipharus*). The correlation of this faunal phase is still problematical.

Siberia *Fallotaspis explicata* Repina and *F. sibirica* Repina appear above *Profallotaspis* within the first Atdabanian archaeocyathan assemblage and within the range of *Rhombocorniculum insolutum*; it is roughly coincident with the top of the range of *Camenella garbowskae* and *Mobergella radiolata* (Khomentovsky & Repina 1965; Repina 1981; Sokolov & Zhuravleva 1983). *Bigotina* and *Bigotinops* may also be found with *Fallotaspis* (Khomentovsky & Repina 1965; Rozanov 1973), though this is not confirmed by Sokolov & Zhuravleva (1983). Fallotaspids are succeeded at the base of the *Pagetiellus anabarus* Zone by olenellids such as *Archaeaspis hupei* Repina, with *Nevadella subgroenlandica* (Rep.), opisthoparians *Pseudoresserops delicatus* Rep., *P. oculatus* Rep., and *Bigotina rara* Rep. appearing

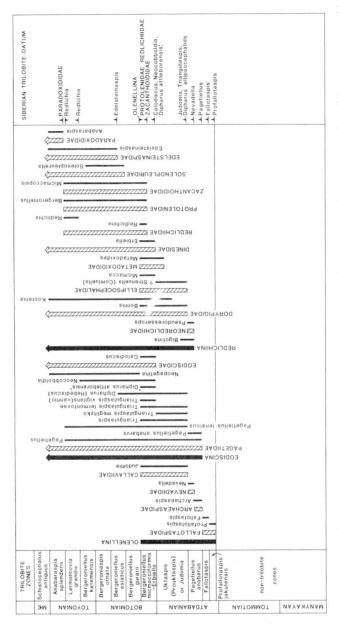

Fig. 7.16. Composite ranges of trilobite species, genera, and higher taxa in strata from the Siberian Platform. Datum points of first appearance, with potential for international correlation, are shown at the right. Compiled from sources in the text.

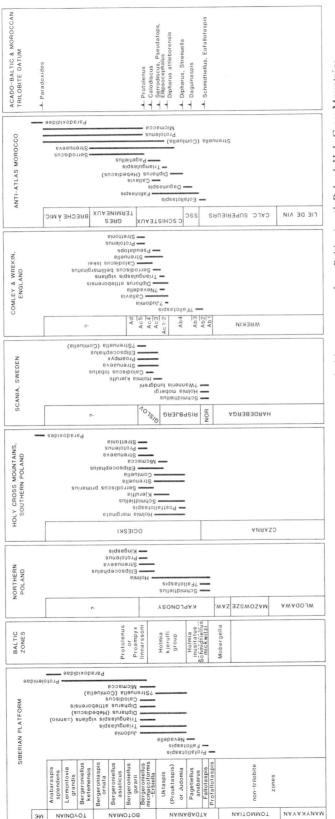

Fig. 7.17. Composite ranges of selected trilobite taxa from the Siberian Platform, compared with some ranges from Baltica (north Poland, Holy Cross Mountains (southern Poland), and Scania), Avalon (Comley and Wrekin), and the Moroccan margin of Gondwana. Datum points of first appearance which are useful for correlation are shown on the right. Note the widespread appearance of the *Dipharus attleborensis*, *Serrodiscus*, and *Calodiscus* pelagic fauna in the middle part of the Lower Cambrian, roughly spanning the interval of demise of olenellids *Judomia*, *Holmia*, and *Callavia*, and the appearance of protolenids. Compiled from sources in the text.

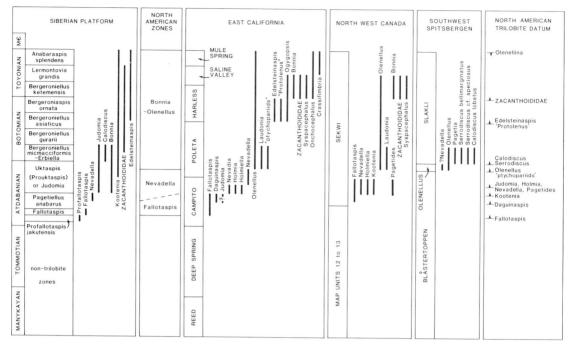

Fig. 7.18. Composite ranges of selected trilobite taxa from the Siberian Platform, compared with some ranges from the North American craton (eastern California, north-west Canada, and south-west Spitsbergen). Datum points of first appearance shown at the right are examples of forms useful for international correlation. Note the homotaxial appearances of *Fallotaspis*, *Kootenia*, *Judomia*, *Calodiscus*, and *Edelsteinaspis*, but diachronous appearances of *Bonnia*, Zacanthoididae, and the prolonged range of Olenellina in North America. Compiled from sources in the text.

in the upper part of the zone. *Pagetiellus anabarus* Lazarenko notably marks the appearance of eodiscids and ranges through the *P. anabarus* Zone, while *Triangullina parvula* Rep. appears in the middle of the zone (Khomentovsky & Repina 1965; Repina 1981; Sokolov & Zhuravleva 1983). The elements of *Rhombocorniculum insolutum* persist into the lower parts of this zone, but *R. cancellatum* and *Lapworthella lucida* appear contemporaneously with *Nevadella* in the upper part (e.g. Meshkova 1969; Rozanov & Sokolov 1984).

Morocco and Spain In the Anti-Atlas, *Fallotaspis tazemmourtensis* Hupé appears above assemblages with *Eofallotaspis* and *Hupetina*, and ranges from trilobite Zones I through V (Hupé 1952, 1960). Zone I contains *Fallotaspis*, *Pararedlichia*, *Bigotinops*, and *Tazzemourtia* (Hupé 1952; Boudda *et al.* in Cowie & Glaessner 1975) and *Lemdadella* (Liñan & Sdzuy 1978). Although archaeocyathans found below Zone I were compared with 'mid-Atdabanian' assemblages of Siberia by Rozanov & Debrenne (1974) this age was disputed by Sdzuy (1978). The first redlichiid in the High Atlas of Morocco

is *Lemdadella* (Sdzuy op. cit.) which also appears with *Bigotina* in the Sierra Morena, Spain (Liñan & Sdzuy 1978); the latter occurrence is above archaeocyathan assemblages thought by Zamarreño & Debrenne (1977) to be 'mid-Atdabanian' and above small shelly problematica (Liñan 1978) that include *Allonnia* and possible *Chancelloria*, *Halkieria*, and *Pelagiella* (my tentative interpretation of figures in Liñan 1978), pls 1, 2. This assemblage suggests a China Zone III to V correlation.

Assemblages of *P. anabarus* or early *Holmia* Zone age may correlate with Hupé Zone II in Morocco, bearing *Fallotaspis* and *Choubertella* (Sokolov & Zhuravleva 1983, fig. 42), and Zone III which contains *Fallotaspis*, *Daguinaspis*, *Resserops*, and *Abadiella*. In Spain and Sardinia, the higher assemblages with the *Dolerolenus* fauna are probably of pre-*Judomia* age (Sdzuy 1971).

Western North America *Fallotaspis* dominates the lowest olenellid or *Fallotaspis* Zone (Fritz 1972; Nelson 1978) but the top of this zone may be diachronous and the upper part may include equivalents of the *Nevadella* Zone (Wiggett 1977; Rozanov & Sokolov 1984).

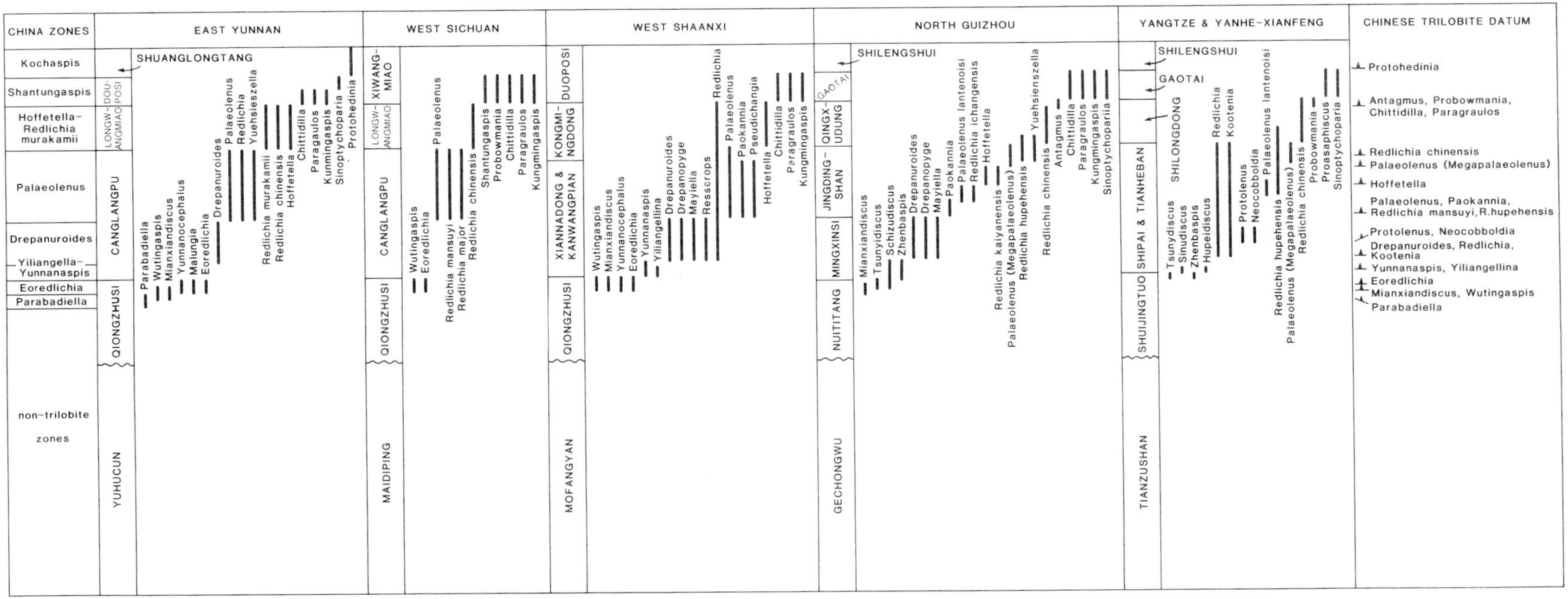

Fig. 7.19 Composite ranges of selected trilobite taxa from the Yangtze Platform sections in eastern Yunnan, western Sichuan, northern Guizhou, Yangtze Gorges, and Yanhe–Xianfeng. Datum points on the right show the first appearances of taxa useful for correlation across the Palaeotethyan region, and a few with international significance.

Earliest occurrences of the genus may be in the Andrews Mountain Member, Campito Formation of the White-Inyo Mountains, eastern California, where occur forms similar to *F. tazemmourtensis* from Morocco (Nelson 1978). This member also yields *Volborthella tenuis* (e.g. Yochelson 1977), which first occurs with *Schmidtiellus mickwitzi* faunas in Scandinavia (Bergström 1981). The upper part of the *Fallotaspis* Zone in California with *Fallotaspis* cf. *longa*, *Daguinaspis*, and *Judomia*(?) may be of *Pagetiellus anabarus* age, while the immediately succeeding assemblage with *Nevadia*, *Holmia*, and *Holmiella* (Nelson 1978) may be of this age or compare with the early *Judomia* Zone. In the Canadian Cordillera, *Fallotaspis*, *Fallotaspis*?, and *Parafallotaspis* occur near the base of the Sekwi Formation and equivalents, in beds below the *Nevadella* Zone bearing *Nevadella* and *Pagetides* (Fritz 1972). The lower assemblage may be of *P. anabarus* age (e.g. Rozanov & Sokolov 1984).

Scandinavia   The olenellids *Schmidtiellus mickwitzii* (Schmidt), *Holmia mobergi* Bergström, *Wanneria? lundgreni* (Moberg) and *Holmiella* sp. are the oldest trilobites found in the *S. mickwitzi* Zone of Scandinavia (Bergström 1973, 1981). This correlates with the base of the *Holmia* Zone on the East European Platform. The assemblage is broadly correlated with beds bearing *Mobergella radiolata*, *M. holsti*, *M. turgida* (Bergström 1981), and *R. insolutum* (Brasier, new data) in the West Baltic area that also contain Talsy-type acritarch assemblages (Moczydłowska & Vidal 1986) but the argument is circular and careful study is required. A major problem is the presence of hiatuses in Baltic sequences used for acritarch biostratigraphy. One possibility is that *Skiagia* acritarchs first appeared with *Mobergella radiolata* in the Lontova–Lukati hiatus, while *S. mickwitzi* trilobite faunas appeared in the upper part of the range of *M. holsti* and *M. turgida*. *Schmidtiellus mickwitzi* in Scania is believed to be concurrent with strata bearing *Rhombocorniculum insolutum* and ?*Mobergella* sp. in the Green Shales of Bornholm, which is a likely source for phosphatized ?*Schmidtiellus* fragments reworked in the basal Middle Cambrian (Poulsen 1978). This suggests correlation between the *S. mickwitzi* Zone of the Scandinavia with the *Fallotaspis* to early *Pagetiellus anabarus* Zones in Siberia.

East European Platform   In Poland are found the problematical arthropods *Livia convexa* Lendzion in the lower *Mobergella* Zone, and *Livia plana* Lendzion plus *Cassubia infercambriensis* (Lendzion) in the upper *Mobergella* Zone, below the first appearance of *Schmidtiellus* sp. in the Vergale Horizon (Urbanek & Rozanov

1983; Lendzion 1983*a*). The latter is associated with *Fallotaspis*? sp. and *Volborthella tenuis* (Volkova *et al.* 1983; Lendzion 1983*a*).

England; south-east Newfoundland   The upper Home Farm Member of Nuneaton, part of the Comley Sandstone of Shropshire, and the Smith Point Limestone of Newfoundland may be of *lenaicus* to ?early *P. anabarus* Zone age, as indicated by their position below *Callavia* faunas and the presence in England of *Camenella baltica*, *Mobergella* sp., *Rhombocorniculum insolutum*, and *Sunnaginia parva*. The absence of trilobite skeletons cannot be due to unsuitable depth but may relate to temperature, since carbonate biomineralization of chitin may have been delayed in cool-water facies.

China   The Badaowan Member of the Qiongzhusi Formation, and equivalents elsewhere, may be of this age (or earlier?) because of the presence of chancelloriids *Allonnia* and *Archiasterella* and their position beneath beds with *Lapworthella cornu* and *Tannuolina multiforata*. The widespread metal-enriched black shales of this time (Xiang *et al.* 1981) indicate conditions that perhaps were . unsuitable for carbonate biomineralization of trilobite chitin.

Early Judomia *age trilobites*

Judomia is a widespread callavid olenellid characteristic of Upper Atdabanian strata in Siberia and presumed equivalents elsewhere. Callaviids (i.e. *Callavia*, *Judomia*) and opisthoparian strenuellids are common elements of faunas of this age, and the eodiscids *Dipharus* (*Hebediscus*) and *Triangulaspis* appear in many regions with this fauna. Late *Judomia* age trilobite assemblages may contain distinctive eodiscids (and microfossils), discussed separately below.

Siberia   The *Judomia* Zone has olenellid *Judomia mattajensis* Laz., *J. tera* Rep. plus the strenuellid *Strenuella* (*Comluella*) sp., eodiscids *Dipharus* (*Hebediscus*) sp., *Pagetiellus lenaicus* (Toll), *Triangulaspis meglitzkii* (Toll) and *T. lermontovae* Laz.; the upper part of the zone with *Neocobboldia dentata* (Lerm.) and *N. paradentata* Rep. (Rozanov & Sokolov 1984) is probably equivalent to the early phase of the *Serrodiscus bellimarginatus* fauna in England and Newfoundland (see below).

North America   *Fallotaspis* was largely replaced by nevadiinids in the *Nevadella* Zone (Fritz 1972). In California, *Fallotaspis* cf. *longa* occurs with *Daguinaspis* and *Judomia*(?) in the upper part of the *Fallotaspis* Zone, below the first appearance of *Holmia* (Nelson 1978). More certain records of *Judomia*, however, are found in the Canadian Cordillera, where the *Nevadella* Zone has olenellids *Nevadella*, *Judomia*, *Bradyfallotaspis*, *Holmiella*, dorypygid *Kootenia*, and eodiscid *Pagetides* near its local base (Fritz 1972). *Lapworthella filigrana* from the *Nevadella* Zone in the Cassiar Mountains (Conway Morris & Fritz 1984) and Mexico to California (McMenamin 1987) suggests correlation with the upper *P. anabarus* to low *Judomia* Zone in Siberia. The supposed *Nevadella* Zone trilobites from the outer detrital belt in the Mackenzie Mountain (Fritz 1973) are more problematical since they contain *Serrodiscus bellimarginatus*-like forms that elsewhere appear at a higher level (see below).

South-east Newfoundland   *Callavia* Zone trilobites first appear above a hiatus in the topmost Smith Point Limestone and through much of the Brigus Formation (Fletcher 1972). Here, the earliest assemblages have olenellid *Callavia broeggeri* (Walcott) and strenuellids *Acanthomicmacca* and *Strenuella* (*Comluella*) below beds with *Dipharus attleborensis* (Shaler & Foerste) and *Strenuella strenua* (Billings). Early *Rhombocorniculum cancellatum* occur 6 m above the base of the Brigus Formation (Dr S. Bengtson, written communication 1986), suggesting that their appearances were roughly coincident, though this requires confirmation. *Callavia* beds with *Nevadella* near North Weymouth, Massachusetts, suggest correlation of the *Nevadella* and *Callavia* Zones (Fritz 1972).

England   The first trilobite is a single olenellid fragment from Ab3 Beds of the Lower Comley Sandstone, Shropshire. Although suggested to be *Fallotaspis*? by Hupé (1952, p. 127), the latter author never saw the specimen and expressed doubts about it (1958, p. 30). Bergström (1973) regarded it as *Kjerulfia*?, which suggests a younger age. It occurs above *Mobergella* cf. *turgida* and *Camenella baltica* in Bed Ab1 and below Ac1 Beds with *Callavia broeggeri*, *Judomia*(?), and *Kjerulfia*(?), and Ac2 beds with *C. broeggeri* and *Nevadella*(?) *cartlandi* (Raw), though these also contain *D. attleborensis* (Raw 1936; Bergström 1973; Rushton 1974; Landing *et al.* 1980). The appearance of *Callavia* is roughly coincident with *Rhombocorniculum cancellatum*, the latter is definitely present in Ac2 and is suspected from Ac1 (in green sandstone clasts reworked into basal Middle Cambrian; Rushton *in* Grieg *et al.* 1968). At Nuneaton, possible *Callavia* fragments first occur at the base of the Purley Shale (Smith & White 1963), associated with undescribed conodonts and *Platysolenites antiquissimus* (Brasier 1986).

Scandinavia   The *Holmia* n. sp. Zone of Bergström (1981, formerly '*Callavia*' n. sp. and now *Holmia inusitata* Ahlberg. 1986) seems to have been a time of relatively barren sandy strata, with only the nominate species. This is here tentatively suggested to span the late *P. anabarus* to early *Judomia* Zone.

Morocco   Zones III to V of Hupé, with *Fallotaspis*, *Daguinaspis*, *Resserops*, *Dipharus* (*Hebediscus*), *Strenuaeva*, and *Comluella* are correlated with the *Judomia* Zone of Siberia by Rozanov & Sokolov (1984).

Serrodiscus bellimarginatus faunas

A distinctive and widespread assemblage of eodiscids, once loosely called the *Dipharus attleborensis* fauna (Robison *et al.* 1977; Landing *et al.* 1980) appears widely at a time of major transgression, probably reflecting a phase of the eustatic sea-level rise. Global elements include pagetiids *Triangulaspis vigilans* (Matthew) (= *annio* Cobbold), *Dipharus attleborensis* (Shaler & Foerste), and blind eodiscids *Serrodiscus bellimarginatus* (Shaler & Foerste), *S. speciosus* (Ford), *Calodiscus lobatus* (Hall) and *C. schucherti* (Matthew). A succession of these important indices can be recognized in Acadian (i.e. English to maritime North American) sequences.

The polymerids vary with time and faunal province, but olenellids are often joined or ousted by opisthoparian trilobites in this interval. In Siberia, south-east Newfoundland, and England, and also Scandinavia and the Baltic, forms such as strenuellids, protolenids, dorypygids, and ellipsocephalids became more important. In the Canadian Cordillera, *Nevadella* was replaced by *Olenellus* without overlap (Fritz 1972) and ptychopariids appeared in California (Nelson 1978). The assemblage occurs in the upper part of the *Callavia* Zone on both sides of the Atlantic, in the lowest Botomian *Bergeroniellus micmacciformis*–*Erbiella* Zone/*Calodiscus*–*Erbiella* Zone of Siberia (Robison *et al.* 1977; Repina 1972, 1981; Sokolov & Zhuravleva 1983), in the *Elliptocephala asaphoides* fauna of the Taconic region of north-eastern USA (e.g. Theokritoff 1981) and in the lower part of the *Bonnia–Olenellus* Zone of Greenland and Spitsbergen (Cowie 1971*a*, 1974).

Small shelly fossils provide important support for this correlation, with comparable first appearances of the *Lapworthella cornu* group, *Microdictyon effusum* group, *Hadimopanella* spp., and *Pelagiella lorenzi*, and the continuing presence of *Rhombocorniculum cancellatum*.

South-eastern Newfoundland and England   The former is the most important section for elucidating the succession of trilobites at this level (Fletcher 1972; pers. comm. 1986), while English occurrences add supporting data. The *Serrodiscus bellimarginatus* fauna appears in the upper part of the *Callavia* Zone (better regarded as a 'Superzone') and the succession appears to be as follows:

1. *Acanthomicmacca* interval: the earliest *Callavia broeggeri* assemblages appear locally above a hiatus in the top of the Smith Point Limestone and in the basal Brigus formation (see above). These contain conservative forms of *Rhombocorniculum cancellatum*.

2. *Dipharus attleborensis* interval: higher beds have *Dipharus attleborensis*, *Acanthomicamacca walcotti* (Matthew), and *Strenuella strenua* and other forms. A similar assemblage occurs in the *Callavia* Sandstone Ac1 Bed and *Callavia* Limestone bed of Comley. Here occur well-developed forms of *Rhombocorniculum cancellatum*.

3. *Serrodiscus bellimarginatus* interval: the preceding trilobites are joined higher up by *Triangulaspis vigilans* and *Serrodiscus bellimarginatus*. A similar assemblage occurs in the *Serrodiscus bellimarginatus* Ac3 Beds of Comley, which also has *Pseudatops reticulatus* (Walcott) and Newfoundland has *Calodiscus lobatus* (= *C. lakei* Rasetti of Ac4 according to Ahlberg 1983). The Ac3 Limestone has *Sunnaginia angulata*, highly developed *R. cancellatum*, *R. spinosus*, *Amphigeisina* sp., and *Hertzina elongata*. The Redlands Limestone at a similar level in Newfoundland has *Microdictyon effusum* and *Hertzina elongata*. Towards the top of this interval in Newfoundland, *D. attleborensis* is absent. The *Strenuella* Limestone Ac4 Beds of Comley have *Lapworthella cornu cornu*, *Microdictyon* cf. *effusum*, and *Pelagiella lorenzi*. In the Antigonish Highlands of Nova Scotia is a similar association with *Acanthomicmacca*, *Callavia broeggeri*, *Comluella* sp., *Ladadiscus* sp., *Strenuella strenua*, *Serrodiscus bellimarginatus*, and *Triangulaspis vigilans* (Landing *et al.* 1980). Their assemblage has SSFs *Sunnaginia imbricata* (or *angulata*), developed *R. cancellatum*, *Amphigeisina?* sp., *Lapworthella* sp. and *Pelagiella* sp.

4. *Strenuella sabulosa* interval: higher beds, above the top of *Callavia* (i.e. marking the base of the *Protolenus* Zone) contain *S. bellimarginatus* and *Strenuella sabulosa* Rushton, but lack *Triangulaspis vigilans*, *Callavia broeggeri*, *Strenuella strenua*, *Acanthomicmacca walcotti*, and *Comluella* spp. A similar assemblage at Nuneaton has *Serrodiscus bellimarginatus*, *S.* cf. *serratus* R. & E. Richter, *Ladadiscus llarenai* (R. & E. Richter), and *Strenuella sabulosa*, about 70 m above the base of the Purley Shale, notably associated with *Pelagiella* cf.

primaeva (Shaler & Foerste) (Rushton 1966). This interval may be represented by a hiatus in the Comley Limestone (Dr T. P. Fletcher, pers. comm. 1986).

5. *Serrodiscus speciosus* interval: higher beds in Newfoundland contain *Serrodiscus speciosus* (Ford) and *Calodiscus lobatus*, above the extinction of *Serrodiscus bellimarginatus* and *Strennuella sabulosa*. This interval also appears to be missing at Comley

6. *Protolenus howleyi* and higher intervals: the latter appears with *Condylopyge* (the earliest known agnostid) and the brachiopod *Botsfordia caelata* (Hall) above the disappearance of *S. speciosus*. *Botsfordia* and *Condylopyge* also appear with the higher *Acidiscus theristes* fauna in Nuneaton (Rushton 1966). The *Protolenus* Limestone Ac5 Beds and the *Lapworthella* Limestone Ad Beds of Comley must span the latest early Cambrian interval. These have small shelly fossils *Lapworthella cornu*, *Rhombocorniculum cancellatum*, *R. spinosus* and *Pelagiella* sp.

Siberia On the Lena river, the *Bergeroniellus micmacciformis–Erbiella* Zone is said to contain *Dipharus* (*Hebediscus*) *attleborensis*, *Triangulaspis vigilans*, *T. lermontovae*, *Pagetiellus lenaicus*, and *Neocobboldia dentata* at the base, in units with Botomian archaeocyathan indices *Botomocyathus zelenovi* Zhur. and *Porocyathus squamosus* (Zhur.) (Rozanov 1973; Repina 1981; Rozanov & Sokolov 1984). Protolenids such as *Bergeroniellus* become important through this zone and provide most of the indices through the Botomian. Other polymerids include species of *Erbiella* (dinesid), *Micmacca* (strenuellid), *Bonnia*, and *Kootenia* (dorypygids) and surviving *Judomia* (e.g. Repina 1981; Sokolov & Zhuravleva 1983). The Judomian–Olenek facies of Siberia also has *Calodiscus schucherti* (Matthew) in the basal *Calodiscus–Erbiella* Zone of the Botomian (Repina 1972). Some authors refer this so-called '*Dipharus attleborensis*' assemblage incorrectly to the Upper Atdabanian (e.g. Robison *et al.* 1977; Landing *et al.* 1980; Bergström & Ahlberg 1981). *Dipharus* (*Hebediscus*) sp. or *Dipharus* (*Hebediscus*) *ponderosus* Lerm. are certainly reported below *Bergeroniellus* and *Erbiella* (Khomentovsky & Repina 1965; Rozanov 1973; Rozanov & Sokolov 1984). Most importantly, the specimens of *D. attleborensis* are not identical to those from the *Callavia* Zone of Newfoundland (Dr T. Fletcher, pers. comm. 1986). The top of the *Judomia* Zone and the *B. micmacciformis–Erbiella* Zone notably contain *Microdictyon* sp., *Hadimopanella knappologica*, *Lapworthella cornu dentata*, and *Rhombocorniculum cancellatum* (Rozanov & Sokolov 1984). The *B. mica-*

macciformis–Erbiella Zone also has *Pelagiella lorenzi* and *Gapparodus bisulcatus* (op cit., Missarzhevsky 1982). This suggests that the base of this zone correlates roughly with the upper part of the *Serrodiscus bellimarginatus* interval in Newfoundland and England.

North America The base of the *Bonnia–Olenellus* Zone is placed at the first appearance of *Olenellus* (Fritz 1972). The lower part usually has olenellids *Olenellus* and *Laudonia*, and ptychopariids (Nelson 1978), including *Proliostracus* (Fritz 1972). In East Greenland, *Calodiscus lobatus* occurs with *Olenellus* and ?*Weymouthia*, in the Bastion Formation and with *Olenellus*, *Bonnia*, *Kootenia*, *Proliostracus*, and *Wanneria* in the Ella Island Formation (Cowie 1971*a*). *Calodiscus lobatus* also occurs with *Wanneria* in the Devil's Cove Formation, at the base of the *Bonnia–Olenellus* Zone in western Newfoundland (North 1972, p. 257).

Spitsbergen and Greenland The Slakli Limestone of Spitsbergen has *Olenellus* cf. *thompsoni* Hall, *Olenellus* spp. indet., *Serrodiscus bellimarginatus*, *S.* cf. *speciosus*, *C. lobatus*, and *Calodiscus* sp. indet. (Major & Winsnes 1955; Cowie 1974; Ahlberg 1983). '*Platyceras primaevum*' probably indicates *Pelagiella* sp. and a similar assemblage from the Blastertoppen Dolomite also yields *Hadimopanella apicata* (Wrona 1982). In North Greenland, an assemblage with *S. bellimarginatus* and *Calodiscus* sp. likewise yields *Hadimopanella apicata* and *Pelagiella* sp. (Peel & Larsen 1984). In East Greenland, the upper Bastion Formation has *Olenellus* and *Calodiscus lobatus* with *Botsfordia caelata*, *Mobergella braastadii*, and '*Discinella micans*' (Poulsen 1932; Cowie 1971*a*).

Taconics, eastern USA The *Elliptocephala asaphoides* fauna has the olenellid *E. asaphoides* Emmons, conocoryphid *Pseudatops reticulatus* plus *Serrodiscus speciosus*, and *Calodiscus lobatus*. These are associated with *Lapworthella cornu schodakensis*, *Microdictyon* cf. *effusum*, and '*Discinella micans*' (e.g. Theokritoff 1964, 1981; Landing 1984).

Scandinavia Beds of the *Holmia kjerulfi* Zone contain *H. kjerulfi* (Linnarsson), *H. sulcata* Bergström, *H. grandis* Kiaer, *Kjerulfia*, rare *Fallotaspis*, *Calodiscus lobatus*, and the first ptychopariids *Ellipsocephalus*, *Strenuaeva*, *Comluella*?, and *Proampyx* (Ahlberg 1981; Bergström 1981; Bergström & Ahlberg 1981; Ahlberg 1983). *Runcinodiscus* cf. *index* Rushton also occurs in the *Holmia* Shale of Mjøsa, Norway (Ahlberg 1983). The Gislöv Formation has *Lapworthella cornu cornu*, *Rhombocorniculum cancellatum*, and *Botsfordia* sp.

East European Platform The upper or *Strenuaeva* Subzone of the *Holmia* Zone in Poland has *Holmia kjerulfi*, *H. grandis*, *Ellipsocephalus* cf. *gripi* (Kautsky), *Strenuaeva primaeva* (Brogger), *Strenuella* aff. *salopiensis* Cobb., *S.* sp. ex. gr. *polonica* Samsonowicz (Lendzion 1983*a*, 1983*b*).

Gorlitz, GDR The *Serrodiscus* slate has *Holmia* cf. *sulcata*, *Kjerulfia* cf. *lata* Kiaer, *Serrodiscus speciosus silesius* R. & E. Richter, *Lusiatops* sp., and *Micmacca* sp. (Schwarzbach 1939; Rozanov 1973; Bergström & Ahlberg 1981).

Central Poland A lower assemblage has *Holmia marginata* Orlowski, *Postfallotaspis spinatus* Orlowski, *Schmidtiellus panowi* (Samsonowicz), *Strenuella* spp., *Micmacca* sp., and *Comluella* spp., suggesting an early *Holmia kjerulfi* Zone age. An upper fauna has *Ellipsocephalus* spp., *Strenuaeva* spp., and *Serrodiscus primarius* Orlowski; the upper part of this assemblage contains *Protolenus* spp. (Orlowski 1985).

Morocco Zone VI of Hupé, with *Callavia*, *Antatlasia*, *Gigantopygus*, *Termierella*, and the pagetiids *Triangulaspis*, *Delgadella*, and *Pagetiellus lenaicus* is probably of this age, as may be part of Zone V (Sdzuy 1971). *Serrodiscus*, *Saukianda*, and *Protolenus* appear at the top (Hupé 1960; Szduy 1971, 1978).

Iberia The *souzai–schneiderei* Band has *Delgadella souzai* (Delgado) (cf. *Dipharus*), *Calodiscus schucherti*, *Triangulaspis fusca* Sdzuy, and *Judomia*? *lotzei* (R. & E. Richter). The *serratus* Band has *Ladadiscus llarenai* R. & E. Richter, *Serrodiscus serratus*, *S. speciosus silesius* R. & E. Richter, and *T.* cf. *fusca* (Sdzuy 1962, 1971).

Protolenus *faunas*

In the Atlantic faunal province of eastern North America, England, Wales, and the rest of Europe, succeeding *Protolenus* faunas contain protolenids and ellipsocephalids but no olenellids. Beds above the range of *Judomia* in Siberia may be of similar age. On the North American craton, however, olenellids survived until the end of the Lower Cambrian but ptychopariid trilobites predominated; *Salterella* provides a possibly useful marker fossil. Near the top of the Lower Cambrian is a widespread *Bathydiscus–Serrodiscus* assemblage (e.g. Robison *et al.* 1977). Since *Protolenus* Zone assemblages are of more interest to the Lower–Middle Cambrian boundary than to the Precambrian–Cambrian boundary problem, they will not be discussed further here.

South China trilobite zones

The succession of mainly redlichiid trilobites through the Lower Cambrian of the Yangtze Platform is reviewed in Xiang *et al.* (1981), Luo *et al.* (1982), and Xing *et al.* (1983) and the zones are illustrated in Fig. 7.19. Trilobites provide a reasonable biostratigraphy in southern China, with comparable Canglangpuian to Longwangmiaoian forms appearing in North China and along the Palaeotethyan Belt into Kashmir (e.g. Shah 1982). Endemism is marked, however, with few pandemic polymerid or pelagic forms reaching this region.

The Qiongzhusian Stage is marked by the appearance of redlichiids *Parabadiella* and *Eoredlichia*, together with primitive-looking pagetiids such as *Mianxiandiscus*. The age of this assemblage is arguably mid- to late Atdabanian, though others have argued for early Atdabanian or older (see Brasier, this volume, pp. 64).

The Canglangpuian Stage contains three main zones. *Protolenus* appears with *Neocobboldia* in the second, *Drepanuroides* Zone, so that a Botomian correlation seems likely.

The final stage of the Lower Cambrian in China is the Lonwangmiaoian, which compares in position with the ?Lower Toyonian, while the succeeding Maozhuangian may be of Toyonian or younger age.

7.4.7. Acritarch stratigraphy

Following extinction of the Vendian microflora, it has been suggested that there was a periodic increase in the diversity of acritarchs, particularly of spinose acanthomorph acritarchs of biostratigraphical value. (Vidal & Knoll 1983). There are problems with this view, however, as will be discussed below. Our knowledge of the natural succession of microfloras is largely limited to subsurface cores in the East European Platform (e.g. Volkova *in* Urbanek & Rozanov 1983) and strata of the Fennoscandian Shield (Vidal 1981) or the North American Platform (Downie 1982; Knoll & Swett 1987) where there are many breaks in deposition and sparse skeletal fossils give minimal yardstick biostratigraphical control.

The earlier assemblages generally comprise simple and long-ranging sphaeromorph acritarchs such as *Leiosphaeridia*. Assemblages of this type occur in the Lublin and Włodawa Formations of the Lublin Slope in Poland, on the southern margin of the East European Platform, where they are associated with the algal frond *Vendotaenia*, the organic tube *Sabellidites*, the first *Platysolenites antiquissimus*, and *Aldanella polonica* (Moczydłowska & Vidal 1986). An almost identical association of *Leiosphaeridia*, *Sabellidites*, *P. antiquissi-*

mus, and *Aldanella kunda* occurs in the Dividal Group of northern Norway (op. cit.). Another similar assemblage, but with *Micrhystridium tornatum* Volkova, is recorded from the Rovno Formation in the western part of the European Platform (Volkova *in* Urbanek & Rozanov 1983) and all may be ascribed to the *Sabellidites cambriensis* Zone. Such poor acritarch assemblages could occur at higher intervals in restricted shallow-water settings, so the biostratigraphical significance of this assemblage is limited.

A slightly more varied assemblage is known from the Lontova Formation and equivalent strata from the *Platysolenites antiquissimus* Zone according to Volkova (op. cit.; Knoll & Swett 1987), with the notable addition of *Micrhystridium tornatum* Volkova, *Granomarginata prima* Naumova, *G. squamacea* Volkova, and the ?prasinophycean cyst *Tasmanites tenellus* Volkova. The Mazowsze Formation of the Lublin Slope has usually been correlated with the Lontova Formation, since it contains a similar restricted skeletal assemblage dominated by *Platysolenites antiquissimus* and Lontova acritarchs (e.g. Volkova *in* Urbanek & Rozanov 1983). Re-sampling by Moczydłowska & Vidal (1986) reveals a surprisingly more varied acritarch assemblage, however, with the notable addition of dense process-bearing *Comosphaeridium* n. spp., *C. strigosum* Jankauskas, keeled *Pterospermella* sp., sculptured *Cymatiosphaera minuta* Jankauskas, plus *Tasmanites bobrowskii* Wazynska. Assemblages with *Comasphaeridium* and *Micrhystridium* in the Hardeberga Quartzite of Scania are also ascribed to a Mazowsze age, as is the poorer assemblage of the Ringsaker Quartzite of Mjøsa, Norway (Moczydłowska & Vidal 1986). Some of these additional elements appear elsewhere in the post-Lontova, Talsy (= Lükati) Horizon of '*Holmia* A' age in the sense of Vidal (1981), though skeletal fossils and stratigraphical position do not support such an age for the Mazowsze Formation. Moczydłowska & Vidal (1986) have, however, extended the '*Holmia* A' Zone down to include the Mazowsze Formation and the *Platysolenites* Zone, inferring a correlation with the Talsy Formation. This action is open to criticism:

(1) there are no *Holmia* trilobites at this level;
(2) they have shifted '*Holmia* A, B, and C' terms downwards, causing confusion;
(3) there are other interpretations.

Firstly, the Masowsze assemblage could be intermediate in age between the Lontova and Talsy Formations, equivalent strata having been removed in the East European Platform and Spitsbergen during the major post-Lontova hiatus (e.g. Brangulis *et al.* 1981; Knoll & Swett 1987). Secondly, The Lontova, Masowsze, and

Talsy assemblages may have been facies controlled and long ranging. The Lontova Formation, for example, may have only a restricted littoral assemblage, lacking the more offshore acanthomorphs (cf. Downie 1982).

A serious blow to acritarch stratigraphy at this level comes from work on the Siberian Platform (Khomentovsky 1986). Lower Vendian to Nemakit–Daldynian strata generally have a Lontova-like assemblage plus *Micrhystridium certum* Trestsh., *Bavlinella faveolata* (Shepeleva), and other Vendian forms. But in the Interior Province, these are joined by *C. strigosum*, *Goniosphaeridium primarium* (Jankauskas), and *Baltisphaeridium cerinum* Volkova, which on the East European Platform appear in the Talsy Formation; and *Skiagia ciliosa* (Volkova) plus *Leiovalia tenera* (Kirjan.), which appear in the Vergale Formation. Either these strata are not Vendian at all or the Talsy and Vergale assemblages are strongly diachronous in their appearance. That the latter is more likely to be the case is suggested by finds of *Baltisphaeridium* and associated spinose microfossils in cherts of the Doushantuo Formation, hundreds of metres below the earliest skeletal faunas in the Dengying Formation of China (Awramik *et al.* 1985).

Distinctive larger acanthomorphs with closed processes (*Baltisphaeridium*) or open ones (*Skiagia*: see Downie 1982), are certainly characteristic of the Talsy Formation, which notably has *S. ornata* (Volkova) and *S. orbiculare* (Volkova). This acritarch assemblage is known from *Mobergella*-bearing beds in the West Baltic and Mjøsa district, from the *Rhombocorniculum insolutum*-bearing Green Shales of Bornholm and generally from *Schmidtiellus mickwitzi* Zone strata in the Balto-Scandinavian area (e.g. Vidal 1981). Many acritarchs that are reported to appear first in the Vergale Horizon of the East European Platform by Volkova *in* Urbanek & Rozanov (1983; '*Holmia B*' *sensu* Vidal 1981) appear with arguably older skeletal assemblages in Scandinavia (Vidal 1981) for reasons that are not yet clear. This means that strata assigned a *Holmia* B age in the sense of Vidal (1981; e.g. the Lingulid Sandstone) cannot be firmly correlated.

Acritarch assemblages described from north-west Scotland and Greenland (Downie 1982) are of middle to late early Cambrian *Bonnia–Olenellus* Zone age, the base of which (as discussed above) may correlate with the *Serrodiscus bellimarginatus* fauna, the *Holmia kjerulfi* Zone, the Vergale Horizon, and the Atdabanian–Botomian transition. Elements that appear in both the Scottish Fucoid beds and the Vergale Horizon include *Skiagia* cf. *insigne* (Fridrichsone), *Ovulum saccatum* Jankauskas, *Aranidium* cf. *izhoricum* Jankauskas, *Multiplicisphaeridium dendroideum* (Jankauskas), *Cymatiosphaera postii* Jankauskas, and *Goniosphaeridium impli-*

catum (Fridrichsone). Similar assemblages occur in the Lower Bastion to Ella Island Formations of East Greenland and the *Holmia* Shale of Norway (Downie 1982).

A difficult situation therefore exists at present. Talsy-type assemblages are now recorded from strata not only with arguably Tommotian skeletal fossils, but much further down into Lower Vendian and Sinian rocks. It will be necessary to study these early records, and the palaeographical and facies pattern of apparent diachronism, before confident progress can be made.

7.5. CONCLUSION

7.5.1. Biostratigraphical assemblages

Taking a broad view, it seems possible to recognize about five successive and widespread fossil assemblages of skeletal fossils, characteristic of intervals up to and including the early Botomian. Each of these contains elements of evolutionary series that provide backbone stratigraphy. But most groups appear abruptly, without known ancestors in the fossil record, in a manner suggestive of major migratory radiations connected with transgressive episodes. The main phase of these cryptogenic migrations was in the *Aldanella attleborensis* interval.

The Anabarites trisulcatus–Protohertzina anabarica *assemblage*

This spans the lower range of the indices, below the appearance of the *Latouchella korobkovi* and *Aldanella attleborensis* groups. Typical members include the organic tubes of *Sabellidites* and phosphatic tubes of *Hyolithellus*, both of which are known from earlier levels. New appearances include hyoliths with simple rounded cross-sections (e.g. *Circotheca*, *Conotheca*, and *Ladatheca*), spinose-walled ?molluscs (e.g. *Cassidina*, *Maikhanella*), and simple molluscs (*Purella*). Examples include the *Purella antiqua* Zone in the upper part of the Manykayan Nemakit–Daldyn Series of northern Siberia, and lower Zone I of the Yangtze Platform, China.

Assemblages with this character but bearing hyoliths with oval cross-sections (e.g. *Turcutheca crasseocochlia*, *Ovalitheca* sp.), anabaritid *Tiksitheca licis*, conulariids (e.g. *Barbitositheca ansata*), may lie in the upper part of this interval (e.g. those in the phosphorite intervals of Iran, Kazakhstan, India, and China) or in the following interval.

The Aldanella attleborensis *assemblage*

The first assemblage is here joined or replaced by an assemblage whose diversity and architecture indicates a period of cryptic evolution. This migrating 'horde' includes sculptured bellerophontacean monoplacophorans (e.g. *Latouchella korobkovi* group), gastropods (e.g. *Aldanella attleborensis* group, *Barskovia*), paterinid brachiopods (e.g. *Aldanotreta*), archaeocyathans of the *sunnaginicus* Zone, tommotiids (e.g. *Sunnaginia imbricata*, *Tommotia* spp., *Eccentrotheca* sp., *Yunnanotheca* sp.), and hyolithelminthes (e.g. *Torellella curva* and *T.* cf. *lentiformis*). Allathecid hyoliths with rounded triangular cross-sections (e.g. *Allatheca*) and anabaritids with more-developed architecture (e.g. *Anabaritellus hexasulcatus*) indicate evolutionary development of the initial skeletal fauna. Assemblages of this type occur in the *sunnaginicus* Zone of Siberia and Zone II (Marker B) of the Yangtze Platform and may correlate with the *Aldanella attleborensis* Zone in south-east Newfoundland and eastern Massachusetts. Early *Platysolenites* Zone assemblages may also be of this age in Baltica.

Assemblages bearing *Torellella lentiformis*, *Lapworthella tortuosa* group, and lipped hyoliths (e.g. *Burithes*) occupy a slightly higher position in Siberia (upper *sunnaginicus* to *regularis* Zones) and Mongolia, and also occur in Newfoundland and England.

The Rhombocorniculum insolutum *Assemblage*

The index appears without forebears, joined by other cryptogenic groups: problematical discs of the *Mobergella radiolata* group, agglutinated *Volborthella tenuis*, and the first olenellid trilobites (e.g. *Fallotaspis*, *Schmidtiellus*). Hyolithelminthes *Torellella biconvexa* or *T. laevigata* and tommotiid *Camenella baltica* indicate this fauna. Large patelliform monoplacophorans (e.g. *Randomia*, *Tannuella*), obolellid brachiopods (e.g. *Obolella*) and bivalves (*Fordilla*) also appear in this interval. Examples include the upper Bonavista to Smith Point Formations of Newfoundland, the upper Home Farm Member of Nuneaton, England, the Green Shales of Bornholm and the *lenaicus* to *pinus/anabarus* Zones in Siberia. This fauna is not clearly recognized from China and may be represented by parts of Zones II and III, the barren interval or the widespread hiatus in between.

The Rhombocorniculum cancellatum *Assemblage*

This spans the lower range of *R. cancellatum*, which is joined by other forms developed from known ancestors, such as *Lapworthella lucida* or *L. filigrana*. The trilobites *Callavia*, *Judomia*, *Dipharus*, and *Triangulaspis* appear here, while bradoriid crustaceans arise from cryptogenic origins. Examples include the Ab3 to Ac2 Beds of Shropshire in England, the lower Brigus Formation of south-east Newfoundland and the upper *Pagetiellus anabarus* to *Judomia* Zones of Siberia. On the Yangtze Platform, *R. cancellatum* appears in Qiongzhusian

strata with *Cambroclavus* and *Allonnia*, and the latter datum (plus *Sinosachites*, marking Zone III) is suggested to be contemporaneous with this assemblage.

The Lapworthella cornu *Assemblage*

Rhombocorniculum cancellatum is here joined by developmental forms such as: *Rhombocorniculum spinosus*, tommotiids of the *Lapworthella cornu* group, (e.g. *L. cornu cornu*, *L. cornu dentata*, *L. cornu shodackensis*, and *L. cornu nigra*), *Sunnaginia angulata*, gastropod *Pelagiella lorenzi*, and lingulid *Botsfordia caelata*. *Lingulella* is also widespread, though it appears earlier in the Avalon region. The problematical 'buttons' of *Hadimopanella knapplogica* and *H. apicata*, and the net-like problematicum *Microdictyon effusum* have a cryptogenic origin. This is a relatively cosmopolitan assemblage, associated in the olenellid realm with blind ?pelagic trilobites such as *Serrodiscus bellimarginatus*, *Calodiscus lobatus*, and *Pseudatops* spp. Examples include Beds Ac3 to Ac4 of the Comley Limestone and the middle Purley Shales of Nuneaton, England, the upper Brigus Formation of south-east Newfoundland, the Gislöv Formation of Scania, and the upper Atdabanian–lower Botomian of Siberia. Zones IV to V on the Yangtze Platform, with *Botsfordia caelata*, *Lapworthella cornu*, and *Pelagiella* spp., are suggested to be of mid to late Atdabanian age.

7.5.2. The bio-event approach to correlation

The writer agrees with Scott (1985) that biozone-based stratigraphy does not make the most of the diverse evidence available for correlation. Comparison of homotaxial first-appearance datum points (i.e. FADs, or biohorizons) of taxa through several sections, provide many more points for correlation, as well as a check on diachronism. Both Walliser (1986) and Kauffman (1986) have extolled the virtues of an even broader, bio-event stratigraphy; i.e. events involving punctuated evolution, population bursts, productivity increases, immigration and emigration, ecostratigraphical changes, regional colonization, mass mortality, and extinction. These bio-events can then be calibrated against unusual sedimentological or geochemical events (e.g. bentonites, carbon isotopes, and magnetic reversals). The methodology is pragmatic rather than ideological, beginning with local correlation and extending outwards to the limits of practicality. It obviously requires thorough documentation of diverse fossil groups through numerous bio-events and geo-events in many sections.

A start has already been made towards bio-event stratigraphy and Anglo-Siberian correlation (Brasier

1986*a*), Anglo-Avalon correlation (Brasier, this volume, pp. 98–100) and Palaeotethyan correlation (Brasier 1986*b*, Brasier & Singh 1987; Brasier, this volume, pp. 66–7) of Precambrian–Cambrian boundary sequences. As argued there, correlation along strike, and within-terrane, should precede attempts to correlate across the strike and between separate blocks.

The potential of the bio-event approach is demonstrated for some of the main sections in the preceding parts of this chapter (Figs 7.11 to 7.19). Shown here are ranges of selected skeletal taxa, correlated by means of their point of first appearance. These points are shown at the right-hand side of each diagram, and though liable to modification, they demonstrate a broadly homotaxial pattern that is inconsistent with significant diachroneity. Clearly, the best insurance against diachronism is to analyse the first appearances of diverse biotic groups and to integrate these data with other events. It must be admitted that most of the named data points were selected because they seem to give a congruent pattern of succession in different areas; there are also incongruent taxa—and incongruent occurrences—which have not been detailed in this preliminary attempt and these will require explanation.

Though the selected data points do not entirely coincide in their order of appearance, they do reveal a general homotaxial pattern of faunal succession, as inspection of the range charts will show. If this is a true reflection of a sequence of biological events, it would seem to confirm that migratory waves of novel taxa took place during the course of the 'Cambrian transgression'.

Acknowledgements

This chapter has benefited greatly from discussions at various times with other experts. I mention particularly Dr J. W. Cowie, Dr A. Yu Rozanov, Dr F. Debrenne, and Dr T. P. Fletcher. I am indebted to Dr Xiang Liwen of the Chinese Academy of Geological Sciences, Beijing, Mr Jiang Zhiwen of the Yunnan Institute of Geology, Kunming, and Mr He Tinggui of Chengdu College of Geology for their kind hospitality and the opportunities to see for myself the Chinese sections and material assisted by the Royal Society and the Scottish Academic Press for their project on the forthcoming book on 'The Cambrian System of China'; to Dr T. P. Fletcher for his loan and donation of comparative material from Newfoundland; to Dr P. Singh and Dr P. Kalia for their provision of material from India; to Dr B. Hamdi for his provision of material from Iran; to Dr S. Bengtson for organizing the seminal Uppsala meeting (for which this review was originally prepared) and for making available the unpublished results of his work in Scania, and on *Microdictyon*; to Dr J. Shergold for his stimulus to begin work on the biostratigraphy and Dr. S. Conway Morris for helpful discussion.

Mr Paul McSherry assisted with the diagrams and Mr John Garner with the plates, while Mr Ali-Reza Ashouri and Mr Ian Alexander helped with micropalaeontological preparations.

REFERENCES

Abaimova, G. P. (1978) [Anabaritids—ancient fossils with a calcareous skeleton. In *New material on the stratigraphy and paleontology of Siberia*, (ed. S. V. Sukhov)], pp. 77–83. Nauka, Novosibirsk. [In Russian.]

Ahlberg, P. (1981). Ptychopariid trilobites in the Lower Cambrian of Scandinavia. *US geol. Surv. Open-File Rep.*, **81–743**, 5–7.

Ahlberg, P. (1983). Redescription of a Lower Cambrian eodiscid trilobite from Norway. *Norsk geol. Tiddskr.*, **63**, 289–90.

Ahlberg, P. (1984). A Lower Cambrian trilobite fauna from Jamtland central Scandinavian Caledonides. *Geol. Fören. Stockholm Förhandl.*, **105**, 349–61.

Ahlberg, P. (1986). Lower Cambrian olenellid trilobites from the Baltic Faunal Province. *Geol. Fören. i Stockholm Förhandl.* **108**, 39–56.

Åhman, E. & Martinsson, A. (1965). Fossiliferous Lower Cambrian at Aspelund on the Skaggenas Peninsula. *Geol. Fören. i Stockholm Förhandl*, **87** 139–151.

Awramik, S. M., McMenamin, D. S., Yin Chongyu, Zhao Ziqiang, Ding Qixiu, & Zhang Shusen. (1985). Prokaryotic and eukaryotic microfossils from a Proterozioc/Phanerozoic transition in China. *Nature*, **315**, 655–8.

Azmi, R. J. (1983). Microfauna and age of the Lower Tal Phosphorite of Mussoorie Syncline, Garwhal Lesser Himalaya, India. *Himalayan Geol.* **11**, 373–409.
Himalaya, India. *Himalayan Geol.* **11**, 373–409.

Azmi, R. J. & Pancholi, V. P. (1983). Early Cambrian (Tommotian) conodonts and other shelly microfauna from the Upper Krol of Mussoorie Syncline, Garwhal Lesser Himalaya with remarks on the Precambrian–Cambrian boundary. *Himalayan Geol.*, **11**, 360–72.

Bengtson, S. (1968). The problematic genus *Mobergella* from the Lower Cambrian of the Baltic area. *Lethaia*, **1**, 325–51.

Bengtson, S. (1970). The Lower Cambrian fossils *Tommotia*. *Lethaia*, **3**, 363–92.

Bengtson, S. (1976). The structure of some Middle Cambrian conodonts and the early evolution of conodont structure and function. *Lethaia*, **9**, 185–206.

Bengtson, S. (1977*a*) Aspects of problematic fossils in the early Palaeozoic. *Acta Univ. Uppsaliensis*, **415**, 1–71.

Bengtson, S. (1977*b*). Early Cambrian button-shaped microfossils from the Siberian Platform. *Palaeontology*, **20**, 751–62.

Bengtson, S. (1980). Redescription of the Lower Cambrian *Lapworthella cornu*. *Geol. Fören. i Stockholm Förhandl.*, **102**, 53–5.

Bengtson, S. (1983). The early history of the Conodonta. *Fossils & Strata*, **15**, 5–19.

Bengtson, S. (1985). Taxonomy of disarticulated fossils. *J. Paleont.*, **59**, 1350–8.

Bengtson, S. (1986). A new Mongolian species of the Lower Cambrian genus *Camenella* and the problems of scleritome-based taxonomy of the Tommotiidae. *Paläont. Z.*, **60**, 45–55.

Bengtson, S. & Conway Morris, S. (1984). A comparative study of Lower Cambrian *Halkieria* and Middle Cambrian *Wiwaxia*. *Lethaia*, **17**, 307–29.

Bengtson, S. & Fletcher, T. P. (1983). The oldest sequence of skeletal fossils in the Lower Cambrian of southeastern Newfoundland. *Can. J. Earth Sci.*, **20**, 525–36.

Bengtson, S. & Missarzhevsky, V. V. (1981). Coeloscleritophora—a major group of enigmatic Cambrian metazoans. *US geol. Surv. Open-File Rep.*, **81–743**, 19–21.

Bengston, S., Missarzhevsky, V. V. & Rozanov, A. Yu. (1984). The Precambrian–Cambrian boundary: a plea for caution. *IGCP Project 29 Circular Newsletter*, June 1984, 14, 15.

Bengtson, S., Matthews, S. C., & Missarzhevsky, V. V. (1986*a*). The Cambrian net-like fossil *Microdictyon*. In *Problematic fossils*, (ed. A. Hoffman & M. H. Nitecki). Oxford Monogr. Geol. & Geophys., 5, Oxford University Press, pp. 97–115.

Bengtson, S., Fedorov, A. B., Missarzhevsky, V. V., Rozanov, A. Yu., Zhegallo, E. A., and Zhuravlev, A. Yu. (1986*b*). *Tumulduria incomperta* and the case for Tommotian trilobites. *Lethaia*, **20**, 361–70.

Berg-Madson, V. (1985). Middle Cambrian biostratigraphy, fauna and facies in southern Baltoscandia. *Acta Univ. Uppsaliensis*, **781**, 1–37.

Bergström, J. (1973). Classification of olenellid trilobites and some Balto-Scandian species. *Norsk geol. Tidsskr.*, **53**, 283–314.

Bergström, J. (1981). Lower Cambrian shelly faunas and biostratigraphy in Scandinavia. *US geol. Surv. Open-File Rep.*, **81–743**, 22–5.

Bergström, J. & Ahlberg, P. (1981). Uppermost Lower Cambrian biostratigraphy in Scania, Sweden. *Geol. Fören. i Stockholm Förhandl.*, **103**, 193–214.

Bischoff, G. C. O. (1976). *Dailyatia*, a new genus of Tommotiidae from Cambrian strata of S.E. Australia (Crustacea. Cirrepedia). *Senckenbergiana Lethaea*, **57**, 1–33.

Boorgard, van den, M. (1983). The occurrence of *Hadimopanella oezgueli* Gedik in the Lancara Formation in NW Spain. *Proc. Kon. Ned. Akad. Wet.*, **B86**, 331–41.

Brangulis, A. P., Kirjyanov, V. V., Mens, K. A., & Rozanov, A. Yu. (1981). Vendian and Cambrian Palaeogeography of the East European Platform. *US geol. Surv. Open-File Rep.*, **81–743**, 26–8.

Brasier, M. D. (1976). Early Cambrian intergrowths of archaeocyathids, *Renalcis* and pseudostromatolites from South Australia. *Palaeontology*, **19**, 223–45.

Brasier, M. D. (1979). The Cambrian radiation event. In *The origin of major invertebrate groups*, (ed. M. R. House), pp. 103–59. Syst. Assoc. Spec. Vol., 12. Academic Press, London.

Brasier, M. D. (1980). The Lower Cambrian transgression and glauconite–phosphate facies in western Europe. *J. geol. Soc. Lond.*, **137**, 695–703.

Brasier, M. D. (1981). Sea-level changes, facies changes and the late Precambrian–early Cambrian evolutionary explosion. *Precambr. Res.*, **17**, 105–23.

Brasier, M. D. (1984). Microfossils and small shelly fossils from the Lower Cambrian *Hyolithes* Limestone at Nuneaton, English Midlands. *Geol. Mag.*, **121**, 299–53.

Brasier, M. D. (1985). Evolutionary and geological events across the Precambrian–Cambrian boundary. *Geol. Today*, **1**, 141–6.

Brasier, M. D. (1986a). The succession of small shelly fossils (especially conoidal microfossils) from English Precambrian–Cambrian boundary beds. *Geol. Mag.*, **123**, 237–56.

Brasier, M. D. (1986b). Precambrian–Cambrian boundary biotas and events. In *Global Bio-Events* (ed. O. Walliser), pp. 109–17. Lecture Notes in Earth Sciences, **8**, Springer-Verlag, Berlin.

Brasier, M. D. & Hewitt, R. A. (1979). Environmental setting of fossiliferous rocks from the uppermost Proterozoic–Lower Cambrian of central England. *Palaeogeogr., Palaeoclimatol., Palaeoecol.*, **27**, 35–57.

Brasier, M. D. & Hewitt, R. A. (1981). Faunal sequence within the Lower Cambrian 'Non-Trilobite Zone' (s.l.) of central England and correlated regions. *US geol. Surv. Open-File Rep.*, **81–743**, 26–8.

Brasier, M. D., Hewitt, R. A., & Brasier, C. J. (1978). On the Late Precambrian–Early Cambrian Hartshill Formation of Warwickshire. *Geol. Mag.*, **115**, 21–36.

Brasier, M. D. & Singh, P. (1987). Microfossils and Precambrian–Cambrian boundary stratigraphy at Maledota, Lesser Himalaya. *Geol. Mag.*, **124**, 323–45.

Cherchi, A. & Schroeder, R. (1985). Middle Cambrian Foraminifera and other microfossils from SW Sardinia. *Boll. Soc. Paleontol. Ital.*, **23**, 149–60.

Cobbold, E. S. (1921). The Cambrian horizons of Comley (Shropshire) and their Brachiopoda, Pteropoda and Gasteropoda etc. *Q. J. geol. Soc. Lond.*, **76**, 325–86.

Cobbold, E. S. & Pocock, R. W. (1934). The Cambrian area of Rushton (Shropshire). *Phil. Trans. R. Soc. Lond.*, **B223**, 305–409.

Conway Morris, S. & Fritz, W. H. (1984). *Lapworthella filigrana* n. sp. (*incertae sedis*) from the Lower Cambrian of the Cassiar Mountains, northern British Columbia, Canada, with comments on the possible levels of competition in the early Cambrian. *Paläont. Z.*, **58**, 197–209.

Cowie, J. W. (1971a). Cambrian of the North American Arctic Regions. In *Cambrian of the New World*, (ed. C. H. Holland), pp. 325–84. John Wiley, London.

Cowie, J. W. (1971b). Lower Cambrian faunal provinces. *Geol. J.*, Spec. Iss., **4**, 31–46.

Cowie, J. W. (1974). The Cambrian of Spitsbergen and Scotland. In *Cambrian of the British Isles, Norden, and Spitsbergen*, (ed. C. H. Holland), pp. 123–56. John Wiley, London.

Cowie, J. W. (1978). I.U.G.S./I.G.C.P. Project 29 Precambrian–Cambrian Boundary Working Group in Cambridge, 1978. *Geol. Mag.*, **115**, 151–2.

Cowie, J. W. (1984). Introduction to papers on the Precambrian–Cambrian boundary. *Geol. Mag.*, **121**, 137–8.

Cowie, J. W. (1985). Continuing work on the Precambrian–Cambrian boundary. *Episodes*, **8**, 93–7.

Cowie, J. W. & Johnson, M. R. W. (1985). Late Precambrian and Cambrian geological time scale. In *The Chronology of the geological record* (ed. N. J. Snelling). *Mem. geol. Soc. Lond.*, **10**, pp. 47–64. Blackwell Scientific, Oxford.

Cowie, J. W. & Glaessner, M. F. (1975). The Precambrian–Cambrian Boundary: A Symposium. *Earth-Sci. Rev.*, **11**, 209–51.

Cowie, J. W., Rushton, A. W. A., & Stubblefield, C. J. (1972). A correlation of Cambrian rocks in the British Isles. *Spec. Rep. geol. Soc. Lond.*, **2**, 42 pp.

Crimes, T. P. & Anderson, M. (1985). Trace fossils from Late Precambrian–early Cambrian strata of southeastern Newfoundland (Canada): temporal and environmental implications. *J. Paleont.*, **59**, 310–43.

Debrenne, F. (1969). Lower Cambrian Archaeocyatha from the Ajax Mine, Beltana, South Australia. *Bull. Brit. Mus. natur. Hist., Geol. Ser.*, **17**, (7), 295–376.

Debrenne, F. & Debrenne, M. (1978). Archaeocyathid fauna from the lowest fossiliferous levels of Tiout (Lower Cambrian, Southern Morocco). *Geol. Mag.*, **115**, 101–19.

Debrenne, F. & James, N. (1981). Reef-associated archaeocyathans from the Lower Cambrian of Labrador and Newfoundland. *Palaeontology*, **24**, 343–78.

Debrenne, F. & Rozanov, A. Yu. (1983). Paleogeographic and stratigraphic distribution of regular Archaeocyatha. *Geobios*, **16**, 727–36.

Debrenne, F. & Vacelet, J. (1984). Archaeocyatha: is the sponge model consistent with their structural organization? *Palaeontographica Americana*, **54**, 358–69.

Debrenne, F., Rozanov, A. Yu., & Webers, G. F. (1984). Upper Cambrian Archaeocyatha from Antarctica. *Geol. Mag.*, **121**, 291–9.

Donovan, S. K. (1987). The fit of the continents in the late Precambrian. *Nature*, **327**, 139–41.

Doré, F. & Reid, R. E. (1965). *Allonnia tripodophora* nov. gen., nov. sp., nouvelle éponge du Cambrien inférieur du Carteret (Manche). *C. R. Soc. géol. Fr.*, 1965, 20–1.

Downie, C. (1982). Lower Cambrian acritarchs from Scotland, Norway, Greenland and Canada. *Trans. R. Soc. Edin.: Earth Sci.*, **72**, 257–85.

Fletcher, T. P. (1972). Geology and Lower to Middle Cambrian trilobite faunas of southwest Avalon, Newfoundland. Unpublished Ph.D. thesis, University of Cambridge, UK.

Firby, J. B. & Durham, J. W. (1974). Molluscan radula from earliest Cambrian. *J. Paleont.*, **48**, 1109–19.

Fonin, V. D. & Smirnova, T. N. (1967). New group of problematic early Cambrian organisms and methods of preparing them. *Paleont. J.*, **2**, 7–18. [English translation.]

Føyn, S. & Glaessner, M. F. (1979). *Platysolenites*, other animal fossils, and the Precambrian–Cambrian transition in Norway. *Norsk geol. Tidsskr.*, **59**, 25–46.

Fritz, W. H. (1972). Lower Cambrian trilobites from the Sekwi Formation type section, Mackenzie Mountains, northwestern Canada. *Bull. geol. Surv. Can.*, **212**, 58 pp., 20 pls.

Fritz, W. H. (1973). Medial Lower Cambrian trilobites from the Mackenzie Mountains, northwestern Canada. *Geol. Surv. Can. Pap.*, **73–24**, 43 pp.

Fritz, W. H. (1980). International Precambrian–Cambrian boundary working group's 1979 field study to Mackenzie Mountains, Northwest Territories, Canada. *Geol. Surv. Can. Pap.*, **80–1A**, 41–5.

Fritz, W. H. & Crimes, T. P. (1985). Lithology, trace fossils, and correlation of Precambrian–Cambrian boundary beds, Cassiar Mountains, north-central British Columbia. *Geol. Surv. Can. Pap.*, **83–13**.

Fritz, W. H. & Yochelson, E. L. (1988). The status of *Salterella* as a Lower Cambrian index fossil. *Can. J. Earth Sci.*, **25**, 403–16.

Gedik, I. (1977). Orta Toroslar'da konodont biyostratigrafisi. *Bul. Türk. Jeol. Kurumu*, **20**, 35–48. [In Turkish with an English summary.]

Gedik, I. (1981). *Hadimopanella* Gedik, 1977 'nin stratigrafik dagilimi ve mikroyapisi konusunda bazi gozlemler. *Karadeniz Tek. Univ. Yer Bilimleri Dergisi, Jeol.*, **1**, 159–63. [In Turkish.]

Glaessner, M. F. (1976). Early Phanerozoic annelid worms and their geological and biological significance. *J. geol. Soc. Lond.*, **132**, 259–75.

Glaessner, M. F. (1978). The oldest Foraminifera. *Bull. Bur. Min. Res.*, **192**, 61–5.

Glaessner, M. F. (1984). *The Dawn of animal life*. Cambridge University Press.

Grabau, A. W. (1900). Palaeontology of the Cambrian terranes of the Boston Basin. *Occ. Pap. Boston natur. Hist. Soc.*, **4**, 601–94.

Grieg, D. C., Wright, J. E., Hains, B. A., & Mitchell, G. H. (1968). Geology of the country around Church Stretton, Craven Arms, Wenlock Edge and Brown Clee. *Mem. geol. Surv. GB*, **166**, 379 pp.

Hamdi, B., Brasier, M. D., & Jiang Zhiwen. A succession of early skeletal microfaunas from Precambrian–Cambrian boundary strata, Elburz Mountains, Iran. *Geol. Mag.*, (submitted).

He Ting-gui, Pei Fang, & Fu Guang-Hong. (1984). Some small shelly fossils from the Lower Cambrian Xinji Formation in Fangcheng County, Henan Province. *Acta Palaeont. Sinica*, **23**, 350–7.

Hill, D. (1972). Archaeocyatha. In *Treatise on invertebrate paleontology*, (ed. C. Teichert), 158 pp. Part E, Vol. 1 Geological Society of America & University of Kansas Press.

Hinz, I. (1983). Zur unterkambrischen Mikrofauna von Comley und Umgebung in Shropshire. Unpublished Diploma Thesis, Bonn University, pp. 1–55.

Hinz, I. (1987). The Lower Cambrian microfauna of Comley and Rushton, Shropshire/England. *Palaeontographica*, **A198**, 41–100.

Hupé, P. (1952). Contribution à l'étude du Cambrian inférieur et du Pré-Cambrian III de l'Anti-Atlas Marocain. *Notes et Mém. Serv. Min. Carte géol. Maroc*, **103**. (Issued 1953).

Hupé, P. (1958). Discussion. In *Les relations entre Précambrien et Cambrien*. Central National de la Recherche Scientifique, Paris.

Hupé, P. (1960). Sur le Cambrien inférieur du Maroc. *21st International Geological Congress*, Norden, at Copenhagen. VIII, 75–85.

Hsü, K. J., Oberhansli, H., Gao, J. Y., Sun Shu, Chen Haihong, & Krahenbühl, U. (1985). 'Strangelove ocean' before the Cambrian explosion. *Nature*, **316**, 809–11.

Jell, P. (1975). The abathochroal eye of *Pagetia*, a new type of trilobite eye. In *Evolution and morphology of the Trilobita, Trilobitoidea and Merostomata*, (ed. A. Martinsson). pp. 33–43. Fossils & Strata, 4.

Jiang Zhiwen (1980). The Meishucun Stage and fauna of the Jinning County, Yunnan. *Bull. Chin. Acad. geol. Sci.*, *Ser. I*, **2**, 75–92. [In Chinese.]

Jiang Zhiwen (1985). On the genus *Tannuolina. Acta Micropalaeont. Sinica*, **2**, 231–8.

Kauffman, E. G. (1986). High resolution event stratigraphy: regional and global Cretaceous bio-events. In *Global bio-events*, (ed. O. H. Walliser), pp. 279–336. Lecture Notes Earth Sci. **8**, Springer-Verlag, Berlin.

Khomentovsky, V. V. (1986). The Vendian System of Siberia and a standard stratigraphic scale. *Geol. Mag.*, **123**, 333–48.

Khomentovsky, V. V. & Repina, L. N. (1965). *Lower Cambrian stratotypes of Siberia*. Akad. Nauk, Moscow.

Kirschvink, J. L. & Rozanov, A. Yu. (1984). Magnetostratigraphy of Lower Cambrian strata from the Siberian Platform: a palaeomagnetic pole and a preliminary polarity time-scale. *Geol. Mag.* **121**, 189–203.

Knoll, A. H. & Swett, K. (1987). Micropalaeontology across the Precambrian–Cambrian boundary in Spitsbergen. *J. Paleont.*, **61**, 898–926.

Kumar, G., Bhatt, D. K., & Raina, B. K. (1987). Skeletal microfauna of Meishucunian and Qiongzhusian (Precambrian–Cambrian boundary) age from the Ganga Valley, Lesser Himalaya, India. *Geol. Mag.*, **124**, 167–71.

Lambert, I. B., Walter, M. R., Zang Wenlong, Lu Songnian, & Ma Guogan (1987). Palaeoenvironment and carbon isotope stratigraphy of the Yangtze Platform. *Nature*, **325**, 140–2.

Landing, E. (1984). Skeleton of lapworthellids and the suprageneric classification of tommotiids (early and middle Cambrian phosphatic problematica). *J. Paleont.*, **58**, 1380–98.

Landing, E. (in press). Lower Cambrian of Eastern Massachusetts: stratigraphy and small shelly fossils. *J. Paleont.*

Landing, E. & Brett, C. E. (1982). Lower Cambrian of eastern Massachusetts: microfaunal sequence and the oldest known borings. *Geol. Soc. Am., Abstr. Prog.*, **14**, p. 35.

Landing, E., Nowlan, G. S., & Fletcher, T. P. (1980). A microfauna associated with early Cambrian trilobites of the *Callavia* Zone, northern Antigonish highlands, Nova Scotia. *Can. J. Earth Sci.*, **17**, 400–18.

Laurie, J. R. (1986). Phosphatic fauna of the early Cambrian Todd River Dolomite, Amadeus Basin, central Australia. *Alcheringa*, **10**, 431–54.

Laurie, J. R. & Shergold, J. H. (1985). Phosphatic organisms and the correlation of Early Cambrian carbonate formations in central Australia. *BMR J. Aust. Geol. Geophys.*, **9**, 83–9.

Lendzion, K. (1983*a*). Biostratygrafia osadów kambru w polskiej części platformy wschodnioeuropejskiej. *Kwartalník Geologiczný*, **27**, 669–94.

Lendzion, K. (1983*b*). [The development of the Cambrian platform deposits in Poland.] Prace Instytutu Geologicznego, Warszawa, **CV**, 55 pp. [In Polish with English summary.]

Liñan, E. (1978). *Bioestratigrafia de la Sierra de Cordoba*. Ph.D. Thesis, Univ. Granada.

Liñan, E. & Sdzuy, K. (1978). A trilobite from the Lower Cambrian of Cordoba (Spain) and its stratigraphical significance. *Senckenbergiana Lethaea*, **59**, 387–99.

Lindström, M. (1977). White sand crosses in turbated siltstone bed, basal Cambrian, Kinnekulle, Sweden. *Geologica et Palaeontologica*, **11**, 1–5.

Lipps, J. H. & Sylvester, A. G. (1968). The enigmatic Cambrian fossil *Volborthella* and its occurrence in California. *J. Paleont.*, **42**, 329–36.

Lochman-Balk, C. (1956). Stratigraphy, paleontology and paleogeography of the *Elliptocephala asaphoides* strata in Cambridge and Hoosick Quadrangles, New York. *Bull. geol. Soc. Am.*, **67**, 1331–96.

Lochman-Balk, C. (1971). The Cambrian of the craton of the United States. In *Cambrian of the New World*, (ed. C. H. Holland), pp. 79–168. John Wiley, London.

Lochman-Balk, C. & Wilson, J. L. (1958). Cambrian biostratigraphy in North America. *J. Paleont.*, **32**, 312–50.

Luo Huilin, Jiang Zhiwen, Wu Xiche, Song Xueliang, & Ouyang Lin (1982). [*The Sinian–Cambrian boundary in Eastern China.*] People's Publishing House, Beijing, China, 265 pp. [In Chinese.]

Luo Huilin *et al*. (1984). *Sinian–Cambrian boundary stratotype section at Meishucun, Jinning, Yunnan, China*. People's Publishing House, Yunnan, China, 154 pp.

McMenamin, M. A. S. (1987). Lower Cambrian trilobites, zonation and correlation of the Puerto Blanco Formation, Sonora, Mexico. *J. Paleont.*, **61**, 738–49.

McMenamin, M. A. S., Awramik, S. M., & Stewart, J. H. (1983). Precambrian–Cambrian transition problem in western North America: Part II. Early Cambrian skeletonized fauna and associated fossils from Sonora, Mexico. *Geology*, **11**, 227–30.

Magaritz, M., Holser, W. T., & Kirschvink, J. L. (1986). Carbon-isotope events across the Precambrian–Cambrian boundary on the Siberian Platform. *Nature*, **320**, 258–9.

Major, H. & Winsnes, T. S. (1955). Cambrian and Ordovician fossils from Sorkapp Land, Spitsbergen. *Norsk Polarinst. Skrift*, **106**, 47 pp.

Mambetov, A. M. (1977). On revision of the genus *Helenia*. *Paleont. J.*, **10**, 90–5.

Martinsson, A. (1974). The Cambrian of Norden. In *Cambrian of the British Isles, Norden and Spitsbergen*, (ed. C. H. Holland), pp. 185–283. John Wiley, London.

Matthews, S. C. (1973). Lapworthellids from the Lower Cambrian Strenuella Limestone at Comley, Shropshire. *Palaeontology*, **16**, 139–48.

Matthews, S. C. & Missarzhevsky, V. V. (1975). Small shelly fossils of late Precambrian and early Cambrian age: a review of recent work. *J. geol. Soc. Lond.*, **131**, 289–304.

Meshkova, N. P. (1969). [To the question of the palaeontological characteristics of the Lower Cambrian sediments of the Siberian Platform. In *Biostratigraphy and fauna of the Lower Cambrian of Siberia and the Far East*, (ed. I. T. Zhuravleva)], pp. 158–74, Nauka, Moscow. [In Russian.]

Missarzhevsky, V. V. (1966). [The first finds of *Lapworthella* in the Lower Cambrian of the Siberian Platform]. *Paleont. Zhurnal*, 1966, 13–18. [In Russian.]

Missarzhevsky, V. V. (1969). [Description of hyolithids, gastropods, hyolithelminths, camenides and forms of an obscure taxonomic position.] In *The Tommotian Stage and the Cambrian lower boundary problem*, (A. Yu. Rozanov *et al.*). Trudy Geol. Inst. Nauka, Moscow, 206, pp. 93–175. [In Russian; English translation, US Dept. of the Interior, 1981.]

Missarzhevsky, V. V. (1973). [Conodont-shaped organisms from the Precambrian and Cambrian boundary strata of the Siberian Platform and Kazakhstan.] In *Palaeontological and biostratigraphical problems in the Lower Cambrian of Siberia and the Far East*, (ed. I. T. Zhuravleva), pp. 53–7, pls 9–10. Nauka Publishing House, Novosibirsk. [In Russian.]

Missarzhevsky, V. V. (1977) [Conodonts(?) and phosphatic problematica from the Cambrian of Mongolia and Siberia]. In *Palaeozoic invertebrata of Mongolia*. Joint Soviet-Mongolian Palaeontological Expedition, Transactions (ed. L. P. Tartarinov *et al.*), pp. 10–19. [In Russian.]

Missarzhevsky, V. V. (1982). [Subdivision and correlation of the Precambrian–Cambrian boundary beds using some groups of the oldest skeletal organisms.] *Byulleten' Moskovskogo Obshchestva Ispytatelei Prirody, Otdelenie Geologii*, **57**, (5), 52–67. [In Russian.]

Missarzhevsky, V. V. (1983). [Stratigraphy of oldest Phanerozoic deposits of Anabar massif.] *Soviet Geol.* **9**, *Stratigr. & Palaeogeogr.*, 62–73. [In Russian.]

Missarzhevsky, V. V. & Mambetov, A. J. (1981). [Stratigraphy and fauna of Cambrian and Precambrian boundary beds of Maly Karatau.] *Trudy Akademii Nauka SSSR, Moscow*, **326**, 1–90. [In Russian.]

Moczydłowska, M. & Vidal, G. (1986). Lower Cambrian acritarch zonation in southern Scandinavia and southeastern Poland. *Geol. Fören. i Stockholm Förhandl.*, **108**, 201–33.

Müller, K. J. (1959). Kambrische Conodonten. *Z. Deutschen Geol. Gesell.*, **111**, 434–85.

Narbonne, G. M., Myrow, P. M., Landing, E., & Anderson, M. (1987). A candidate stratotype for the Precambrian–Cambrian boundary, Fortune Head, Burin Peninsula, southeastern Newfoundland. *Can. J. Earth Sci.*, **24**, 1277–93.

Nelson, C. (1978). Late Precambrian–Early Cambrian stratigraphic and faunal succession of eastern California and the Precambrian–Cambrian boundary. *Geol. Mag.*, **115**, 121–6.

North, F. K. (1971). The Cambrian of Canada and Alaska. In *Cambrian of the New World*, (ed. C. H. Holland), pp. 219–324. John Wiley, London.

Notholt, A. G. & Brasier, M. D. (1986). Chapter 7. Regional review: Europe. In *Proterozoic and Cambrian phosphorites*, (eds P. J. Cook & J. H. Shergold). Cambridge University Press.

Nowlan, G. S., Narbonne, G. M., & Fritz, W. H. (1985). Small shelly fossils and trace fossils near the Precambrian–Cambrian boundary in the Yukon Territory, Canada. *Lethaia*, **18**, 233–56.

Odin, G. S. *et al.* (1983). Numerical dating of Precambrian–Cambrian boundary. *Nature*, **301**, 21–3.

Orlowski, S. (1985). Lower Cambrian and its trilobites in the Holy Cross Mts. *Acta geol. Polonica*, **35**, 231–50.

Palmer, A. R. (1969). Cambrian trilobite distributions in North America and their bearing on the Cambrian palaeogeography of Newfoundland. In *North Atlantic geology and continental drift*, (ed. G. M. Kay), pp. 139–44. Mem. Am. Assoc. Pet. Geol., 12.

Palmer, A. R. (1971). The Cambrian of the Great Basin and adjacent areas, western United States. In *Cambrian of the New World*, (ed. C. H. Holland), pp. 1–78. John Wiley, London.

Palmer, A. R. (1974). Search for the Cambrian World. *Am. Scient.*, **62**, 216–24.

Palmer, A. R. & Rozanov, A. Yu. (1976). Archaeocyatha from New Jersey: evidence for an intra-Cambrian unconformity in the north-central Appalachians. *Geology*, **4**, 773–4.

Parrish, J. T., Ziegler, A. M., Scotese, C. R., Humphreville, R. G., & Kirschvink, J. L. (1986). Proterozoic and Cambrian phosphorites—specialist studies: Early Cambrian palaeogeography, palaeoceanography and phosphorites. In *Proterozoic and Cambrian phosphorites*, (eds P. J. Cook & J. H. Shergold), pp. 280–94. Cambridge University Press.

Peel, J. S. & Larsen, N. H. (1984). *Hadimopanella apicata* from the Lower Cambrian of western North Greenland. *Rapp. Grønlands geol. Unders.*, **121**, 89–96.

Piper, J. D. (1982). The Precambrian palaeomagnetic record: the case for the Proterozoic supercontinent. *Earth planet. Sci. Lett.*, **59**, 61–89.

Pojeta Jr, J. & Runnegar, B. (1976). The palaeontology of rostroconch mollusks and the early history of the phylum Mollusca. *Prof. Pap. US geol. Surv.*, **968**, 88 pp., 54 pls.

Poulsen, C. (1932). The Lower Cambrian faunas of East Greenland. *Medd. Grønland*, **87**, (6).

Poulsen, C. (1967). Fossils from the Lower Cambrian of Bornholm. *Kong. Danske Vidensk. Selsk. Mat.–Fys. Medd.* **36**, (2), 48 pp.

Poulsen, V. (1978). The Precambrian–Cambrian boundary in parts of Scandinavia and Greenland. *Geol. Mag.*, **115**, 131–6.

Qian Yi, & Zhang Shi-Ben (1983). Small shelly fossils from the Xihaoping Member of the Tongying Formation in Fangxian county of Hubei Province and their stratigraphical significance. *Acta palaeont. Sin.*, **22**, 82–94. [In Chinese with English summary.]

Qian Yi, Yu Wen, Liu Diyong, & Wang Zonghe (1985). Restudy on the Precambrian–Cambrian boundary section at Meishucun of Jinning, Yunnan. *Kexue Tongbao*, **30**, 1086–90. [In English.]

Raw, F. (1936). Mesonacidae of Comley in Shropshire, with a discussion of classification within the family. *Q. J. geol. Soc. Lond.*, **92**, 236–93.

Repina, L. N. (1972). [Trilobites from horizons of the Lower Cambrian sections of the River Sukharika (Igarka Region). In *Problems of biostratigraphy and palaeontology of the Lower Cambrian of Siberia* (ed. I. T. Zhuravleva)], pp. 184–216. [In Russian.]

Repina, L. N. (1981). Trilobite biostratigraphy of the Lower Cambrian stages in Siberia. *US geol. Surv. Open-File Rep.*, **81–743**, 173–80.

Robison, R. A. (1972). Mode of life of agnostid trilobites. *24th Int. geol. Congr.*, Montreal, **7**, 33–40.

Robison, R. A., Rosova, A. V., Rowell, A. J., & Fletcher, T. P. (1977). Cambrian boundaries and divisions. *Lethaia*, **10**, 257–62.

Rozanov, A. Yu. (1973). [*Regularities in the morphological evolution of regular archaeocyathans and the problems of the Lower Cambrian stage division.*] Nauka, Moscow. [In Russian.]

Rozanov, A. Yu. (1984). The Precambrian–Cambrian boundary in Siberia. *Episodes*, **7**, 20–24.

Rozanov, A. Yu. & Debrenne, F. (1974). Age of archaeocyathid assemblages. *Am. J. Sci.*, **274**, 833–48.

Rozanov, A. Yu. *et al.* (1969). [*The Tommotian Stage and the Cambrian lower boundary problem.*] Trudy Geol. Inst., Nauka, Moscow, 206. [In Russian; English translation, US Dept. of the Interior, 1981.]

Rozanov, A. Yu. & Sokolov, B. S. (1984). [*Lower Cambrian stage subdivision. Stratigraphy.*] Akademii Nauk SSSR: Iztdatelstov 'Nauka', Moscow. [In Russian.]

Rushton, A. W. A. (1966). The Cambrian trilobites from the Purley Shales of Warwickshire. *Palaeontogr. Soc. Monogr.*, **511**.

Rushton, A. W. A. (1974). The Cambrian of Wales and England. In *Cambrian of the British Isles, Norden and Spitsbergen*, (ed. C. H. Holland), pp. 43–122. John Wiley, London.

Rushton, A. W. A. (1978). Appendix 3. Description of the macrofossils from the Dolwen Formation. *Bull. geol. Surv. GB*, **61**, 46–8.

Schwarzbach, M. (1939). Die Oberlausitzer *Protolenus*-fauna; Weitere Funde aus dem schlesischen Kambrium und ihre allgemeine Bedeutung. *Jb. Preuss. geol. Landesanst. Berlin, für 1938*, **59**, 769–85.

Scott, G. (1985). Homotaxy and biostratigraphical theory. *Palaeontology*, **28**, 777–82.

Sdzuy, K. (1962). Trilobiten aus dem Unter-Kambrium der Sierra Morena (S-Spanien). *Senckenbergiana Lethaea*, **43**, 181–229.

Sdzuy, K. (1969). Unter- und mittelkambrische Porifera (Chancelloriida und Hexactinellida). *Paläont. Z.*, **43**, 115–47.

Sdzuy, K. (1971). Acerca de la correlacion del Cambrico Inferior en la Peninsula Iberica. *Geologica Economica, Madrid* **2**, 769–82.

Sdzuy, K. (1978). The Precambrian–Cambrian boundary beds in Morocco (preliminary report). *Geol. Mag.*, **115**, 83–94.

Sedgwick, A. & Murchison, R. I. (1835). On the Silurian and Cambrian Systems, exhibiting the order in which the older sedimentary strata succeeded each other in England and Wales. *London & Edinburgh phil. Mag.*, **7**, 483–5.

Sepkoski Jr, J. J. (1979). A kinetic model of Phanerozoic taxonomic diversity. II. Early Phanerozoic families and multiple equilibria. *Paleobiology*, **5**, 222–51.

Shah, S. (1982). Cambrian stratigraphy of Kashmir and its boundary problems. *Precambr. Res.*, **17**, 87–98.

Shergold, J. H. & Brasier, M. D. (1986). Biochronology of Proterozoic and Cambrian phosphorites. In *Proterozoic and Cambrian phosphorites*, (eds P. J. Cook & J. H. Shergold), pp. 295–326, Cambridge University Press.

Signor, P. W. & Mount, J. F. (1986). Lower Cambrian stratigraphic paleontology of the White-Inyo Mountains of Eastern California and Esmeralda County, Nevada. In *Natural history of the White-Inyo Range, eastern California and western Nevada and high altitude physiology*, (eds C. A. Hall Jr. & D. J. Young), pp. 6–15. Univ. California, White Mountain Res. Stat. Symp., Vol. 1.

Signor, P. W., Mount, J. M., & Onken, B. R. (1987). A pre-trilobite shelly fauna from the White-Inyo region of eastern California and western Nevada. *J. Paleontol.*, **61**, 425–38.

Singh, P. & Shukla, D. S. (1981). Fossils from the Lower Tal: their age and its bearing on the stratigraphy of Lesser Himalaya. *Geosci. J.*, **11**, 157–76.

Smith, J. D. D. & White, D. E. (1963). Cambrian trilobites from the Purley Shales. *Palaeontology*, **6**, 397–407.

Sokolov, B. S. & Fedonkin, M. A. (1984). The Vendian as the terminal system of the Precambrian. *Episodes*, **7**, 12–19.

Sokolov, B. S. & Zhuravleva, I. T. (1983). [*Lower Cambrian stage subdivision of Siberia. Atlas of fossils.*] Trans. Inst. Geol. & Geophys., 558. Akademii Nauk SSSR: Izdatelstvo 'Nauka', Moscow. [In Russian.]

Taylor, M. E. (1966). Precambrian mollusc-like fossils from Inyo County, California. *Science*, **153**, 198–201.

Theokritoff, G. (1964). Taconic stratigraphy in northern Washington County, New York. *Bull. geol. Soc. Am.*, **75**, 171–90.

Theokritoff, G. (1979). Early Cambrian provincialism and biogeographic boundaries in the North American region. *Lethaia*, **12**, 281–95.

Theokritoff, G. (1981). Early Cambrian faunas of eastern New York State—taphonomy and ecology. *US geol. Surv. Open-File Rep.*, **81–743**, 228–30.

Tucker, M. E. (1985). Carbon isotope excursions in Precambrian/Cambrian boundary beds, Morocco. *Nature*, **319**, 48–9.

Urbanek, A. & Rozanov, A. Yu. (1983). *Upper Cambrian and Cambrian palaeontology of the East-European Platform.* Wydawnictwa Geologiczne, Warszawa.Val'kov, A. K. (1982). [*Biostratigraphy of the Lower Cambrian of the eastern Siberian Platform.*] Nauka, Moscow, 91 pp. [In Russian.]

Val'kov, A. K. (1982). [*Biostratigraphy of the Lower Cambrian of the eastern Siberian Platform.*] Nauka, Moscow, 91 pp. [In Russian.]

Vidal, G. (1981). Lower Cambrian acritarch stratigraphy in Scandinavia. *Geol. Fören. i Stockholm Förhandl.* **103**, 183–92.

Vidal, G. & Knoll, A. H. (1983). Proterozoic plankton. *Mem. geol. Soc. Am.*, **161**, 265–77.

Voronin, Yu. I. *et al.* (1982). [*The Precambrian/Cambrian boundary in the geosynclinal areas (the reference section of Salany-Gol, MPR).*] Trans. Joint Soviet–Mongolian Palaeont. Expedition, 18. Nauka Izdatelstvo, Moscow. [In Russian.]

Walliser, O. H. (1986). Towards a more critical approach to bio-events. In *Global bio-events*, (ed. O. H. Walliser), pp. 5–16. Lecture Notes in Earth Sci., 8, Springer-Verlag, Berlin.

Whittington, H. B. (1979). Early arthropods, their appendages and relationships. In *The origin of major invertebrate groups*, (ed. M. R. House). Systematics Assoc. Spec. Vol., 12, pp. 253–68. Academic Press, London.

Wiggett, G. J. (1977). Late Proterozoic–Early Cambrian biostratigraphy, correlation, and paleoenvironments, White-Inyo facies, California–Nevada. *Spec. Rep. California Div. Min. Geol.*, **129**, 87–92.

Wrona, R. (1982). Early Cambrian phosphatic microfossils from southern Spitsbergen (Horsnund region). *Palaeont. Polonica*, **43**, 9–16.

Wrona, R. (1987). Cambrian micro fossil *Hadimopanella* Gedik from glacial erratics in West Antarctica. *Paleont. Polonica*, **49**, 39–48.

Xiang Liwen *et al.* (1981). *Stratigraphy of China, Volume 4. The Cambrian System of China*. Geological Publishing House, Beijing.

Xing Yusheng & Luo Huilin (1984). Precambrian–Cambrian boundary candidate, Meishucun, Jinning, Yunnan, China. *Geol Mag.*, **121**, 143–54.

Xing Yusheng, Ding Qixiu, Luo Huilin, He Ting-gui, & Wang Yangeng (1983). [The Sinian–Cambrian boundary of China.] *Bull. Inst. Geol., Chin. Acad. geol. Sci.*, **10**. [In Chinese.]

Xing Yusheng, Ding Qixui, Luo Huilin, He Ting-gui, & Wang Yangeng (1984). The Sinian–Cambrian boundary of China and its related problems. Geol. Mag., **121**, 155–70.

Yang Xianhe, He Yuanxiang, & Deng Shouhe (1983). [On the Sinian–Cambrian boundary and the small shelly fossil assemblages in Nanjiang area, Sichuan.] *Bull. Chengdu Inst. Geol. Min. Res., Chin. Acad. geol. Sci.*, **4**, 91–105. [In Chinese]

Yin Jicheng, Ding Lianfang, He Tinggui, Li Shilin, & Shen Lijuan (1980). *The Palaeontology and sedimentary environment of the Sinian System in Emei–Ganluo area, Sichuan*. Sichuan, The People's Publishing House, 231 pp.

Yochelson, E. I. (1977). Agmata, a proposed extinct phylum of early Cambrian age. *J. Paleont.*, **51**, 437–54.

Yochelson, E. I. (1981). A survey of *Salterella* (Phylum Agmata). *Open File Report. U.S. geol. Surv.*, **81–743**, 244–8.

Yochelson, E. I. (1983) *Salterella* (Early Cambrian: Agmata) from the Scottish Highlands. *Palaeontology*, **26**, 253–60.

Yochelson, E. I., Henningsmoen, G., & Griffin, W. L. (1977). The early Cambrian genus *Volborthella* in southern Norway. *Norsk geol. Tiddskr.*, **57**, 133–51.

Yue Zhao (1987). The discovery of *Tannuolina* and *Lapworthella* from Lower Cambrian in Meishucun (Yunnan) and Maidiping (Sichuan) sections. *Prof. Pap. Stratigr. Palaeont.*, **16**, 173–80. [In Chinese].

Zamarreño, I. & Debrenne, F. (1977). Sédimentologie et biologie des constructions organogénés du Cambrien inférieur du Sud de l'Espagñe. *Mémoires BRGM.*, **89**, 49–61.

Zhou Benhe & Xiao Ligong (1984). Early Cambrian monoplacophorans and gastropods from Huainan and Huoqiu Counties, Anhui Province. *Prof. Pap. Stratigr. Palaeont.*, **13**, 125–40. Geological Publishing House, Beijing.

Zhuravlev, A. Yu. (1986). Evolution of archaeocyathans and palaeobiogeography of the early Cambrian. *Geol. Mag.*, **123**, 377–85.

Zhuravleva, I. T. (1960). [*Archaeocyatha of the Siberian Platform*.] Akad. Nauka, Moscow. [In Russian.]

Zhuravleva, I. T. (1970). Porifera, Sphinctozoa, Archaeocyathi—their connections. *Symp. zool. Soc. Lond.*, **25**, 41–59.

Zhuravleva, I. T. (1974). [Biology of Archaeocyatha. In *Stratigraphical studies, jubilee book in honour of academician B. S. Sokolov.*], pp. 107–29. Izdatelstvo. Nauka, Moscow. [In Russian.]

Zhuravleva, I. T. (1975). [Description of the palaeontological characteristics of the Nemakit–Daldyn Horizon and its possible equivalents in the territory of the Siberian Platform. In *Equivalents of the Vendian Complex in Siberia*, (Eds B. S. Sokolov & V. V. Khomentovsky)], pp. 62–100. Trudy Inst. Geol. Geofis., Akad, Nauk, SSSR, Moscow, 232. [In Russian.]

Ziegler, A. M., Scotese, C. R., McKerrow, W. S., Johnson, M. E., & Bambach, R. K. (1979). Paleozoic paleogeography. *Ann. Rev. Earth planet. Sci.*, **7**, 473–502.

Trace fossils

T. P. Crimes

8.1. INTRODUCTION

It is becoming increasingly recognized that trace fossils may be of considerable importance in correlating latest Precambrian and early Cambrian strata. The best preserved sequences spanning the Precambrian–Cambrian boundary are mostly composed of shallow-water clastic sediments, and it is precisely these lithologies in which trace fossils are most abundant. In contrast, small shelly fossils are almost totally restricted to carbonate rocks, and microfossils have a similar preference.

In the past, the main objections to using trace fossils for correlation have been their marked facies control and long time-ranges. It has, however, been shown recently that the facies control of trace fossils is much less marked in the Lower Cambrian than later (Crimes & Anderson 1985). The long time-ranges of most trace fossils limit their usefulness for correlation in Phanerozoic sequences, but the rapid evolution which took place at about the Precambrian–Cambrian boundary created a number of short-ranging aberrant forms which have potential for correlation. More importantly, at about this time, most of the long-ranging forms evolved, and it is their first appearance which provides the main criterion for correlation.

The possibilities of using trace fossils for correlation at this level were reviewed briefly by Crimes (1975, pp. 115–16), and more comprehensively by Alpert (1977). Further contributions have been made by Fedonkin (1979, 1980a,b) and Keller & Rozanov (1979). There has been much additional work in the last five years, and the current situation has been discussed by Crimes (1987).

Trace fossils have been described from sections spanning the Precambrian–Cambrian boundary in:

(1) Argentina,
(2) Australia,
(3) Canada,
(4) China,
(5) England,
(6) Greenland,
(7) India,
(8) Norway,
(9) Poland,
(10) South-west Africa (Namibia),
(11) Spain,
(12) Sweden,
(13) USA,
(14) USSR.

Trace-fossil-bearing sequences of this age are known to exist in many other countries, but are awaiting detailed description (e.g. Mexico).

The purpose of this review is to describe the trace-fossil occurrences in the well-documented sections and to discuss their usefulness for correlation. The absence of a formally agreed type section, and therefore an inability to define what is 'Precambrian' and what 'Cambrian', could make discussion of correlation between the various sections verbally cumbersome. The stratigraphical terms most widely used at present to cover this boundary interval are those employed for the Russian Platform. This is also an area where much trace-fossil work has already been accomplished. This review will therefore utilize the Russian subdivisions and use 'Vendian' for the latest Precambrian, 'Tommotian' for a transitional unit, and 'Atdabanian' for the lowest Cambrian. This is purely for convenience of description, and without prejudice to the Chinese boundary candidate and the stratigraphical terminology applied to it.

8.2. TRACE FOSSILS AND CORRELATION IN THE PRECAMBRIAN–CAMBRIAN BOUNDARY INTERVAL

A contrast between the paucity of trace fossils in the Precambrian and their abundance and diversity in the Cambrian was recognized by Seilacher (1956). He reviewed the world-wide literature and showed that while a few trace fossils, such as *Gordia* (= *Helminthoidichnites*) and *Planolites* occur in undoubted Precambrian rocks, many more are present in the Cambrian (e.g. *Corophioides*, *Dictyodora*, *Diplocraterion*, *Phycodes*, *Rhizocorallium*, *Scolicia*, *Skolithos*, and *Teichichnus*). He used this contrast to demonstrate an explosive evolution at the base of the Cambrian System and concluded that this trace-fossil change marks the beginning of the Cambrian Period better than any other time indicator.

Crimes & Crossley (1968) used the occurrence of *Oldhamia* and *Monocraterion* (= *Histioderma*) to infer that the Cahore Group of County Wexford, Eire, was early to middle Cambrian in age, rather than Precambrian as had been previously considered (e.g. Tremlett 1959). Similarly, in County Dublin, Eire, Crimes (1976) concluded that the occurrence of *Arenicolites*, *Granularia*, *Oldhamia*, and *Teichichnus* in the Bray Group indicated an early mid-Cambrian, rather than Precambrian, age. Both of these inferences have since been confirmed by the discovery of microfossils by the Geological Survey of Ireland.

Bergström (1970) used the occurrence of *Rusophycus parallelum* Bergström to indicate that the Hardeberga Quartzite of Sweden should be considered as Cambrian, rather than Eocambrian or Precambrian.

Daily (1972) found *Diplocraterion* and *Cruziana* to be common below the earliest trilobites and archaeocyathans in Australia, and considered such occurrences to indicate an early Cambrian rather than a Precambrian age.

In Argentina, Aceñolaza & Durand (1973) used the presence of trace fossils such as *Cochlichnus*, *Diplichnites*, *Helminthopsis*, and *Oldhamia* to assign an Eocambrian or Cambrian age to rocks previously considered to be Precambrian.

Crimes (1975, pp. 115–16) briefly reviewed the literature, and concluded that trace fossils are more common than body fossils in many late Precambrian–Lower Cambrian sequences. He suggested that this, together with the rapid evolution of the animals at that time, means that trace fossils should at least be considered in defining a base to the Cambrian System, and might, in fact, be most valuable in adequately locating that base in many sections.

Following a comprehensive review, Alpert (1977) concluded from the occurrences of trace fossils in sections spanning the Precambrian–Cambrian boundary, that each ichnogenus falls into one of three groups. Group 1 included those whose incoming was taken to indicate an early Cambrian age (e.g. *Cruziana*, *Diplocraterion*, *Phycodes*, *Plagiogmus*, *Rusophycus*, and *Teichichnus*). Group 2 included those whose ranges extended across the boundary and whose occurrence could not therefore be used to delineate it (e.g. *Curvolithus*, *Didymaulichnus*, *Gordia* (= *Helminthoidichnites*), Planolites, Scolicia, and *Skolithos*), while Group 3 included trace fossils known only from the late Precambrian (e.g. *Bunyerichnus*, *Buchholzbrunnichnus*, and *Archaeichnium*).

Alpert therefore suggested that the basal Cambrian boundary should be place at, or just below, the lowest horizon containing trilobite trace fossils or other trace fossils indicating a Cambrian age.

An extensive review of the literature, together with personal observations of many sections world-wide, led Crimes (1987) to recognize three zones with regard to the incoming of trace fossils, as shown in Table 8.1.

An analysis of the order of incoming of trace fossils in eleven sections world-wide, suggested that they appeared normally in approximately the order given below, but it must be appreciated that this is only a provisional ranking and will be subject to modification as detailed work on other sections becomes available.

(18)	*Cruziana*	youngest	
(17)	*Plagiogmus*		Atdabanian
(16)	*Rusophycus*		– – – – –?– – – – –
(15)	*Taphrhelminthopsis*		
(14)	*Diplocraterion*		
(13)	*Teichichnus*		
(12)	*Bergaueria*		Tommotian
(11)	*Phycodes*		
(10)	*Diplichnites*		
(9)	*Monomorphichnus*		– – – – –?– – – – –
(8)	*Scolicia s.l.*		
(7)	*Cochlichnus*		
(6)	*Skolithos*		
(5)	*Gordia*		Vendian
(4)	*Neonereites*		
(3)	*Arenicolites*		
(2)	*Didymaulichnus*		
(1)	*Planolites*	oldest	

Table 8.1. The three zones of Crimes (1987) for the incoming of trace fossils.

		Trace fossils probably restricted to this zone	Trace fossils first appearing in this zone
Zone I	(Upper Vendian)	*Bilinichnus, Harlaniella, Intrites, Nenoxites, Palaeopascichnus, Vendichnus, Vimerites.*	*Arenicolites, Cochlichnus, Didymaulichnus, Gordia, Neonereites, Planolites, Scolicia, Skolithos.*
Zone II	(Lower Tommotian)		*Diplocraterion, Phycodes, Teichichnus, Treptichnus*
Zone III	(Upper Tommotian– Lower Atdabanian)	*Astropolichnus, Plagiogmus, Taphrhelminthopsis circularis* Crimes *et al.*	*Cruziana, Diplichnites, Dimorphichnus, Rusophycus*

The trace fossils present in Vendian strata are illustrated in Fig. 8.1, and some of those which occur first in Tommotian and Atdabanian strata below the earliest trilobites are shown in Fig. 8.2.

This analysis can form the basis for discussion of the stratigraphical range and correlation of many of the Precambrian–Cambrian sequences which will now be described, but more data are still required and only a complete ichnofauna should be used to attempt correlations. Deductions based on only one or two ichnogenera might well be misleading.

8.3. TRACE-FOSSIL-BEARING PRECAMBRIAN–CAMBRIAN SEQUENCES

In the last fifteen years trace fossils have been recorded from many Precambrian–Cambrian sequences (Fig. 8.3). These will now be reviewed and correlations suggested, starting in the west in the Americas and proceeding eastwards to end at the proposed stratotype section in China.

8.3.1. Cassiar Mountains, in British Columbia, Canada

A measured section over 1 000 m thick has been described by Fritz & Crimes (1985). The Stelkuz Formation (609 + m) is overlain by the Boya Formation (400 m), while the earliest trilobites (*Fallotaspis* zone) are found in the succeeding Rosella Formation. The first appearance in this section of the different trace fossils is indicated in Table 8.2.

Two of the trace fossils in the lower part of the Stelkuz Formation first appear in Zone I, thereby suggesting that this part of the succession may be of Vendian or younger age. The middle part of the formation includes the first appearance of some more Zone I types (*Didymaulichnus, Neonereites*, and *Skolithos*) but has, in addition, *Taphrhelminthopsis circularis* Crimes *et al.* This Zone III trace fossil only occurs high up in other successions and suggests a Tommotian age. The occurrence of another Zone III trace fossil, *Plagiogmus*, in the upper part of the formation strongly suggests an age at least as young as late Tommotian. The abundance and diversity of trace fossils increases in the Boya Forma-

Table 8.2. The first appearance of different ichnogenera in the Cassiar Mountains, British Columbia, Canada.

Boya Formation (400 m)

Upper part	*Cruziana, Rusophycus*
Lower part	*Monomorphichnus, Phycodes, Treptichnus*

Stelkuz Formation (609 + m)

Upper part	*Diplocraterion, Plagiogmus, Teichichnus*
Middle part	*Didymaulichnus, Neonereites, Skolithos, Taphrhelminthopsis*
Lower part	*Chondrites, Helminthopsis, Gordia, Planolites*

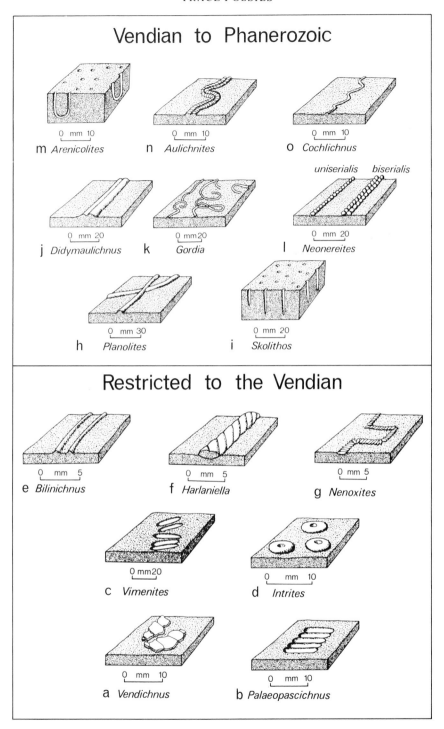

Fig. 8.1. The trace fossils present in late Precambrian (Vendian) strata.

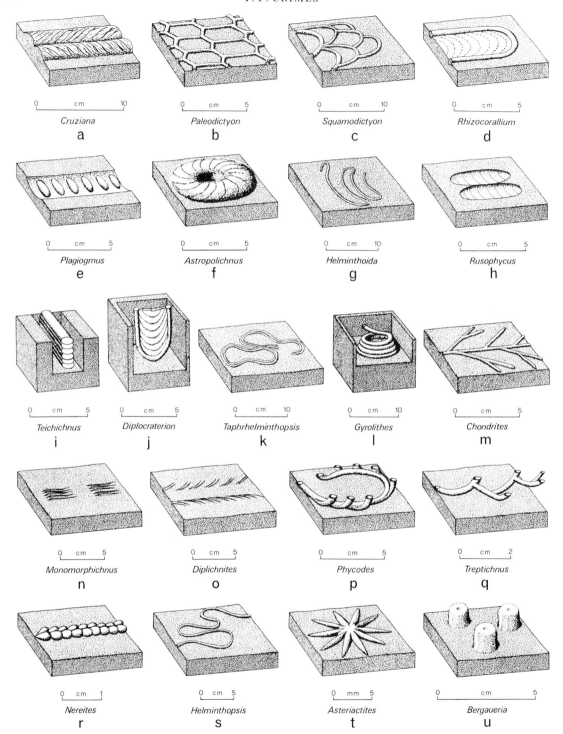

Fig. 8.2. Some of the trace fossils which first appear before the earliest trilobites in Tommotian and Atdabanian strata.

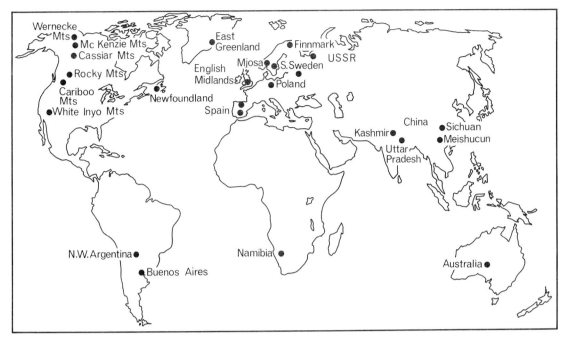

Fig. 8.3. Location of main sections yielding late Precambrian and Lower Cambrian trace fossils.

tion, and the occurrence of *Cruziana* and *Rusophycus* indicates that its upper part is very probably of early Atdabanian age. This is consistent with the first trilobites being present in the overlying Rosella Formation.

8.3.2. Cariboo Mountains, British Columbia

Young (1972) found trace fossils below the lowest trilobites (*Fallotaspis*) in the Cariboo Mountains. However, only *Chondrites*, ?*Didymaulichnus*, *Planolites*, and ?*Scolicia* were recorded and these are insufficient to permit any correlation.

8.3.3. Rocky Mountains, British Columbia

The Rocky Mountains section described by Young (1972) has a more abundant and diverse ichnofauna than that in the Cariboo Mountains—which are only 95 km away but were separated from it by the Rocky Mountain trench.

The first appearance of trace fossils in this section is shown in Table 8.3. More collecting in this area is necessary but the >1 200 m-thick Miette Group has only the Zone I *Didymaulichnus*, while the 1 300 m-thick McNaughton Formation has yielded the Zone I trace fossil *Planolites* in its lower part, and the Zone III forms *Cruziana*, *Diplichnites*, and *Rusophycus* in its middle part and again near the top, suggesting that, while the lower beds might be Vendian, a horizon high in the Tommotian or low in the Atdabanian is reached by about the middle of the McNaughton Formation. Indeed, the occurrence of *Cruziana* would favour an Atdabanian rather than a Tommotian age. The occurrence of *Nevadella* at the top of the formation infers that the highest strata at least, must be Atdabanian. Therefore this section would seem to be younger than suggested by Young (1972, fig. 2) who regarded almost all the McNaughton Formation as Precambrian. It is likely that the Vendian–Tommotian boundary is low down or at the base of the McNaughton Formation.

8.3.4. McKenzie Mountains, Northwestern Territories, Canada

Trace fossils have been described from the Sekwi Brook area, where they occur in an unnamed Late Proterozoic formation (Hofmann, 1981). They consist of *Gordia* and *Torrawangea*, both of which have been described from strata of similar age elsewhere. A more extensive ichnofauna has been cited by Fritz (1980) and the first appearance of the trace fossils in the section is indicated

Table 8.3. The first appearance of different ichnogenera in the Rocky Mountains, British Columbia, Canada.

McNaughton Formation (1 300 m)	
Upper part	*Rusophycus, Skolithos*
Middle part	*Cruziana, Diplichnites, Monomorphichnus*
Lower part	*Planolites*
Miette Group (1 200 m +)	
Upper part	*Didymaulichnus*

in Table. 8.4, which utilizes the Canadian Geological Survey's informal stratigraphical terminology based on 'map units'.

The first record of *Scolicia* and *Neonereites* this high up in this section, together with the absence of *Didymaulichnus* below *Phycodes*, indicates that further collecting is necessary here, but the occurrence of the Zone II trace fossils *Phycodes* and *Treptichnus* in the middle part of Map Unit 12 suggests that the Vendian–Tommotian boundary is probably at about this level, as indicated by Fritz (op. cit., figs. 7–13). The occurrence of *Rusophycus* in the overlying Map Unit 13 indicates an age at least as young as late Tommotian and this is consistent with the record of Atdabanian microfossils at the top of the unit (Fritz op. cit., p. 44).

8.3.5. Wernecke Mountains, Yukon

In this area Nowlan *et al.* (1985) have reported *Palaeophycus* from an horizon described as 'Map Unit 11', but *Cruziana* and *Rusophycus* from near the base of the overlying Vampire Formation. A disconformity separates the two units. The Vampire Formation contains a small shelly fauna in its lowest metre, which they tentatively suggest might be correlated with the pre-Tommotian Nemakit–Daldyn Horizon of Siberia. The occurrence of the Zone III traces *Cruziana* and *Rusophycus* at this level suggests, however, a late Tommotian or an early Atdabanian age. Clearly more

work is necessary in this area to allow firm conclusions, but the *Rusophycus* (Nowlan *et al.* op. cit., fig. 4a) appears to resemble *R. avalonensis* Crimes & Anderson, reported recently from a probable high Tommotian–Atdabanian level in Newfoundland (Crimes & Anderson 1985) and a similar horizon in North Spain (Seilacher 1970, fig. 7–3: reported as *Cruziana fasciculata* but see Crimes & Anderson, op. cit. p. 331).

At present therefore, a Tommotian to early Atdabanian age is preferred for the part of the Vampire Formation exposed here but, in the absence of further evidence, the age of the disconformably underlying 'Map Unit 11' must remain uncertain.

8.3.6. Newfoundland

An extensive collection of trace fossils has been described from below the lowest occurrence of trilobites in Newfoundland (Crimes & Anderson 1985). They occur in the 915 m-thick Chapel Island Formation and >47 m-thick Random Formation; their distribution is shown in Table 8.5.

The discovery of *Harlaniella* near the base of Member 1 indicates a Vendian age for at least part of this unit (Bengtson & Fletcher 1983), and this is consistent with the first appearance of the Zone I trace fossils *Gordia* and *Planolites*.

Member 2 has yielded an abundant and diverse ichnofauna, including the Zone 1 trace fossils *Arenico-*

Table 8.4. The first appearance of different ichnogenera in the McKenzie Mountains, Northwest Territories, Canada.

Map Unit 13 (150+ m)	
Lower part	*Bergaueria, Rusophycus, Scolicia, Teichichnus*
Map Unit 12 (c.900 m)	
Upper part	*Neonereites*
Middle part	*Gyrolithes, Didymaulichnus, Phycodes, Treptichnus*
Lower part	*Planolites*

Table 8.5. The first appearance of different ichnogenera in the Burin Peninsula, Newfoundland.

Random Formation (47+ m)	
	Cruziana, Diplocraterion, Paleodictyon, Rusophycus, Scolicia, Squamodictyon
Chapel Island Formation (c.915 m)	
Member 5 (c.150 m) II	*Palaeophycus*
Member 3 (150 m)	*Astropolichnus, Helminthoida, Helminthopsis, Hormosiroidea, Taphrhelminthopsis*
Member 2 (50 m)	*Arenicolites, Bergaueria, Cochlichnus, Curvolithus, Didymaulichnus, Gyrolithes, Monomorphichnus, Neonereites, Nereites, Phycodes, Protopaleodictyon, Skolithos.*
Member 1 (400 m)	*Buthotrephis, Gordia, Harlaniella, Planolites.*

lites, Cochlichnus, Didymaulichnus, Neonereites, and *Skolithos*, and also the Zone II *Phycodes*. The diversity, and occurrence of *Phycodes*, suggests that the Vendian–Tommotian boundary may be high in the 400 m-thick Member 1, rather than at the base of Member 2.

Member 3 also has a high abundance and diversity of trace fossils, and has additionally yielded the Zone III *Astropolichnus* and *Taphrhelminthopsis*, although the latter was not positively identified as the more diagnostic *T. circularis* which is probably restricted to the Upper Tommotian (Crimes & Anderson 1985, p. 334).

The Random Formation, which commences 315 m above the top of Member 3, is the next unit to yield an extensive ichnofauna, and includes the Zone III traces *Cruziana* and *Rusophycus*. The first trilobites are in the top of the Smith Point Limestone which is separated from the Random Formation by 30 m of sediments and a stratigraphical break. Therefore it would seem that the Random Formation is of late Tommotian to early Atdabanian age.

8.3.7. White-Inyo Mountains, California

The trace fossils from the Precambrian–Cambrian boundary interval in the White-Inyo Mountains have been described by Alpert (1976, 1977).

The lowest reported trace fossil was *Planolites* which was found in the Reed Dolomite (Table 8.6). The Upper Member of the overlying Deep Spring Formation contains the Zone II trace fossil *Phycodes* and the Zone III *Diplichnites* and *Rusophycus*, as well as *Monocraterion*, which is relatively complex and has so far only been reported at relatively high levels. This ichnofauna strongly suggests a late Tommotian age for at least the Upper Member. The relative paucity of trace fossils in the lower part of the Deep Spring Formation and, more particularly in the Reed Dolomite, is probably largely due to the absence of suitable facies, with most of the rock being carbonates. Therefore the Vendian–Tommotian boundary is probably well below the Upper Member of the Deep Spring Formation and may be near its base or in the underlying Reed Dolomite.

The presence of *Cruziana* in the lower part of the Campito Formation indicates a horizon at least as high as the Upper Tommotian and probably into the Lower Atdabanian, and this is supported by the occurrences of *Fallotaspis* in the upper part of that unit.

8.3.8. Argentina

Trace fossils are abundant and diverse in the Puncoviscana and laterally equivalent Sancho Formations of

Table 8.6. The first appearance of different ichnogenera in the White-Inyo Mountains, California.

Campito Formation (736 m) (*Fallotaspis* zone)	*Arthrophycus, Bergaueria, Cruziana, Dactyloidites, Teichichnus, Zoophycos*
Campito Formation (414 m) (below *Fallotaspis* zone)	*Belorhaphe, Cruziana, Cochlichnus, Helminthopsis*
Deep Spring Formation (500 m)	*Diplichnites, Monocraterion, Monomorphichnus, Phycodes, Scolicia, Skolithos, Rusophycus*
Reed Dolomite (650 m)	*Planolites*

north-western Argentina. These units have yielded no body fossils but are considered to range in age across the Precambrian–Cambrian boundary, with the trace fossils used as evidence for an early Cambrian age for the upper part of these successions (Aceñolaza & Durand 1973; Aceñolaza 1978; Aceñolaza & Toselli 1981). Measured sections are not available, but these formations have together yielded the following trace fossils: *Cochlichnus*, *Dimorphichnus*, *Diplichnites*, *Glockeria*, *Gordia*, *Helminthopsis*, *Nereites*, *Oldhamia*, *Planolites*, *Protichnites*, *Protovirgularia*, *Tasmanadia*, and *Torrawangea*.

Some of the trace fossils (e.g. *Diplichnites*, *Dimorphichnus*, *Glockeria*, and *Oldhamia*) indicate a horizon at least as high as Upper Tommotian and the occurrences may well span the Precambrian–Cambrian boundary. This ichnofauna is also of particular interest because of the occurrence of forms typical of both shallow and deep water. For example, *Dimorphichnus*, *Diplichnites*, *Protichnites*, and *Protovirgularia* are usually restricted to shallow water, while *Glockeria*, *Helminthopsis*, *Nereites*, and *Oldhamia* are typical of deep water. In the absence of detailed sections and facies analyses, it is not possible to say if there was an alternation of facies or whether it might be another example of the mixing of shallow- and deep-water forms in a tidal sequence, as has been reported from Newfoundland (Crimes & Anderson 1985).

In the Province of Buenos Aires, Poiré *et al.* (1984) have recently reported *Didymaulichnus* and *Palaeophycus* from the Sierras Bayas Formation, which they refer to the Precambrian, and *Arthrophycus*, *Corophioides*, *Crossopodia*, *Cruziana*, *Dimorphichnus*, *Isopodichnus*, *Nereites*, and *Palaeophycus* from the overlying Balcarce Formation which has also yielded *Plagiogmus* (Del Valle pers. comm. 1986) and *Phycodes* aff. *pedum* (Regalia & Herrera 1981), and is tentatively ascribed to the Cambrian. The occurrence of *Cruziana* and *Dimorphichnus* suggests a horizon at least as high as Upper Tommotian and more probably Atdabanian or younger, for part of the formation.

8.3.9. Greenland

In East Greenland there is a succession of approximately 500 m between the late Precambrian tillite and the upper part of the Lower Cambrian. No trace fossils or body fossils, except algal stromatolites, have been found in the two lowermost units—the 0–300 m-thick Canyon Formation or the 0–25 m-thick Spiral Creek Formation. The earliest trace fossils are in the unconformably overlying Kloftelv Formation (70–75 m thick), where Cowie & Spencer (1970) reported *Skolithos linearis* Haldeman in loose boulders. Trace fossils are,

however, more abundant and diverse in the succeeding Lower Bastion Formation (50–55 m-thick), where they found *Monomorphichnus* (= Arthropod scratch-marks, op. cit., pl. 2a,b), *Phycodes*, *Planolites*, *Scolicia*, and *Skolithos*. They also mention (op. cit., p. 94) a 'possible arthropod walking track (*Diplichnites*)'. The Upper Bastion Formation has the first body fossils, including brachiopods, gastropods, hyoliths, bradoriids, and eodiscids plus olenellid trilobites. Trace fossils appear again in the overlying Ella Island Formation, and *Cruziana*, *Diplichnites*, *Plagiogmus*, and *Skolithos* are reported, together with varied body fossils.

The occurrence of the Zone II trace fossil *Phycodes* and Zone III *Diplichnites* suggests that the Lower Bastion formation is at least as high as Lower Tommotian and is very probably higher. This is consistent with the diverse body-fossil fauna in the overlying Upper Bastion formation, implying an Atdabanian age. The base of the Tommotian is probably therefore in or below the Kloftelv Formation.

8.3.10. Europe

Finnmark, North Norway

In Finnmark, Banks (1970) has described a trace-fossil-bearing sequence which lies between the late Precambrian tillites and fossiliferous Lower Cambrian in the Tanafjord area.

The stratigraphy and relevant trace-fossil occurrences are summarized in Table 8.7.

The trace fossils from the Innerelv and Manndraperelv Members are all known to range back to the late Precambrian (Vendian). The overlying Lower Breivik Member, however, includes the Zone I *Cochlichnus*, together with the Zone II *Phycodes* and *Treptichnus*. *Rusophycus* is mentioned (Banks 1970, p. 28), but described only as 'paired sets of claw impressions' and not figured. This may be a primitive form of *Rusophycus* but this is not clear from the information given. The occurrence of *Phycodes* and *Treptichnus* does, however, incidate a horizon at least as high as the Lower Tommotian. *Rusophycus* would suggest a younger age. The occurrence of the Zone III traces *Cruziana*, *Dimorphichnus*, *Plagiogmus*, and *Rusophycus* in the Lower Duolbasgaissa Member indicates a late Tommotian or more probably an early Atdabanian age. This is consistent with occurrences of the trilobite *Holmia* in about the middle of the Upper Duolbasgaissa Member.

The trace fossils therefore suggest that the Vendian–Tommotian boundary is at least as low as the base of the Breivik Member and may well be lower. The Breivik Member is probably early Tommotian and the Lower Duolgasgaissa Member late Tommotian to early Atda-

Table 8.7. The first appearance of ichnogenera in Finnmark, North Norway.

Digermul group

Upper Duolbasgaissa Member (300 m)	*Rhizocorallium, Syringomorpha*
Lower Duolbasgaissa Member (210 m)	*Cruziana, Dimorphichnus, Diplocraterion,*
	Plagiogmus, Rusophycus
Upper Breivik Member (350 m)	*Gyrolithes, Teichichnus*
Lower Breivik Member (250 m)	*Cochlichnus, Phycodes, Treptichnus*
Manndraperelv Member (190 m)	*Curvolithus, Planolites*
Innerelv Member (275 m)	*Arenicolites, Skolithos*

banian in age. The Tommotian–Atdabanian boundary would then probably be in the Lower Duolbasgaissa Member.

South Norway

In the Mjøsa district of South Norway, *Diplocraterion, Monocraterion,* and *Skolithos* occur within the Ringsaker Quartzite (Føyn & Glaessner 1979, fig. 3), above which there is a stratigraphical break before beds with *Platysolenites* are encountered. The trace fossils indicate a Tommotian age for the quartzite.

Sweden

The Hardeberga Quartzite in South Sweden has *Diplocraterion* and *Rusophycus*, and these were used by Bergström (1970) to postulate an early Cambrian age. *Rusophycus* is typically a Zone III trace fossil, while *Diplocraterion* is most commonly first encountered high up in Zone II. The available evidence suggests a late Tommotian age.

Plagiogmus has also been reported from Öland and nearby areas in erratic blocks. These apparently came from 'Lower Cambrian strata', possibly at about the same level as horizons with *Mobergella* (Jaeger & Martinsson 1980). *Mobergella* is normally found in Upper Tommotian strata, and such an age is consistent with the record of *Plagiogmus*.

England

The Precambrian–Cambrian sequences in the English Midlands have been described by Brasier *et al.* (1978), and Brasier & Hewitt (1979). The preservation of some

of the trace fossils makes positive identification difficult, but the essential elements are included in Table 8.8.

All of the trace fossils occurring in the two lowest members (Park Hill and Tuttle Hill) are first encountered in Vendian strata elsewhere, but the record of the Zone III *Rusophycus* in the overlying Jees Member indicates a level at least as high as Upper Tommotian. *Teichichnus*, however, is first reported in the overlying Home Farm Member and this typically appears in the Lower Tommotian. Exposure in this area is largely restricted to quarries, and collecting is not always easy: further work is necessary before any viable conclusions can be made. At this stage it seems that the Vendian/Tommotian boundary is to be placed at the base of, or more probably below, the Jees Member.

Spain

The best exposed and most complete section spanning the Precambrian–Cambrian boundary, so far described from Spain, is probably on the North Coast, in Asturias. There, Crimes *et al.* (1977) have recorded a 1 600 m-thick succession in which the earliest trilobites are only a few metres from the top and are suggested to be of low early Cambrian (early Atdabanian) age. The section and its trace fossils are summarized in Table 8.9.

The Vendian–Tommotian boundary may lie between the Asma Beds, which have yielded only *Arenicolites*, and the upper part of the overlying Palomar Beds which include the Zone II *Diplocraterion*.

The Zone III *Rusophycus* and *Astropolichnus* occur, respectively, in the Rubia and Castañal Beds, while the first trilobites are near the top of the Cayetano Beds. The

Table 8.8. The first appearance of ichnogenera in the English Midlands.

Hartshill Formation

Home Farm Member	*Teichichnus*
Jees Member	*Rusophycus* (= *Isopodichnus*)
Tuttle Hill Member	*Didymaulichnus, Gordia*
Park Hill Member	*Arenicolites, Neonereites, Planolites*

Table 8.9. First appearance of ichnogenera in the Asturian coast section (North Spain).

Candana Quartzite	
Cayetano Beds (390 m)	*Bergaueria, Cochlichnus, Taphrhelminthopsis*
Castañal Beds (360 m)	*Astropolichnus, Monomorphichnus*
Rubia Beds (470 m)	*Rusophycus*
Palomar Beds (250 m)	*Diplocraterion, Planolites, Skolithos*
Asma Beds (130 m)	*Arenicolites*

succession from within the Rubia Beds to the middle of the Cayetano Beds is probably late Tommotian in age, with the highest beds being Atdabanian.

To the south, in the Cantabrian Mountains, Crimes *et al.* (1977) have described an extensive suite of trace fossils from two sections in the Herreria Sandstone.

The first, between Los Barios de Luna and Irede, is some 860 m thick unconformably overlying Precambrian rocks of the Mora Formation (Crimes *et al.* op. cit., fig. 2). The top 55 m of the section has yielded a *Dolerolenus* fauna, but the occurrence of *Arenicolites, Cruziana, Diplocraterion, Rusophycus,* and *Skolithos* in the lowest few metres suggests that this is an incomplete sequence probably commencing in the Upper Tommotian and ending in the Atdabanian.

The second section, in the vicinity of the Porma Dam, consists of 550 m of sandstones and subordinate shales with an unexposed base. *Arenicolites, Cruziana, Diplichnites, Diplocraterion, Rusophycus,* and *Skolithos* all occur in the lowest 150 m of the section, thereby suggesting that it commences in the Upper Tommotian or higher. About 150 m from the top, trace fossils are abundant and diverse, including *Arenicolites, Cruziana, Diplichnites, Diplocraterion, Phycodes, Plagiogmus, Planolites, Rusophycus, Skolithos,* and *Teichichnus*. This extensive ichnofauna with many elements from Zone III, which have already appeared much lower, suggests an Atdabanian age. In the absence of trilobites an early Atdabanian age is most likely.

The correlation of the Cantabrian sections with the thicker sequence on the north coast has been much discussed, and the possibility that the former are incomplete has been raised (see Van den Bosch 1969). The trace fossils suggest that the coastal section may well range much lower, thereby implying a greater break at the underlying unconformity in the Cantabrian Mountains.

Trace fossils have also been reported from Precambrian–Cambrian boundary sequences in south-west and central Spain (Brasier *et al.* 1979; Liñán *et al.* 1984).

Three sections have been studied. The best is in the Valley of the Rio Uso in the Toledo Mountains, and this has been proposed as the Spanish candidate for the global Precambrian–Cambrian stratotypes (Liñán *et al.*

op. cit., p. 225). The section includes three main units: the Upper Alcudian, yielding algal stromatolites; the overlying Pusa Shales with *Monomorphichnus* and possibly *Phycodes* plus *Treptichnus*; and the Azorejo Sandstone with *Astropolichnus, Didymaulichnus, Diplocraterion, Diplichnites, Gordia, Monomorphichnus, Plagiogmus, Planolites, Psammichnites, Scolicia,* and *Rusophycus*. The occurrence of such a diverse ichnofauna with the inclusion of the Zone III trace fossils *Astropolichnus, Diplichnites, Plagiogmus,* and *Rusophycus* indicates a level at least as high as Upper Tommotian for this unit. The Tommotian–Vendian boundary may well be in or at the base of the Pusa Shales.

A second section, in the Sierra de Cordoba, has *Neonereites*? low down and *Monomorphichnus, Phycodes,* and *Rusophycus* over 1 000 m higher in the Julia Member. These traces suggest a late Tommotian age.

In a third section, in the Sierra de Guadalupe, *Cochlichnus* occurs first, but *Phycodes, Planolites,* and *Treptichnus* are present about 400 m higher and suggest at least an early Tommotian age for the higher beds.

USSR and East European Platform

The study of trace fossils at about the Precambrian–Cambrian boundary has received much attention in the USSR. Some of the information is available only in Russian and the following account is based on English translations and data from the Russian literature kindly provided by Dr M. A. Fedonkin. For simplicity, the information from the various sections has been combined. The areas studied are the Baltic region, the Dnestr River (Podolia, USSR), Poland, and the southeast White Sea region. Most of the information comes from papers by Fedonkin (1976, 1977, 1978, 1979, 1980a,b, 1981), Keller & Rozanov (1979), and Urbanek & Rozanov (1983), with further assistance from Fedonkin (pers. comm. 1981). Regrettably, trace fossils are virtually absent from the carbonate-dominated sequences in Siberia, and information is not yet available from the Mongolian sections.

The distribution of the trace fossils within the Russian stratigraphical units is shown in Table 8.10.

The lowest subdivision (Vendian) is normally accepted as Precambrian in age and includes some

Table 8.10. First appearance of ichnogenera in the USSR and East European Platform. Composite table derived from sections in the Baltic, Dnestr River (Podolia, USSR), Poland and south-east White Sea region.

Lower Atdabanian (Talsy)	*Cruziana, Dimorphichnus, Diplichnites, Gordia, Halopoa, Rhizocorallium, Sagitichnus*
Upper Tommotian (Lontova)	*Chondrites, Diplocraterion, Rusophycus*
Lower Tommotian (Rovno)	*Bergaueria, Phycodes, Teichichnus, Treptichnus*
Vendian	*Aulichnites* (cf. *Scolicia*), *Bilinichnus, Cochlichnus, Didymaulichnus, Harlaniella, Intrites, Neonereites, Nenoxites, Palaeopascichnus, Planolites, Skolithos, Suzmites, Vendichnus, Vimenites*

unusual trace fossils which appear to be restricted to about this horizon (e.g. *Vendichnus*, *Nenoxites*, and *Intrites*), and others which first appear at this level but continue into the Cambrian (e.g. *Aulichnites, Cochlichnus,* and *Didymaulichnus*). Most of these are surface traces produced either on the ocean floor or by burrowing between beds (cf. Fedonkin, 1980*b*).

The Lower Tommotian sees the first appearance of some more complex burrow systems including *Phycodes, Teichichnus,* and *Treptichnus*; *Diplocraterion* has not been reported so far. It appears at about this level in some sections elsewhere, but the earliest forms are very small and simple. It appears here in the Upper Tommotian, together with the first arthropod (probably trilobite) trace—*Rusophycus*. This is a resting trace which precedes the furrowing trace (*Cruziana*) in almost all sections, world-wide. The Lower Atdabanian includes *Cruziana*, other trilobite traces such as *Diplichnites* and *Dimorphichnus*, and a variety of other forms such as *Halopoa* and *Rhizocorallium*. *Gordia* is also reported at this level for the first time, but it occurs much lower in many other sections.

The two most significant changes in the ichnofauna are the incoming of relatively complex feeding-burrow systems at the base of the Lower Tommotian, and the incoming of *Cruziana* and certain other trilobite trace fossils at the base of the Atdabanian.

8.3.11. Namibia, South-west Africa

Trace fossils have been described from a sequence probably spanning the Precambrian–Cambrian boundary in Namibia, by Germs (1972) and Crimes & Germs (1982).

Two sections have been investigated, one lying to the south of Osis Farm and the other to the north (Crimes & Germs op. cit., text-fig. 2).

The succession is subdivided into three Subgroups which are, in ascending sequence, the Kuibis, Schwarz-rand, and Fish River Subgroups (Crimes & Germs op. cit.). The Kuibis Subgroup has yielded only ?*Skolithos* and *Bergaueria*. The form of the latter trace fossil closely resembles *Intrites* from the Vendian of the USSR and can probably be included in that ichnogenus. The overlying Schwarzrand Subgroup to the north of Osis Farm yielded *Planolites, Skolithos*, the radiating trace fossil *Brooksella*, and very small aberrant forms of *Diplocraterion* (Crimes & Germs op. cit.). To the south of Osis Farm, Germs (1972, fig. 1-6) figured a bilobed trace fossil which can be recognized as *Didymaulichnus* sp., and a trace fossil consisting of a single line of pellets (plate 1-5, 2-1), as in *Neonereites uniserialis* Seilacher. The top of the Schwarzrand Subgroup included *Neonereites uniserialis*, *N. biserialis* Seilacher, and *Phycodes pedum* Seilacher. The upper limit of the 'Ediacara' fauna is about 100 m below this last horizon and separated from it by a stratigraphical break.

The upper part of Fish River Subgroup yielded *Skolithos, Phycodes* cf. *pedum*, and the very strange *Enigmatichnus africani* to Crimes & Germs (op. cit.). The only trace fossil recorded from the lower part of this unit is from the Breckhorn Formation and this was described by Germs (1972, plate 2-9) as worm tracks (*Helminthoidichnites?*). From the photograph it seems to be bilobed and circling, and it closely resembles *Taphrhelminthopsis circularis*.

The Kuibis Subgroup has a variety of soft-bodied fossils of Ediacarian type, together with the trace fossil *Intrites* (= *Bergaueria*), and can therefore be inferred to be of Vendian age. The trace fossils from the overlying Schwarzrand Subgroup are a rather strange collection with some small aberrant forms (e.g. *Diplocraterion* and *Nereites*), and there are none that could be used to satisfactorily indicate a Tommotian age—until the occurrence of *Phycodes pedum* in the Nomtsas Formation at the top and above a disconformity. The Vendian–Tommotian boundary may therefore be at the base of the Nomtsas Formation. The ichnofauna is neither abundant nor diverse enough to indicate the existence of Atdabanian rocks within this section.

8.3.12. India

Trace fossils in Precambrian and Lower Cambrian strata have been found recently in the Himalayas in Uttar Pradesh and Kashmir, India.

In north-west Kashmir, the Lolab Formation has yielded three horizons with trace fossils below the earliest Lower Cambrian trilobites which are here Redlichiids (Raina *et al.* 1983).

The lowest trace-fossil-bearing horizon is approximately 3175 m below the trilobites, and has *Bergaueria* and *Planolites*. About 1400 m below the trilobites, *Arenicolites* (recorded as Ichnogenus A, probably *Arenicolites*), *Planolites*, and *Taphrhelminthopsis* occur. Trace fossils are, however, more abundant and diverse near the top of the formation and Shah (1982) and Shah & Sudan (1983) report *Monomorphichnus*, *Phycodes*, *Planolites*, *Rusophycus*, and *Skolithos*.

The lowest beds may therefore be Vendian in age, but the occurrence of *Taphrhelminthopsis* suggests a possible Tommotian horizon by the middle of the unit. The occurrence of the Zone II *Phycodes* and Zone III *Rusophycus* near the top confirms a Tommotian age, with the highest beds probably at least Upper Tommotian. Nearby, a similar section is exposed in the Hapatnar Valley, and there the middle of the Lolab (Khaiyar) Formation has provided *Astropolichnus* (= *Astropolithon*), *Bergaueria*, *Bifasciculus*, and *Monomorphichnus* (Kumar *et al.* 1984, p. 215). The presence of *Astropolichnus* suggests a late Tommotian to early Atdabanian age, but further collecting is required.

Bhargava & Srikantia (1982) have described an excellent example of *Taphrhelminthopsis circularis* from rocks of unknown age near Hangalpan, Kashmir. This is typical of Zone III and strongly suggests a late Tommotian to early Atdabanian age for these rocks.

The Tal Formation of the Lesser Himalayas, Utar Pradesh, has yielded a wide variety of trace fossils (Kumar *et al.* 1983; Singh & Rai 1983). Most of these come from the Argillaceous Member, which is in about the middle of the formation. The suite of trace fossils includes: *Astropolichnus* (= *Astropolithon*), *Aulichnites*, *Bifungites*, *Chondrites*, *Crossopodia*, *Cruziana*, *Curvolithos*, *Cylindrichnus*, *Dimorphichnus*, *Diplichnites*, *Halopoa*, *Merostomichnites*, *Monocraterion*, *Monomorphichnus*, *Phycodes*, *Plagiogmus*, *Planolites*, *Protichnites*, *Rosselia*, *Scolicia*, *Skolithos*, *Suzmites*, *Taphrhelminthopsis*, and *Tasmanadia*. The quality of reproduction of the illustrations does not allow any checking of the identifications, but clearly this is an abundant and diverse ichnofauna, and the inclusion of Zone III forms such as *Astropolichnus*, *Cruziana*, *Dimorphichnus*, *Plagiogmus*, and *Taphrhelminthopsis* indicates an age at least as high as late Tommotian and more probably Atdabanian. The lower part of the c.750 m-thick Tal Formation is, however, presumably Tommotian. The underlying Krol Formation has yielded only a few vertical burrows and is possibly of Vendian age.

In the Spiti Basin, north-west Himalayas, the Parahio Valley section has *Diplichnites*, *Dimorphichnus*, *Gyrochorte*, *Monomorphichnus*, *Phycodes*, *Planolites*, and *Scolicia* in the upper part of the Kunzam La Formation (Kumar *et al.* 1984). The occurrence of the trilobite traces *Diplichnites* and *Dimorphichnus* suggests a horizon at least as high as Upper Tommotian and more probably Atdabanian.

8.3.13. Australia

Many of the Precambrian–Cambrian boundary sequences in Australia contain abundant trace fossils, but detailed descriptions and measured sections are generally lacking.

Daily (1972, 1973) has described trace-fossil occurrences in the Mount Scott Ranges. The lowest unit—the Pound Quartzite—has an Ediacarian fauna and is unconformably overlain by the 130 m-thick Uratanna Formation and the 60 m-thick Parachilna Formation. The Uratanna Formation has *Rusophycus* about 100 m from its base, while the Parachilna Formation has *Diplocraterion*, *Phycodes pedum*, and *Plagiogmus*. This suite of trace fossils, including the Zone III *Plagiogmus* and *Rusophycus*, indicates a late Tommotian–early Atdabanian age.

At some localities the Uratanna Formation is absent, and *Diplocraterion* burrows from the Parachilna Formation have apparently penetrated the underlying, then unlithified, Pound Quartzite. This suggests that the top of the Pound Quartzite is no older than late Tommotian and may even be early Atdabanian.

Jenkins *et al.* (1983) record marks resembling *Monomorphichnus* in the Ediacara Member of the Rawnsley Quartzite in the Flinders Ranges.

No measured sections are available from the Pound Quartzite, but Glaessner (1969) has described some trace fossils from it, without giving precise horizons. He figures two pelleted trails, one (op. cit., fig. 5a) was later recognized as *Neonereites* (Glaessner 1984, p. 66), while the other (op. cit., fig. 5b) conforms to *Nereites*, as he infers (p. 382). There are also simple sinuous trails (op. cit., fig. 5e) referrable to *Planolites*, and tightly meandering trails resembling *Helminthoida crassa* Schafhäutl (op. cit., fig. 5c,d).

Neonereites and *Planolites* first appear in the Vendian, and the former seems to be particularly common at
this level; *Nereites* also occurs low down in the Nama
Group in rocks of probable Vendian age. The trace
fossils therefore suggest a Vendian age, but in the
absence of measured sections it is not known where the
Vendian–Tommotian boundary might lie.

Trace fossils are recorded as more abundant and
diverse in the upper part of the Arumbera Formation in
the Amadeus Basin, Central Australia (Glaessner 1969,
pp. 382–90), but no measured sections are given. The
following have been described: *Cochlichnus* (recorded as
Gordia), *Didymaulichnus miettensis* (recorded as molluscan trails), *Diplichnites*, *Phycodes pedum*, *Plagiogmus*,
Planolites, and *Rusophycus*. *Plagiogmus* and *Rusophycus* are both Zone III traces and indicate a late
Tommotian–early Atdabanian age.

There is a need for much more work in this very
promising area, but the trace-fossil evidence suggests
that some of the sequences may be slightly younger than
has been commonly thought, with even the top of the
Pound Quartzite possibly ranging into the Atdabanian.

8.3.14. China

In China, trace fossils have been found recently in
several Precambrian–Cambrian boundary sections. The
one which commands the most interest and research is
the proposed global boundary stratotype at Meishucun,
in Yunnan Province near Kunming. Trace fossils from
this section have been described by Jiang *et al.* (1982)
and Crimes & Jiang (1986). Their stratigraphical
distribution is illustrated in Fig. 8.4. and their palaeoenvironmental setting in Fig. 8.5.

The earliest occurrences are in Unit 3 of the
Zhongyicun Member: *Arenicolites*, *Asteriactites*, *Neonereites biserialis*, *N. uniserialis*, and *Sellaulichnus
meishucuniensis* Jiang Zhiwen. *Neonereites* is common in
Vendian strata and *Arenicolites* is also known to occur
this early. Radiating traces, of which *Asteriactites* is an
example, also occur at very low levels (e.g. *Brooksella* in
the Schwarzrand Subgroup, Namibia). This ichnofauna
is therefore consistent with, but does not prove, a
Vendian age. Units 4 and 5 have, as yet, yielded no trace
fossils, but Unit 6 has *Cochlichnus*, *Monomorphichnus*,
Neonereites biserialis, and *N. uniserialis*, while Unit 7
has *Didymaulichnus miettensis* Young below *Cruziana*
and *Rusophycus*. The last two trace fossils indicate a
horizon at least as high as Upper Tommotian and more
probably Lower Atdabanian. This suggests that the
Tommotian–Atdabanian boundary may be within Unit
7. Units 4–6 could then be Tommotian. This is the most
likely inference on the evidence currently available, but

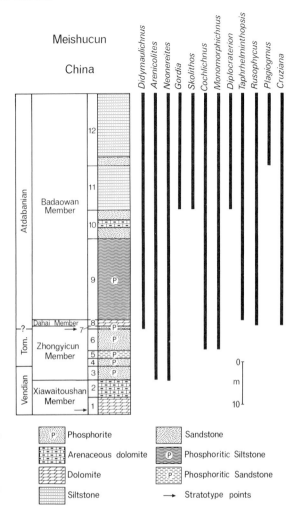

Fig. 8.4. Vertical distribution of trace fossils in the proposed
Precambrian–Cambrian boundary stratotype sec-
tion at Meishucun, Yunnan, China.

with the common occurrence of phosphorites in the
sequence, the possibility of a break cannot be ruled out.
Certainly, with the combined thicknesses of units
4–6—only 8.3 m—the sequence appears to be greatly
condensed.

Trace fossils have not been found in the overlying
Dahai Member but within the Badaowan Member,
Didymaulichnus and *Taphrhelminthopsis* occur in Unit 9,
while Unit 11 has *Arenicolites*, *T. circularis*, *Diplocraterion*, *Gordia molassica* (Heer), and *Skolithos*. *T. circularis*
normally indicates a late Tommotian to early Atdabanian age.

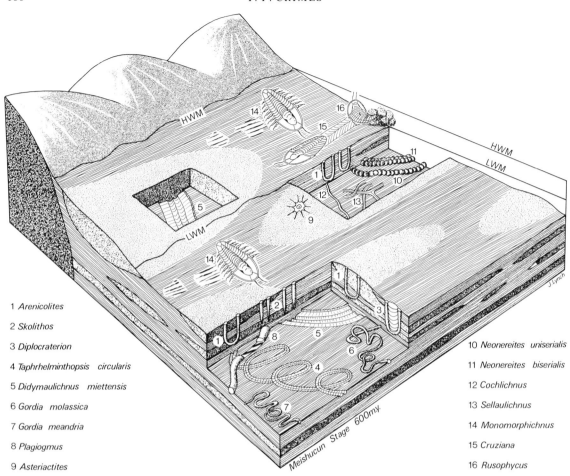

1 *Arenicolites*

2 *Skolithos*

3 *Diplocraterion*

4 *Taphrhelminthopsis circularis*

5 *Didymaulichnus miettensis*

6 *Gordia molassica*

7 *Gordia meandria*

8 *Plagiogmus*

9 *Asteriactites*

10 *Neonereites uniserialis*

11 *Neonereites biserialis*

12 *Cochlichnus*

13 *Sellaulichnus*

14 *Monomorphichnus*

15 *Cruziana*

16 *Rusophycus*

Fig. 8.5. Environmental setting of the trace fossils recorded from the section at Meishucun, Yunnan, China.

Two levels have been suggested for the Precambrian–Cambrian boundary reference point in this section. The first is 0.8 m above the base of the Xiawaitoushan Member (Xing & Luo 1984) and might well correspond with a Vendian age. The second, now preferred point, is at the base of Unit 7 (Cowie 1985). This latter horizon is quite possibly much higher, and the trace fossils suggest that it might be at about the Tommotian–Atdabanian boundary. If this is correct, it would mean that the boundary would be placed after the considerable increase in trace-fossil abundance and diversity ('explosive evolution') but probably before the first trilobites. This might provide a more readily correlatable horizon than one below the diversity increase (i.e. the Vendian–Tommotian boundary). There would seem therefore to be no fundamental reason to reject this section and

boundary point because it might be at a higher level than initially intended when the Precambrian–Cambrian Working Group were looking for a level at 'diversity explosion'.

More work would, however, seem to be necessary on this section hopefully to eliminate the possibility of stratigraphical breaks.

Trace fossils have also been recorded from a boundary sequence in the Emei–Ganluo region of Sichuan (Yang *et al.* 1982). In this section, the base of the Maidiping Member of the Hongchunping Formation corresponds approximately with the lower-boundary reference point. The Maidiping Member has yielded only *Planolites*. The base of the overlying Jiulaodong Formation is approximately equivalent to the upper preferred boundary reference point. The Juilaodong

Formation is divided into Lower, Middle, and Upper Members. The Lower contains 'arthropod scratch marks', *Planolites*, *Asteriactites*?, and *Chondrites*, the Middle has *Neonereites uniserialis*, recorded as *Parascalaritubus* (Yang *et al.* op. cit., pls 1-1, 1-5, 2-2), *Planolites*, *Scolicia*, *Skolithos*, and a circling trace described as *Multilaqueichnus ganluoensis* Yang & Yin, which appears to be very similar to *Taphrhelminthopsis circularis*, while the upper has so far provided only *Monomorphichnus* and *Astropolichnus*? (= *Astropolithon*?), but also has trilobites.

8.4. DISCUSSION

Trace fossils appear to have advantages for correlation at this level:

(1) they are relatively common in clastic sequences;
(2) they are easily found and collected;
(3) most are relatively easy to identify to ichnogeneric level;
(4) they do not occur as derived fossils;
(5) they seem to have a regular order of first appearance at widespread localities;
(6) a few have a restricted stratigraphical range.

The main examples of the usefulness of trace fossils for correlation discussed in this paper are summarized in Fig. 8.6. The main instances where these correlations are at variance with some previous work are:

(1) Rocky Mountains, Canada;
(2) Wernecke Mountains, Canada;
(3) White-Inyo Mountains, USA;
(4) Rio Uso, Spain;
(5) Meishucun, China.

In the Rocky Mountains, the general practice was to consider the Miette Group as Precambrian and the overlying Gog Group as Cambrian (Aitken 1969). Later, Young (1972) placed the Precambrian–Cambrian boundary high up in the McNaughton Formation, that is about two-thirds of the way up the Gog Group. Fritz (1979), however, regarded the Gog Group as Cambrian and claimed a break in the sequence below it, with the underlying Miette Group considered as Precambrian. The present work would place the Vendian–Tommotian boundary low down in the Gog Group or at its base, therefore supporting Fritz's contention.

In the Wernecke Mountains, Nowlan *et al.* (1985) suggested that the Vampire Formation extends down

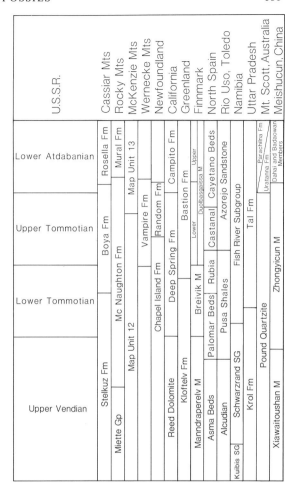

Fig. 8.6. Correlation chart for late Precambrian and Lower Cambrian sequences, based on trace-fossil evidence.

into the Vendian, whereas this study suggests that trace fossils from its base (*Cruziana* and *Rusophycus*) are relatively young and may even indicate an Atdabanian age. The main evidence provided by Nowlan *et al.* (1985) was small shelly faunas from near the base of the Vampire Formation. They claim that these correlate best with the low-diversity assemblage of Vendian age, in terms of number of taxa and in having species and genera in common. They do, however, admit that the taxa identified to the generic and specific level are known to extend 'at least a short distance into the Cambrian'. There is also the possibility that, since most of the shelly faunas are broken and they were deposited immediately above a disconformity, they may have been derived from earlier strata. However, Nowlan *et al.* (op.

cit.) point out that these fossils have not been seen at widespread outcrops of the underlying strata. This section is an important one by virtue of the rare association of trace fossils and small shelly fossils, and it clearly merits further work. On the evidence currently available, however, the trace fossils strongly suggest that the disconformity at the base of the Vampire Formation may be very significant, with the beds immediately above it being of Tommotian or even Atdabanian age.

In the White-Inyo Mountains, at least four different levels have been proposed for the Precambrian–Cambrian boundary (Nelson 1978). Alpert (1977) used trace fossils to suggest that it should be placed at the base of the Upper Member of the Deep Spring Formation. This level is marked by the incoming of *Diplichnites* and *Rusophycus*, both of which first appear high up in the Tommotian or in the Lower Atdabanian, elsewhere. The Vendian–Tommotian boundary therefore appears to be lower—probably at the base of the Deep Spring Formation or in the underlying Reed Dolomite. Support for a lower base is provided by the recent discovery of Tommotian small shelly fossils in the Lower Member of the Deep Spring Formation in Esmeralda County, Nevada (Mount *et al.* 1983). This would also suggest that the Vendian–Tommotian boundary should be placed at or below the base of the Deep Spring Formation.

In the Rio Uso section of the Toledo Mountains, Spain, Brasier *et al.* (1979) placed the Vendian–Tommotian boundary at the contact between the Pusa Shales and overlying Azorejo Sandstone. This was mainly based on the supposed record of *Chuaria* in the Pusa Shales. However, the ichnofaunal diversity in the Azorejo Sandstone suggests an age at least as high as late Tommotian for this unit and this, together with the record of *Monomorphichnus lineatus* Crimes and possible *Diplocraterion* in the Pusa Shales, suggests that the Vendian–Tommotian boundary should be placed within or at the base of that unit. Support for this is provided by Liñán *et al.* (1984), who found the Upper Vendian palynomorph *Bavlinella faveolata* near the base of the Pusa Shales, and also concluded (p. 224) that the

Chuaria were 'barely identifiable forms more probably belonging to other groups'. They therefore also place the Vendian–Tommotian boundary low down in the Pusa Shales.

At Meishucun in China, Xing & Luo (1984 p. 109) correlated the lower part of the Zhongyicun Member with the upper part of the Rovno Horizon or the Lontova Horizon of the Tommotian Stage of Siberia. Thus the base of the Tommotian would be at or slightly above, the base of the Zhongyicun Member. They also consider that the Tommotian–Atdabanian boundary is at least as high as the top of the Badaowan Member. Thus the Tommotian Stage would correlate with most or all of the Zhongyicun Member, the Dahai Member, and the Badaowan Member. Russian work on the shelly faunas (see Cowie 1985, p. 96) claims that the age of the Zhongyicun Member may be younger Tommotian and perhaps equivalent to the overlying Atdabanian. The trace-fossil evidence presented here, however, agrees with Xing & Luo (1984) in placing the Vendian–Tommotian boundary close to the base of the Zhongyicun Member, but strongly suggests that the Tommotian–Atdabanian boundary is no higher than the upper part of the Zhongyicun Member. This agrees with the Russian work which indicates that the upper part of the Zhongyicun Member and the overlying Dahai and Badaowan Members are Atdabanian. These conclusions carry with them the implication that the Tommotian Stage is only represented by a thin sequence at Meishucun. This is consistent with the fact that the Zhongyicun Member is dominated by phosphorites which themselves usually indicate slow accumulation. The inference that the Meishucun section is condensed and in part younger than has been supposed, should not lead to its rejection as the global stratotype, but a very detailed examination for stratigraphical breaks should be made.

Acknowledgement

I am grateful to Joe Lynch for redrawing the diagrams.

REFERENCES

Aceñolaza, F. G. (1978). El Paleozoico inferior de Argentina segun sus trazas fosiles. *Ameghiniana*, **15**, 15–64.

Aceñolaza, F. G. & Durand, F. (1973). Trazas fósiles del basamento cristalino del noroeste Argentino. *Bol. Asociacion geol. Cordoba*, **2**, 45–55.

Aceñolaza, F. G. & Toselli, A. J. (1981). *Geologia del noroeste Argentino*. Univ. Nacional de Tucuman, Publ. 1287.

Aitken, J. D. (1969). Documentation of the sub-Cambrian unconformity, Rocky Mountains, Main Ranges, Alberta. *Can. J. Earth Sci.*, **6**, 193–200.

Alpert, S. P. (1976). Trilobite and star-like trace fossils from the White-Inyo Mountains, California. *J. Paleont.*, **50**, 226–39.

Alpert, S. P. (1977). Trace fossils and the basal Cambrian boundary. In *Trace fossils 2*, (eds T. P. Crimes & J. C. Harper), pp. 1–8. Geol. J. Spec. Iss., 9. Seel House Press, Liverpool.

Banks, N. L. (1970). Trace fossils from the Late Precambrian and Lower Cambrian of Finnmark, Norway. In *Trace fossils*, (eds T. P. Crimes & J. C. Harper), pp. 19–34. Geol. J. Spec. Iss., 3. Seel House Press, Liverpool.

Bengtson, S. & Fletcher, T. P. (1983). The oldest sequence of skeletal fossils in the Lower Cambrian of southeastern Newfoundland. *Can. J. Earth Sci.*, **20**, 525–36.

Bergström, J. (1970). *Rusophycus* as an indication of early Cambrian age. In *Trace fossils*, (eds T. P. Crimes and J. C. Harper), pp. 35–42. Geol. J. Spec. Iss., 3. Seel House Press, Liverpool.

Bhargava, D. N. & Srikantia, S. V. (1982). *Taphrohelminthopsis circularis* from Cambrian sediments of southeast Kashmir valley. *J. geol. Soc. India*, **23**, 406–7.

Bosch, W. J. Van den (1969). Geology of the Luna-Sil region, Cantabrian mountains (NW Spain). *Leiden geol. Meded.*, **44**, 137 pp.

Brasier, M. D. & Hewitt, R. A. (1979). Environmental setting of fossiliferous rocks from the uppermost Proterozoic–Lower Cambrian of Central England. *Palaeogeogr., Palaeoclimatol., Palaeoecol.*, **27**, 35–57.

Brasier, M. D., Hewitt, R. A., & Brasier, C. J. (1978). On the Late Precambrian–Early Cambrian Hartshill Formation of Warwickshire. *Geol. Mag.*, **115**, 21–36.

Brasier, M. D., Perejon, A., & San José, M. A. (1979). Discovery of an important fossiliferous Precambrian–Cambrian sequence in Spain. *Estud. geol. Inst. Invest. geol. 'Lucas Mallada'.* **35**, 379–83.

Cowie, J. W. (1985). Continuing work on the Precambrian–Cambrian boundary. *Episodes*, **8**, 93–8.

Cowie, J. W. & Spencer, A. M. (1970). Trace fossils from the Late Precambrian–Lower Cambrian of East Greenland. In *Trace fossils*, (eds T. P. Crimes & J. C. Harper), pp. 91–100. Geol. J. Spec. Iss., 3. Seel House Press, Liverpool.

Crimes, T. P. (1975). The stratigraphical significance of trace fossils. In *The study of trace fossils*, (ed. R. W. Frey), pp. 109–30. Springer-Verlag, New York.

Crimes, T. P. (1976). Trace fossils from the Bray Group (Cambrian) at Howth, Co. Dublin. *Geol. Surv. Ireland, Bull.*, **2**, 53–67.

Crimes, T. P. (1987). Trace fossils and the Precambrian–Cambrian boundary. *Geol. Mag.*, **124**, 97–119.

Crimes, T. P. & Anderson, M. M. (1985). Trace fossils from late Precambrian–early Cambrian strata of southeastern Newfoundland (Canada): temporal and environmental implications. *J. Paleont.*, **59**, 310–43.

Crimes, T. P. & Crossley, J. D. (1968). The stratigraphy, sedimentology, ichnology and structure of the Lower Palaeozoic rocks of part of northeastern Co. Wexford. *Proc. R. Irish Acad.*, **67B**, 185–215.

Crimes, T. P. & Germs, G. J. B. (1982). Trace fossils from the Nama Group (Precambrian–Cambrian) of southwest Africa (Namibia). *J. Paleont.*, **56**, 890–907.

Crimes, T. P. & Jiang Zhiwen (1986). Trace fossils from the Precambrian–Cambrian boundary candidate at Meishucun, Jinning, Yunnan, China. *Geol. Mag.*, **123**, 641–9.

Crimes, T. P., Legg, I., Marcos, A., & Arboleya, M. (1977). ?Late Precambrian–low Lower Cambrian trace fossils from Spain. In *Trace fossils 2*, (eds T. P. Crimes & J. C. Harper), pp. 91–138. Geol. J. Spec. Iss., 9. Seel House Press, Liverpool.

Daily, B. (1972). The base of the Cambrian and the first Cambrian faunas. *Centre for Precambrian Res., Univ. Adelaide, S. Australia. Spec. Pap.*, **1**, 13–37.

Daily, B. (1973). Discovery and significance of basal Cambrian Uratanna Formation, Mt. Scott Range, Flinders Ranges, South Australia. *Search*, **4**, 202–5.

Fedonkin, M. A. (1976). Traces of multicellular animals from the Valdai Series. *Izvestiya Akad. Nauk. SSSR, Ser. geol.*, **4**, 129–132.

Fedonkin, M. A. (1977). Precambrian–Cambrian ichnocoenoses of the east European platform. In *Trace fossils 2*, (eds T. P. Crimes & J. C. Harper), pp. 138–94. Geol. J. Spec. Iss., 9. Seel House Press, Liverpool.

Fedonkin, M. A. (1978). Ancient trace fossils and the ways of behavioural evolution of mud eaters. *Palaeont. J.*, **12**, 106–12.

Fedonkin, M. A. (1979). Paleoichnology of Precambrian and Early Cambrian. In *Paleontology of Precambrian and Early Cambrian*, (ed. B. S. Sokolov), pp. 183–92. Akad. Nauk. SSSR, Leningrad.

Fedonkin, M. A. (1980a). Fossil traces of Precambrian Metazoa. *Isvestiya Akademii Nauk SSSR, Ser. geol.*, **1**, 39–46.

Fedonkin, M. A. (1980b). Early stages of evolution of Metazoa on the basis of palaeoichnological data. *Isvestiya Akademii Nauk SSSR, Ser. geol.*, **2**, 226–33.

Fedonkin, M. A. (1981). Belomorskaya biota Venda [White Sea biota of the Vendian.] *Trudy geol. Inst. Akad. Nauk SSSR*, **342**, 99 pp.

Føyn, S. & Glaessner, M. F. (1979). *Platysolenites*, other animal fossils, and the Precambrian–Cambrian transition in Norway. *Norsk. geol. Tidsskr.*, **59**, 25–46.

Fritz, W. H. (1979). Cambrian stratigraphy in the Northern Rocky Mountains, British Columbia. *Current Research, Part B, geol. Surv. Can., Pap.*, **79–1B**, 99–109.

Fritz, W. H. (1980). International Precambrian–Cambrian boundary working group's 1979 field study to Mackenzie Mountains, Northwest Territories, Canada. *Current Research, Part A, geol. Surv. Can., Pap.*, **80–1A**, 41–5.

Fritz, W. H. & Crimes, T. P. (1985). Lithology, trace fossils and correlation of Precambrian–Cambrian boundary beds, Cassiar Mountains, north-central British Columbia, Canada. *Geol. Surv. Can. Pap.*, **83–13**, 24 pp.

Germs, G. J. B. (1972). Trace fossils from the Nama Group, southwest Africa. *J. Paleont.*, **46**, 864–70.

Glaessner, M. F. (1969). Trace fossils from the Precambrian and basal Cambrian. *Lethaia*, **2**, 369–93.

Glaessner, M. F. (1984). *The dawn of animal life*. Cambridge University Press.

Hofmann, H. J. (1981). First record of a Late Proterozoic faunal assemblage in the North American Cordillera. *Lethaia*, **14**, 303–10.

Jaeger, H. & Martinsson, A. (1980). The early Cambrian trace fossil *Plagiogmus* in its type area. *Geol. Foren. i Stockholm Forhandl.*, **102**, 117–26.

Jenkins, R. J. F., Ford, C. H., & Gehling, J. G. (1983). The Ediacara Member of the Rawnsley Quartzite: the context of the Ediacara assemblage (late Precambrian, Flinders Ranges). *J. geol. Soc. Aust.* **30**, 101–19.

Jiang Zhiwen, Luo Huilin, & Zang Shishan (1982). Trace fossils of the Meishucun Stage (Lowermost Cambrian) from the Meishucun section in China. *Geol. Revue*, **28**, 7–13.

Keller, B. M. & Rozanov, A. Yu (1979). *Upper Precambrian and Cambrian Paleontology of East-Eurapian [sic] platform*. Academy of Sciences of the USSR, Moscow.

Kumar, G., Raina, B. K., Bhatt, D. K., & Jangpangi, B. S. (1983). Lower Cambrian body- and trace-fossils from the Tal Formation, Garhwal Synform, Uttar Pradesh, India. *J. palaeont. Soc. India*, **28**, 106–11.

Kumar, G., Raina, B. K., Bhargava, O. N., Maithy, P. K., & Babu, R. (1984). The Precambrian–Cambrian boundary problem and its prospects, Northwest Himalaya, India. *Geol. Mag.*, **121**, 211–19.

Liñán, E., Palacios, T., & Peréjon, A. (1984). Precambrian–Cambrian boundary and correlation from southwestern and central part of Spain. *Geol. Mag.*, **121**, 221–8.

Mount, J. F., Gevirtzman, D. A., & Signor, P. W. (1983). Precambrian–Cambrian transition problem in western North America: Part I. Tommotian fauna in the southwestern Great Basin and its implications for the base of the Cambrian system. *Geology*, **11**, 224–6.

Nelson, C. A. (1978). Late Precambrian–Early Cambrian stratigraphic and faunal succession of eastern California and the Precambrian–Cambrian boundary. *Geol. Mag.*, **115**, 121–6.

Nowlan, G. S. Narbonne, G. M., & Fritz, W. H. (1985). Small shelly fossils and trace fossils near the Precambrian–Cambrian boundary in the Yukon Territory, Canada. *Lethaia*, **18**, 233–56.

Poiré, D. G., Del Valle, A., & Regalia, G. M. (1984). Trazas fosiles en Cuarcitas de la Formacion sierras bayas (Precambrico) y su comparicion con las de la Formacion Balcarce (Cambro-Ordovicico), Sierras septentrionales de la Provincia de Buenos Aires, *Noveno congreso Geol. Argentino S.C. de Bariloche, 1984. Actas*, **4**, 249–66.

Raina, B. K., Kumar, G., Bhargava, O. N., & Sharma, V. P. (1983). ?Precambrian–low Lower Cambrian ichnofossils from the Lolab valley, Kashmir, Himalaya, India. *J. palaeont. Soc. India*, **28**, 91–4.

Regalia, G. M. & Herrera, H. H. (1981). *Phycodes* aff *pedum* (traza fósil) en estratos cuarciticos de San Manuel, Sierras Septentrionales de la provincia de Buenos Aires, Republica Argentina. *Rev. Asoc. Geol. Argentina*, **36**, 257–61.

Seilacher, A. (1956). Der Beginn des Kambriums als biologische Wende. *Neues Jb. Geol. Paläont. Abh.*, **103**, 155–80.

Seilacher, A. (1970). *Cruziana* stratigraphy of 'non-fossiliferous' Palaeozoic sandstones. In *Trace fossils*, (eds T. P. Crimes & J. C. Harper), pp. 447–76. Geol. J. Spec. Iss., 3. Seel House Press, Liverpool.

Shah, S. K. (1982). Cambrian stratigraphy of Kashmir and its boundary problems. *Precambr. Res.*, **17**, 87–98.

Shah, S. K. & Sudan, C. S. (1983). Trace fossils from the Cambrian of Kashmir and their stratigraphic significance. *J. geol. Soc. India*, **24**, 194–202.

Singh, I. B. & Rai, V. (1983). Fauna and biogenic structures in Krol-Tal succession (Vendian–Early Cambrian). Lesser Himalaya: their biostratigraphic and palaeoecological significance. *J. palaeont. Soc. India*, **28**, 67–90.

Tremlett, W. E. (1959). The Pre-Cambrian rocks of southern Co. Wicklow (Ireland). *Geol Mag.*, **96,** 58–68.

Urbanek, A. & Rozanov, A. Yu. (1983). *Upper Precambrian and Cambrian palaeontology of the east-European platform.* Wydawynictwa Geol. Warszawa.

Xing Yusheng & Luo Huilin (1984). Precambrian–Cambrian boundary candidate, Meishucun, Jinning, Yunnan, China. *Geol. Mag.*, **121,** 143–54.

Yang Zunyi, Yin Jicheng, & He Ting-gui (1982). Early Cambrian trace fossils from the Emei–Ganluo region, Sichuan, and other localities. *Geological Revue*, **28,** 291–8.

Young, F. G. (1972). Early Cambrian and older trace fossils from the southern Cordillera of Canada. *Can. J. Earth Sci.*, **9,** 1–17.

9

Chronometry

J. W. Cowie & W. B. Harland

9.1. INTRODUCTION

The age in units of duration of the Precambrian–Cambrian boundary has been the subject of investigation and speculation since the late nineteenth century, when it was bound up with estimates of the age of the Earth. For example, H. S. Williams (1893, p. 295) was one of many who used the 'standard time scale of geochronology, on the basis of the Eocene period for a time unit or geochrone . . .', from which he estimated the time elapsed since the beginning of the Cambrian Period as being equal to 57 geochrons. His Eocene included Palaeocene. Estimates, in the unit (a year), were made by thickness and rate of sedimentation between 1860 and 1909, for a minimum age of the Earth which varied between 3 and 1526 million years (Ma), with the rates of sedimentation used varying between 100 and 8600 years to deposit one foot (from nineteen authors quoted by Holmes 1913). The first radiometric determinations of Cambrian rocks, i.e. Swedish 'Kolm' (Upper Cambrian) gave a value of 395 to 405 Ma (millions of years), and Virginia basalts of 427–453 Ma in age were used on Holmes' time-scale of 1937 with a Precambrian–Cambrian boundary of 470 Ma BP. Since then, many estimates have been made, for example, in Ma:

It is obvious, on the one hand, that traditional dates have persisted in the literature without rigorous reinvestigation, and, on the other hand that modern thorough investigations give results ranging from 530 to 600 Ma. These depend upon the different weights given by different workers to various determinations.

Uncertainty about the chronometric age of the Precambrian-Cambrian boundary is now so great that any decision on where the boundary will be stratigraphically standardized (by international agreement through the International Commission on Stratigraphy, ICS), hardly affects the issue, even if an age is chosen which immediately precedes the earliest trilobites or immediately post-dates the Ediacara faunas.

We can shed no further light on this problem: nevertheless it should be addressed in this review book. Accordingly we thought that it would be most useful to list selected published age determinations, in numerical order of radiometric age in years, with identifying characteristics and without tabulated comment. Clearly, any list is subjective, in so far as it depends on the accidents of availability and the decision to discard aberrant (or apparently aberrant) values. Even so, in this list we hope to avoid the subjective element inherent in the selection and rejection process generally used by

510	Holmes 1947 (B scale)
600	Holmes 1959
570 (579 when recalculated)	Harland *et al.* 1964
574	Lambert 1971
570	von Eysinga 1975
570	Glaessner 1977
574	Armstrong 1978
530	Odin *et al.* 1982
590	Harland *et al.* 1982
590	Keller & Krasnobaev 1983
570	Palmer 1983
530–600	Cowie & Johnson 1985
570	Snelling 1985
540–550	Conway Morris, this book, Chapter 2.
536	Cope & Gibbons 1987

practitioners. The list is intended for inspection and reference. It is neither intended as a data base nor for averaging out to produce a consensus.

In the 1985 volume on 'Geochronology of the Geological Record', the paper submitted by Cowie & Johnson on the 'Late Precambrian and Cambrian geological time-scale' was subsequently reviewed by Odin, *et al.* in a useful but partial, committed, authoritarian paper in the same volume. As Cowie was unable to make a reply in the 1985 volume, a few notes are included below. Cowie & Johnson (1985) were deliberately objective and sceptical, the indeterminate nature of the results at this stratigraphical level merited this approach. The problem remains that there are large areas of the world with important sequences of late Precambrian and Lower Cambrian rocks without any available radiometric dates. The challenge remains to collect, where possible, more new, reliable chronometric data to calibrate the good chronostratigraphical data and successions, whether the latter were previously known or are newly described.

9.2. CLASSIFICATION OF CHRONOSTRATIGRAPHICAL DIVISIONS BEFORE AND AFTER THE PRECAMBRIAN–CAMBRIAN BOUNDARY

In Table 9.1 we have used a chronostratigraphical scale, after Harland *et al.* 1982, modified in some respects, as explained below. We have also adopted the same three-letter abbreviations or contractions for these terms, from Appendix 3 of the above work. Thus, tabulating the sequence around the boundary in question, as set out in Harland *et al.*, we have the classification scheme of Table 9.1.

One of the problems, which has become increasingly evident, is that the Atdabanian and Tommotian divisions set out in detail from Siberia and widely accepted, have proved difficult to use elsewhere. Therefore we have to question whether they represent chronostratigraphic divisions applicable to the global scale. Without prejudging the question, we suggest an alternative scheme which also has some degree of historical precedence.

'Aldanian' is a name that may combine approximately the Atdabanian and Tommotian divisions. Soviet colleagues suggest that Nemakit–Daldyn rocks could possibly represent an earliest Cambrian age, earlier than Tommotian as generally conceived. There may, therefore, be a need for an earliest Cambrian division. For this purpose we apply the older name of 'Etcheminian' as defined by Matthew (1899) and proposed now to accommodate Lower Cambrian strata between the first appearance of *Phycodes pedum* and trilobite-bearing rocks in Newfoundland, especially in that subdivision.

It is, at present, fairly generally agreed that the three most likely sections competing for a Global Stratotype Section and Point (GSSP) are in Siberia (USSR),

Table 9.1. Chronostratigraphical scale, after Harland *et al.* 1982, with three-letter abbreviations and contractions

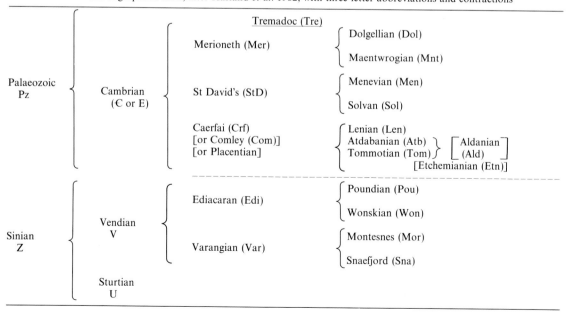

Yunnan (China), and Newfoundland (Canada). It is, however, not impossible that the most complete fossiliferous sequence may yet prove to be in Newfoundland (or possibly in Iran). If the best sequence is in Newfoundland, then this would strengthen the claim for Etcheminian as the earliest Cambrian age.

Since neither the initial Tommotian boundary nor the distinction between Atdabanian and Tommotian boundary nor the distinction between Atdabanian and Tommotian is clear and widely correlatable outside the USSR, it may be that the ages Etcheminian, and/or Meishucunian, and/or Aldanian could successfully form the basis of a new chronostratigraphical classification of earliest Cambrian times.

9.3. COMMENTS ON PROCEDURE

Table 9.2 lists items as pre-, syn-, and post-. The logic of the radiometric calibration of chronostratigraphy is not easy to handle. The best case is when the chronostratigraphical age of the radiometrically dated rock is known: for example, by fossil content. In such case 'syn-X' should be the unambiguous entry.

However, if the rock with fossils (for example) is stratigraphically overlain by the radiometrically determined rock 'Y', or either intruded or underlain by it, there may be some uncertainty. It is not necessarily correct to write the chronometric age of X as 'pre-Y', 'pre-Y or post-Y' respectively, because in each case it is possible that the duration of the chronostratigraphical division X also brackets Y, so that the safest procedure is to write 'pre/syn-Y' or 'syn-Y'. In most cases this possibility is unlikely. Where we have such apparently conflicting data to contend with, it is possible that the resolution of the problem lies in a more rigorous treatment of the 'bracketing' procedure in column 6 of Table 9.2. An alternative way of expressing this kind of relationship is 'pre-*part* X' (or 'post-*part* X') instead of 'pre-X' (*or* 'syn-X' *or* 'post-X').

Needless to say, this logic is entirely independent of other considerations, such as the degree to which the apparent radiometric age approximates to the true chronometric age. The implications of this logical difficulty are that the best case of 'syn-X' is by dating of authigenic minerals in sedimentary rocks which are also dated on correlated biostratigraphy, magnetostratigraphy, and any other available means.

Conversely, bracketing by determined rocks which are no more than contiguous (intrusions, unconformable relationships, etc.) are more suspect, and this may be causing some of the present difficulties. Finally, there is the problem of thermal resetting and the effects of aqueous circulation which affect both the cases.

The great uncertainty that attends the chronometric age of the Precambrian–Cambrian boundary obscures the much lesser, but still evident, uncertainty in biostratigraphical correlation of rocks at this transition. Even so, the principle of the standard Global Stratotype Section and Point, the GSSP (Cowie *et al.* 1986) is not compromised. We still need a reference boundary point, however difficult it proves to correlate with other sections. The likelihood that there may be uncertainty in correlation, therefore, in no way limits the value of being precise about a standard. The alternative would be to hold on to one more element of uncertainty. It is not desirable to have another uncertainty (lack of a GSSP) added to the inevitable uncertainty of correlation.

In studying the chronometry of the Precambrian–Cambrian boundary, it is necessary to examine the dating of the latest Precambrian as well as the lowermost Cambrian. It may seem unnecessary to list ages that seem to be so long in time from the boundary either way. The reason, however, is to allow some estimation of the duration of known history before or after the boundary (sequence stratigraphy), as an added factor in assessing its most likely age.

We may observe that many estimates of the age of the Precambrian–Cambrian boundary have been closely related to studies of the *Phanerozoic* time-scale and were made with little reference to Precambrian rocks.

There is now a significant list of known and well-ordered events within the last one or two hundred million years prior to the Phanerozoic threshold, to enable us to use many late Precambrian dates in the effort to constrain the boundary value.

9.4. COMMENTS ON PARTICULAR DETERMINATIONS IN TABLE 9.2

1. Entry No. 2: 460 ± 9 Ma, see also Compston & Zhang 1983. This date quoted by Cowie & Johnson (1985, p. 52) is an example of dates in the 400–500 Ma bracket—which illustrate the difficulties caused by variation in results, with differing grain size, in the Chinese dates based on illitic shales. This topic is discussed by Odin *et al.* (1985, pp. 69–70). It is basically an argument 'that there exists a variable proportion of detrital minerals in the bulk acid insoluble fraction' and 'that the age of the diagenetic components is ambiguous'. Entries 53, 56, and 57 show ages in the Chinese early Cambrian strata of 580–585 Ma, and these seem to be generally corroborated by Entry 49 of 575 Ma for the younger trilobitic strata and Entry 37 of 568 Ma (whole-rock) for a similar stratigraphical level. Odin *et al.* (1985, p. 70) comment 'that the age of 568 ± 12 Ma is quoted as a U–Pb whole-rock analytical age and is taken as an

Table 9.2.   List of relevant radiometric age determinations in numerical order of millions of years (Ma).

Radio metric age Ma ±	Method/ material	Rock unit sampled (without rank)	Location/ region	pre-, syn-, post- (abbrev. from Table 9.1)	References (primary source precedes secondary)
1. 452 ±31	Rb/Sr	Strettonian of Longmyndian	Salop, UK	syn-? PЄ	Bath 1974
2. 460 ±9	Rb/Sr	Tienzhusan of Dengying	Yangtze Gorges China	syn-Є	Zhang *et al.* 1982
3. 467 ±30	Rb/Sr whole-rock	Harbour Main & Bull Arm Volc.	SE Newfoundland, Canada	syn-PЄ (V)	Fairbairn *et al.* 1966 recalc. McCartney 1966
4. 505 ±48 to 657 ±29	K/Ar	granodiorite	Flemish Cap Avalon Pen. east Newf.	pre-part PE	King *et al.* 1985
5. 526 ±13	Rb/Sr	granite	Musquash N. Brunswick Canada	syn-PЄ (V)	Poole 1980*b*
6. 529 ±6	Rb/Sr illitic shales	Strettonian of Longmyndian	Salop, UK	syn? PЄ	Bath 1974
7. 529 ±9	Rb/Sr whole-rock	granite	Mandar, S. Sinai	syn-??Є	Bielski 1982
8. 531 ±5	U/Pb Pb/Pb zircon	granophyre	Ercall, Salop, UK	pre-late Tom.	Compston *et al.* 1984
9. 533 ±12	Rb/Sr	granophyre	Ercall, Salop, UK	pre-late Tom.	Beckinsale *et al.* 1984*c*
10. 534 ±10	U/Pb zircon	syenite	Jbel Boho, Morocco	??	Choubert 1952 Odin *et al.* 1985
11. 535 ±5	U/Pb monazite	granite	'Armorica', France	??	Odin *et al.* 1985
12. 536 ±14	U/Pb zircon	paragneiss	'Armorica', France	??	Odin *et al.* 1985
13. 537 ±29	Rb/Sr whole-rock	Harbour Main & Bull Arm Volc.	SE Newfoundland, Canada	syn-PЄ	Fairbairn *et al.* 1966 recalc. McCartney *et al.* 1966
14. 540 ±10	U/Pb, Pb/Pb monazite	granite	Vire-Carolles	post-PE (V)	Odin *et al.* 1983
15. 540 ±62	Rb/Sr whole-rock	granite	'Armorica', France	post-PE(V) pre-?Atb	Odin *et al.* 1985
16. 540 ±5	U/Pb, Rb/Sr monazite, sphene, zircon	granite (migmatites)	Vire-Carolles Normandy, France	post-PE(V) pre-?Atb	Peucat 1982
17. 540 ±52	Rb/Sr	diorite	Charnwood, UK	post-Edc pre-Etn	Cribb 1975
18. 541 ±5	U/Pb zircon	granite vein	'Armorica', France	post-PE(V) pre-?Atb	Odin *et al.* (1985) Peucat 1982
19. 542 ±12	Rb/Sr	gneiss Mona Complex	North Wales, UK	pre-Arenig (Ord.)	Beckinsale *et al.* 1984
20. 542 ±31	Rb/Sr whole-rock	granite	Vire-Carolles Normandy, France	post-PЄ pre-Atb	Odin *et al.* 1985
21. 546 ±22	Rb/Sr	diorite	Charnwood, UK	post-Edc	Cribb 1975
22. 547 ±6	Rb/Sr whole-rock	ignimbrite	Lezardrie Ash, UK	post-PЄ pre-Atb	Odin *et al.* 1982

Table 9.2—*continued*

Radio metric age Ma ±	Method/ material	Rock unit sampled (without rank)	Location/ region	pre-, syn-, post- (abbrev. from Table 9.1)	References (primary source precedes secondary)
23. 548 ±5	Rb/Sr	volcanics	Sinai–Jordan	post-? PЄ pre-?	Bielski 1982
24. 549 ±19	Rb/Sr	?Mona Complex	Sarn intrusion, N. Wales, UK	pre-Arenig	Beckinsale *et al.* 1984*a,b*
25. 553 ±10	Rb/Sr whole-rock	Northbridge granite	nr. Hoppin Hill, Mass., USA	syn-Tom pre Є-Tom	Fairbairn *et al.* 1967 recalc. Gale 1982
26. 553 ±17	K/Ar	Pestrotsvet	Aldan R., Yakutia, USSR	syn-Tom	Cowie & Cribb 1978
27. 553 ±10	Rb/Sr whole-rock	granite-gneiss	Hoppin Hill, Mass., USA	pre-LЄ shales ?late Atb	Gale 1982
28. 554	Pb/Pb monazite	granite	Vire-Carolles, Normandy, France	post-?PЄ/ ?Є	Pasteels & Doré 1982
29. 554 ±9.50	Rb/Sr whole-rock	ignimbrite	Port-Scarff, Loquiry, France	post-PЄ	Odin *et al.* 1982
30. 555 ±25	U/Pb sphene	granite vein	'Armorica', France	??	Odin *et al.* 1985
31. 558 ±8	Rb/Sr whole-rock	Uriconian	Salop, England, UK	syn-PЄ	Beckinsale *et al.* 1984*a*
32. 558.10 −16.74	K/Ar	—	Ashinsk, S. Urals, USSR	syn-PЄ	Cowie *in* Harland *et al.* 1964
33. 560	Rb/Sr	granite	Cape Breton I., Nova Scotia, Canada	pre-Є post-? PЄ	Cormier 1972 Clarke *et al.* 1980
34. 610 to 560	K/Ar glauconite	Valdai Group	USSR	syn-PЄ (V)	Cowie & Johnson 1985
35. 561.50 −16.85	K/Ar biotite	granite	Vire-Carolles, Normandy, France	post-PЄ pre-Atb	Cowie *in* Harland *et al.* 1964
36. 563 ±2.50	U/Pb zircon	Quarzazate Volcanics	Morocco	post-PЄ pre-Atb	Odin *et al.* 1982
37. 568 ±12	U/Pb whole-rock	Shuijintuo	East Yangtze Gorges, China	syn-Є	Zhang *et al.* 1984 Xing *et al.* 1984
38. 569 ±12	Rb/Sr	Shuijintuo	Guizhou Prov., S. China	syn-Є	Zhang *et al.* 1984 Xing *et al.* 1984
39. 570 ±4	Rb/Sr illite	Shuijintuo equivalent	East Yangtze Gorges, China	syn-Є	Zhang *et al.* 1982
40. 510 −610	K/Ar illite	Volhyn Basalts	Russian Platform, USSR	syn-PЄ (V)	Cowie & Johnson 1985
41. 572 ±14	Rb/Sr illite	Shuijintuo	East Yangtze Gorges, China	syn-Є	Zhang *et al.* 1984
42. 572 ±4	Rb/Sr illite	Shuijintuo	East Yangtze Gorges, China	syn-Є	Zhang *et al.* 1984
43. 573 ±7	Rb/Sr illite	Shuijintuo	East Yangtze Gorges, China	syn-Є	Zhang *et al.*1984 Xing *et al.* 1984
44. 573 ±32	U/Pb illite	Shuijintuo	East Yangtze, Gorges, China	syn-Є	Zhang *et al.* 1984
45. 573 ±32	Pb/Pb; U/Pb black shales	Shuijintuo	East Yangtze Gorges, China	syn-Є	Xing *et al.* 1984

Table 9.2—*continued*

Radio metric age Ma ±	Method/ material	Rock unit sampled (without rank)	Location/ region	pre-, syn-, post- (abbrev. from Table 9.1)	References (primary source precedes secondary)
46. 574 ±11	Rb/Sr	Shunacadie pluton	Cape Breton I., Nova Scotia, Canada	syn-? PЄ	Poole 1980a
47. 574 ±20	Rb/Sr illite	Shuijintuo	East Yangtze Gorges, China 1982	syn-Є	Ma, pers. comm.
48. 574.61 ±15.00	K/Ar	?	Dniepr River, USSR	syn-PЄ	Armstrong in Cohee 1978
49. 575	Rb/Sr; U/Pb	Shuijintuo	East Yangtze Gorges, China	syn-Є	Compston pers. comm.
50. 578 ±15	U/Pb zircon	rhyolite Ouarzazate Gp.	Bou Ourhioul, High Atlas, Morocco	?	Jéury et al. 1974
51. 578 ±17	K/Ar	Pestrotsvet	Aldan River Yakutia, USSR	syn-Є (Tom)	Cowie & Cribb 1978
52. 579 −17	K/Ar	Laminarites Beds	USSR	syn-PЄ	Cowie in Harland 1964
53. 580 ±4	Rb/Sr illite	Badaowan	East Yunnan, China	syn-Є (Tom)	Xing et al. 1984
54. 580 ±20	Rb/Sr	Volcanics intruded by granite	SE Newfoundland, Canada	syn-PЄ	Dallmeyer et al. 1981
55. 580 ±25	Rb/Sr	Doushantuo	East Yangtze Gorges, China	syn-PЄ (Z)	Zhang et al. 1982
56. 580 ±8	Rb/Sr	Dahai Mb. Yuhucun Fm	East Yunnan, China	syn-Є	Luo, pers. comm. 1984
57. 585 ±8	Rb/Sr	Badaowan	East Yunnan, China	syn-Є (?Atb)	Xing et al. 1984
58. 585 ±15	Rb/Sr	granite	Holyrood, Avalon Penin. SE Newfoundland	post-PЄ (V) pre-Є (?Atb)	Gale 1982
59. 586 +3 −2	rhyolite	Harbour Main Group volcanics	Newfoundland, Canada	syn-PЄ	Krogh et al. 1983
60. 587 ±10	K/Ar	Yudoma	River Yudoma, Siberia, USSR	syn-?PЄ (?V)	Cowie & Cribb 1978
61. 587 ±18	K/Ar glauconite	Yudoma	River Yudoma, Siberia, USSR	syn-?PЄ (?V)	Cowie & Cribb 1978
62. 580 ±8	Rb/Sr illite	Badaowan	East Yunnan, China	syn-Є	Xue 1984 Xing et al. 1984
63. 585 ±15	Rb/Sr illite	Badaowan	East Yunnan, China	syn-Є	Xue 1984 Xing et al. 1984
64. 585 ±15	Rb/Sr	Holyrood granite	Avalon Penin., Newfoundland, Canada	post-PЄ	Poole 1980a,b
65. 587 ±17	Rb/Sr illite	Badaowan	East Yunnan, China	syn-Є	Xing et al. 1984 Luo pers. comm. 1984 Zhang et al. 1984
66. 588 ±13	Rb/Sr illite	Badaowan	East Yunnan, China	syn-Є	Zhang et al. 1984 Xue 1984 Luo pers. comm. 1984

Table 9.2—*continued*

Radiometric age Ma ±	Method/ material	Rock unit sampled (without rank)	Location/ region	pre-, syn-, post- (abbrev. from Table 9.1)	References (primary source precedes secondary)
67. 590 −18	K/Ar glauconite	Blue Clay	Leningrad area, USSR	post-PꞒ pre-Ꞓ (?Tom)	Keller & Krasnobaev 1983
68. 590 ±11	Rb/Sr	Holyrood granite	Avalon Penin, Newfoundland	syn-PꞒ	Dallmeyer *et al.* 1981
69. 590 ±30	U/Pb zircon	Love Cove volcanics	SE Newfoundland, Canada	syn-PꞒ	Dallmeyer *et al.* 1981
70. 590	K/Ar muscovite	granite pebbles, Coldbrook Volc.	New Brunswick Canada	syn-PꞒ	Stockwell *et al.* 1963 Leech *et al.* 1963
71. 592 ±16	Rb/Sr whole-rock	Lontova Clays	USSR, Russian Platform	post-PꞒ pre-?Tom	Keller & Krasnobaev 1983
72. 594 ±11	Rb/Sr whole-rock	Holyrood granite	Newfoundland	post-PꞒ pre-Ꞓ(Atb)	Cowie *in* Harland *et al.* 1964
73. 595	Rb/Sr	Narragansett granites	Rhode I. USA	syn-PꞒ	Skehan & Murray 1980
74. 599 −18	K/Ar	Karatau Series	Kazakhstan, USSR	syn-PꞒ	Cowie *in* Harland *et al.* 19864
75. 600	U/Pb zircon	Love Cove volcanics	SE Newfoundland, Canada	syn-PꞒ	O'Driscoll & Gibbons 1980
76. 602 ±3	U/Th/Pb zircon	rhyolite Mattapan Volc.	Massachusetts, USA	syn-PꞒ (Var)	Kaye & Zartman 1980
77. 602 ±15	Rb/Sr illite	Dengying Fm. Tientzushan Mb.	East Yangtze Gorges, China	syn-Ꞓ	Zhang *et al.* 1982 & 1984 Xing *et al.* 1984
78. 602 ±8	Rb/Sr illite	Tientzushan Mb. Dengying Fm	East Yangtze Gorges, China	syn-Ꞓ (Tom)	Xing *et al.* 1984
79. 603 ±14	Rb/Sr	granite intruding volcanics	Narragansett Rhode I., USA	syn-PꞒ	Smith & Giletti 1978
80. 603 ±31	Rb/Sr illite	Badaowan	East Yunnan, China	syn-Ꞓ	Xing *et al.* 1982 Luo pers. comm. 1984
81. 603 ±34	Rb/Sr	granite	Coedana, N. Wales, UK	?	Beckinsale & Thorpe 1979
82. 605	Rb/Sr whole-rock	granite	Vire-Carolles France	post-? PꞒ pre-?Ꞓ(Atb)	Odin 1982
83. 606 +4 −3	ignimbrite	Harbour Main Gp (James Cove)	SE Newfoundland, Canada	syn-PꞒ	Krogh *et al.* 1983
84. 606 +4 −3	U/Pb zircon	volcanics	Burin Penin., Newfoundland, Canada	pre-Ꞓ	Krogh *et al.* 1983
85. 607 ±11	Rb/Sr	Holyrood granite	Avalon Penin., Newfoundland	post-? V syn-V pre-Ꞓ(Atb)	Frith & Poole 1972
86. 607 +8 −4	Rb/Sr	Marystown Group volcanics	Burin Penin., Newfoundland	syn-PꞒ	Krogh *et al.* 1983
87. 608 ±15	Rb/Sr	Datangpo	East Yangtze Gorges, China	syn-PꞒ (Z)	Zhang *et al.* 1982
88. 608 ±25	Rb/Sr	Marystown volcanics	Burin Penin., Newfoundland	syn-PꞒ	Dallmeyer *et al.* 1981

Table 9.2—*continued*

Radio metric age Ma ±	Method/ material	Rock unit sampled (without rank)	Location/ region	pre-, syn-, post- (abbrev. from Table 9.1)	References (primary source precedes secondary)
89. 610	Rb/Sr	Holyrood granite	Avalon Penin., Newfoundland	syn-PЄ	King *et al.* 1985
90. 610	K/Ar glauconite	just above Laplandian tillite	USSR	post-PЄ (Var) syn-? PЄ (Var)	Chumakov & semikhatov 1979
91. 612 ±36	Rb/Sr	Badaowan	East Yunnan, China	syn-Є	Xing *et al.* 1984
92. 613 ±23	Rb/Sr	Shuijintuo	East Yangtze Gorges, China	syn-PЄ(Z)	Ma *et al.* 1980 Xing *et al.* 1984
93. 614 ±20	U/Pb uraninite	?	Shinkolobwe Katanga, Congo	syn-PЄ(Z)	Cowie *in* Harland *et al.* 1984
94. 614 ±18	Rb/Sr	Brimatuo Mb. Dengying Fm	East Yangtze, China	syn-PЄ(Z)	Zhang *et al.* 1982
95. 615 ±20	K/Ar granodiorite	submarine (at present) intrusion	Flemish Cap east of Avalon, Newfoundland	syn-PЄ pre-PЄ	Pelletier 1971; recalc. King *et al.* 1985
96. 615 ±37	Rb/Sr	Musquash pluton	Ludgate Lake New Brunswick, Canada	syn-PЄ(V)	Olszewski & Gaudette *in* Poole 1980*b*
97. 620 ±20	U/Pb zircon	metam. pyroclastics	North Carolina USA	syn-PЄ (Edc)	Cowie & Johnson 1985
98. 640 −620	Rb/Sr	Pound	Flinders Ranges, South Australia	syn-PЄ(Edc)	Jenkins 1981
99. 620 ±2	Rb/Sr	Holyrood granite	Avalon Penin., Newfoundland	post-PЄ(V) pre-Є(Atb)	Krogh *et al.* 1983
100. 623 ±2	U/Pb zircon	volcanics	Burin Penin., Newfoundland	syn-PЄ	Krogh *et al.* 1983
101. 677 ±72	glauconite	rhyolite	Wrekin, Salop., England, UK	pre-Є(Etn)	Fitch *et al.* 1969, recalc. Patchett *et al.* 1980
102. 632 ±32	glauconite	rhyolite	Wrekin, Salop., England, UK	pre-Є(Etn)	Fitch *et al.* 1969, recalc. Patchett *et al.* 1980
103. 630 ±15	Rb/Sr	Dedham granodiorite	Massachusetts, USA	syn-?Є(Etn) pre-?Є(Etn)	Kaye & Zartman 1980
104. 640	K/Ar muscovite	granite pebbles, Coldbrook Group volcanics	New Brunswick, Canada	syn-PЄ(V)	Leech *et al.* 1963
105. 653 ±23	Rb/Sr	Nyborg Fm (intertillite)	Finnmark, Norway	syn-PЄ (mid. Var)	Pringle 1973
106. 660 −665	K/Ar glauconite	just below Laplandian tillite	Russian Platform, USSR	syn-PЄ (Var) pre-PЄ (Var)	Chumakov & Semikhatov 1979 & 1981
107. 672 ±70	shales	above glacial deposits	Kimberley, Australia	syn-PЄ (?Sturtian & Marinoan)	Cloud & Glaessner 1982
108. 670 ±84	shales	above glacial deposits	Kimberley, Australia	syn-PЄ (?Sturtian & Marinoan)	Cloud & Glaessner 1982
109. 672 ±84	Rb/Sr whole-rocks shales	correl. with Lower Brachina Fm	South Australia	syn-PЄ (Edc)	Cloud & Glaessner 1982

Table 9.2—*continued*

Radio metric age Ma ±	Method/ material	Rock unit sampled (without rank)	Location/ region	pre-, syn-, post- (abbrev. from Table 9.1)	References (primary source precedes secondary)
110. 691 ±29	Rb/Sr	Doushantuo	East Yangtze Gorges, China	syn-P€(Z)	Ma *et al.* pers. comm. 1982; Cowie & Johnson 1985
111. 700 −740	Rb/Sr	Nantuo (glacial epoch)	East Yangtze Gorges, China	syn-P€ (Edc)	Ma *et al.* 1980
112. 700 ±5	Rb/Sr	Doushantuo	East Yangtze Gorges, China	syn-P€(Z)	Zhang *et al.* 1982
113. 727 ±9	Rb/Sr	Doushantuo	East Yangtze Gorges, China	syn-P€(Z)	Zhang *et al.* 1982
114. 728 ±27	Rb/Sr	Datangpo (pre-Nantuo)	East Yangtze Gorges, China	syn-P€(Z)	Ma *et al.*, pers. comm. 1982
115. 750 ±53	Rb/Sr	?	Adelaide Geosyncline, Australia	syn-P€ (Riphean–Vendian boundary)	Preiss & Forbes 1981
116. 750 ±80	Rb/Sr	Coldbrook Group	New Brunswick, Canada	syn-P€(V)	Cormier 1969
117. 751 −833	U/Pb zircon	granodiorite	Flemish Cap., W. of Avalon, Newfoundland	syn-P€	King *et al.* 1985

important criterion of the coherence and meaningfulness of the results'. Odin *et al.* (1985, p. 70) also give their opinion, however, that it would '. . . be an error to interpret the apparent ages from China as giving the age of deposition . . .'. Further discussion and data are clearly needed.

2. Entries 8 and 9 have benefited from a recent publication by Cope & Gibbons (1987)—recent extension of the Ercall Quarry has revealed an unconformable contact (*not*, as previous suggested, an intrusive relationship) between the Ercall granophyre and overlying Lower Cambrian sediments. Like the Chinese Entries 37 and 49, Rb–Sr determinations are apparently corroborated by U–Pb methods. In the case of the Ercall granophyre there are veining, intense alteration, secondary mineralization, and lined microfractures that suggest (Cope & Gibbons 1987, p. 59) limited fluid movement and alteration which might have significantly affected the initial $^{87}Sr/^{86}Sr$ ratio. Full publication is now important for assessing the U–Pb dates on zircons. Compston *et al.* (1984), in an abstract only, also mentioned an older date of 565 ± 7 Ma (considered by Compston *et al.* to be 'spurious'), while the 531 ± 5 Ma date was interpreted as 'well-substantiated'). Work under the International Commission on Stratigraphy on

the Precambrian–Cambrian Boundary is progressing rapidly, but the global stratigraphical correlation is quite incomplete and subject to change at the time of writing. The Tommotian Stage of Siberia is not generally accepted as a global stage and it may be that underlying strata may be assigned to the Cambrian system. Speculative estimates of the duration of earliest Cambrian time, pre-Wrekin Quartzite, are at present premature in global stratigraphical terms and are not based on fully reliable radiometric data.

3. Entries 14, 15, 16, 18, 20, 29, and 30 have been discussed at great length by Odin *et al.* (1985) and Cowie & Johnson (1985). The stratigraphical evidence remains equivocal. The question of rejuvenation was raised first by the veteran French geologists: Choubert & Faure-Muret (1980, pp. 85 & 150)—'Cadomides II were rejuvenated to 560–540 Ma as were their numerous granites'. However, Odin *et al.* (1985, p. 65) dismiss rejuvenation as a very uncertain hypothesis in the light of recent results. Further sophisticated comments by Odin *et al.* (1985) involve the Icartian gneiss being 'allocated' to the early Cambrian Atdabanian Stage as a *reductio ad absurdum* in a site of known radiometric dates giving 2000 Ma for the gneiss (Calver & Vidal 1978; Auvray *et al.* 1980).

The absence of a fully published account of an earliest Cambrian faunal assemblage in the Massif Armoricain is unfortunate and limits stratigraphical interpretation (trace fossils may prove to be a substitute or supplement): their absence may not indicate a stage in the evolution of life, but may be due to an inappropriate facies and is not comparable with the situation in Siberia. The age of the Upper Brioverian remains uncertain and may be late Precambrian or early Cambrian (Mansuy & Vidal (1983) gave good fossil evidence that the Middle–Lower Brioverian should be assigned to a Precambrian age). There is no fundamental reason that the position and nature of the flysch facies of the Upper Brioverian below an unconformity (peneplain) unequivocally excludes it from a Phanerozoic early Cambrian age, even though such arguments were generally used in the past for Precambrian versus Cambrian stratigraphical allocations (Lipalian interval).

4. Entry 25: the Northbridge granite, in common with other Precambrian granites of the Avalon Zone, is close to sediments of early Cambrian age. The granite is either syn-Etcheminian *or* Precambrian to Etcheminian in age and may be indicating an age close to 550 Ma, possibly near the putative Precambrian–Cambrian boundary on the basis of thickness of overlying sediments. Entry 25 was recalculated by Gale (1982). Entry 103: the Dedham Granodiorite at 630 ± 15 Ma may also be syn-?Cambrian (Etcheminian) or pre-?E (?Etcheminian). It appears to be much older radiometrically, and when considered together with Entry 25 and other maritime North American intrusions, indicates the great uncertainty prevailing. It was *not* recalculated.

9.5. CONCLUSIONS

Our objective in this chapter has been to summarize the data relevant to the chronometry of the initial Cambrian boundary, in order to clarify the problem. It is possible that a very thorough study of available data might significantly constrain the age of the boundary. However, it seems currently that decisive or critical determinations are not consistent and so throw doubt on our interpretation of the observational data. Attempts can be made to draw a conclusion by eliminating, for one reason or another, some determinations that conflict with others. However, we are not entirely satisfied that the state of the art can yet give us confidence to do so. These circumstances question many of the assumptions that are commonly made and it has been one purpose of this chapter to encourage radical thinking until a

preponderance of data reinforces one side or another.

The need for radical reappraisal means that it is not necessarily sensible to average-out data if large bodies of data are unreliable. Since a brief review of this matter (Harland 1983), new data have reinforced the claims for both older and younger ages.

For inspection of the list (Table 9.2 pp. 189–94 this volume) it seems to us that the following determinations might prove to be critical:

1. Rb-Sr whole-rock determinations of the *Yunnan* earliest Cambrian rocks. High (old) ages for the boundary are supported by many determinations. However, all could be distorted by significant detrital clay-mineral fractions. Nevertheless, where uranium–lead determinations give supporting evidence, these results may not be discarded lightly.

2. On the other hand, there are a number of plutonic rocks which seem to constrain the age of the boundary the other way. Some of these may have unclear structural relationships with the targeted rocks, but some do somewhat constrain the age, for example:

a. Hoppin Hill and related (e.g. Northbridge) granites in maritime North America; (Entry 25, Table 9.2) 553 ± 10 Ma

b. Ercall granophyre, Salop, UK (Entries 8 & 9, Table 9.2) 531 ± 5 & 533 ± 12 Ma

These data cannot be ignored by those claiming a higher (older) age. The possibility must be borne in mind that we have not sufficiently understood the aspect of magmatism sampled by these methods. There appeared to be two groups of determinations, namely those greater than 590 and those less than 560 Ma, and the traditional ages within the span (e.g. 570 Ma) seemed to have little direct support. Therefore a wholesale discarding of the higher ages or the lower ones might lead, respectively, to an age of less than 560 or more than 590 Ma. Inspection of our list in Table 9.2 shows that the gap is not substantiated. Therefore we have adopted *570 Ma* as a provisional date, knowing that it is presumably not the true age but may avoid a greater error. Conversely, we may return to the *Geological Society of London Memoir* **10** where Cowie & Johnson (1985, p. 61) gave a range of 530–600 Ma to indicate the uncertainty. If we argue for 570 Ma, it is only to provide an interim value as a convention, pending a proper solution.

At the recent conference in St John's Newfoundland, of the Working Group on the Precambrian–Cambrian

Boundary (IUGS), in August 1987, Benus reported a date of 565 ± 3 Ma (U – Pb in zircons) for an ash bed at Mistaken Point containing Precambrian metazoans. The stratigraphical position of this bed is more than 7000 metres below the putative Precambrian–Cambrian boundary horizon (Narbonne 1988, *in press*).

This new result, which still requires confirmation by the publication details, does appear to support a Precambrian–Cambrian radiometric age nearer to the younger limit of 530 Ma mentioned above. These younger limit determinations are not, however, closely defined biostratigraphically, and seem to be coming from the North Atlantic Ocean margins which may have a common tectonostratigraphical history. This is only a small part of the world (comparatively), but one where modern research is giving very interesting results.

REFERENCES

Armstrong, R. L. (1978). Pre-Cenozoic Phanerozoic time scale–computer file of critical dates and consequences of new and in-progress decay-constant revisions. In *Contributions to the geological time scale*, (eds Cohee *et al.*) Am. Assoc. Pet. Geol. Stud. Geol. No. 6, 73–92.

Auvray, B., Macé, J., Vidal, P., & Van Der Voo, R. (1980). *J. geol. Soc. Lond.*, **137**, 207–10.

Bath, A. H. (1974). New isotopic age data on rocks from the Long Mynd, Shropshire. *J. geol. Soc. Lond.*, **130**, 567–74.

Beckinsale, R. D. & Thorpe, R. S. (1979). Rubidium–strontium whole-rock isochron evidence for the age of metamorphism and magnetism in the Mona Complex of Anglesey. *J. geol. Soc. Lond.*, **136**, 433–9.

Beckinsale, R. D., Thorpe, R. S., Wadge, A. J., Evans, J. A., Molyneux, S., & Reedman, A. J. (1984*a*). Rb–Sr whole rock isochron ages for igneous and metamorphic rocks from N. Wales related to calibration of the time scale. *Bull. Liais. Inf. IGCP Proj.*, **196**, (2), p. 43.

Beckinsale, R. D., Evans, J. A., Thorpe, R. S., Gibbons, W., & Harmon, R. S. (1984*b*). Rb–Sr whole-rock isochron ages. $\delta^{18}O$ values and geochemical data for the Sarn Igneous Complex and Parwyd Gneisses of the Mona Complex of Llêyn, North Wales. *J. geol. Soc. Lond.*, (in press).

Beckinsale, R. D., Thorpe, R. S., Wadge, A. J., Evans, J. A., Molyneux, S. & Reedman, A. J. (1984*c*). Rb–Sr whole-rock isochron ages for igneous and metamorphic rocks from North Wales and the Welsh Borderland related to the Phanerozoic time scale. *J. geol. Soc. Lond.*, 17 pp, (in press?).

Bielski, M. (1982). On the problem of the Lower Cambrian age in the Arabian–Nubian Massif 948. In *Numerical dating in stratigraphy* (ed. G. S. Odin), John Wiley, Chichester, pp. 943–8.

Calvez, J. Y. & Vidal, P. (1978). *Contrib. Miner. Petrol.*, **65**, 395–9.

Choubert, G. (1952). Le volcan géorgien de la région d'Alououm (Anti-Atlas). *C. R. Acad. Sci., Paris, Série D*, **234** 350–2.

Choubert, G. & Faure-Muret, A. (1980). The Precambrian in North Peri-Atlantic and South Mediteranean mobile zones: general results. *Earth-Sci. Rev.*, **16**, 85–219.

Chumakov, G. & Semikhatov, M. A. (1979). Precambrian stratigraphic scale of the USSR. *Geol. Mag.* **116**, (6), 419–27.

Chumakov, N. M. & Semikhatov, M. A. (1981). Riphean and Vendian of the USSR. *Precambr. Res.* **15**, 229–53.

Clarke, D. B., Barr, S. M., & Donohoe, H. V. (1980). Granitoid and other plutonic rocks of Nova Scotia. In *Proceedings 'The Caledonides in the USA'*, (ed. D. R. Wones), pp. 107–11. *Dept. geol. Sci. Virginia Polytech. Inst. State Univ. Mem.*, 2.

Cloud, P. E. & Glaessner, M. F. (1982). The Ediacarian Period and System: metazoa inherit the Earth. *Science*, **218**, 783–92.

Compston, W. & Zi-Chau Zhang (1983). The numerical age of the base of the Cambrian. *Geol. Soc. Aust. Abstr.*, **9**, p. 235.

Compston, W., Williams, I. S., Obradovich, J. D., & Patchett, J. D. (1984). Zircon U–Pb age of the Ercall granophyre. *Terra Cognita Spec. Iss.*, Abstracts of ECOG VIII **40.**

Cope, J. C. W. & Gibbons, W. (1987). New evidence for the relative age of the Ercall Granophyre and its bearing on the Precambrian–Cambrian boundary in southern Britain. *Geol. J.*, **22**, 53–60.

Cormier, R. F. (1969). Radiometric dating of the Coldbrook Group of southern New Brunswick, Canada. *Can. J. Earth Sci.*, **6**, 393–8.

Cormier, R. F. (1972). Radiometric ages of granitic rocks, Cape Breton Island, Nova Scotia. *Can. J. Earch Sci.*, **9**, 1074–86.

Cowie, J. W. (1964). The Cambrian period in the Phanerozoic time-scale. In *Q. J. geol. Soc. Lond.*, **120s** (eds W. B. Harland *et al.*), pp. 255–8.

Cowie, J. W. & Cribb, S. J. (1978). The Cambrian System. In *Contributions to the geological time scale*, (G. V. Cohee *et al.*), pp. 355–62. Am Assoc. Pet. Geol. Stud. Geol., No. 6.

Cowie, J. W. & Johnson, M. R. W. (1985). Late Precambrian and Cambrian geological time-scale. In *The Chronology of the geological records* (ed. N. J. Snelling), Geol. Soc. Lond. Mem. 10, Blackwell Scientific, Oxford, 47–64.

Cowie, J. W., Ziegler, W. Boucot, A. J., Bassett, M. G., & Remane, J. (1986). Guidelines and Statutes of the International Commission on Stratigraphy (ICS). *Cour. Forsch- Inst. Senckenberg* **83**, 1–14.

Cribb, S. J. (1975). Rubidium–strontium ages and strontium isotope ratios from the igneous rocks of Leicestershire. *J. geol. Soc. Lond.*, **131**, 203–12.

Dallmeyer, R. D., Odom, A. L., O'Driscoll, C. F., & Hussey, E. M. (1981). Geochronology of the Swift Current granite and host volcanic rocks of the Love Cove Group, south western Avalon Zone, Newfoundland: evidence of a late Proterozoic volcanic–subvolcanic association. *Can. J. Earth Sci.*, **18**, 699–707.

Eysinga, von, D. (1975). Stratigraphic wall chart, Elsevier, Amsterdam.

Fairbairn, H. W., Bottino, M. L., Pinson, W. H. & Hurley, P. M. (1967). Whole-rock age and initial $^{87}Sr/^{86}Sr$ of volcanics underlying fossiliferous Lower Cambrian in the Atlantic provinces of Canada. *Can. J. Earth Sci.*, **3**, 509–21.

Fitch, F. J., Miller, J. A., Evans, A. L., Grasty, R. L., & Meneisy, M. Y. (1969). Isotope age determinations on rocks from Wales and the Welsh Borders. In *The Precambrian and Lower Palaeozoic rocks of Wales*, (ed. A. Wood), pp. 23–46. University of Wales Press, Cardiff.

Frith, R. A. & Poole, W. H. (1972). Late Precambrian rocks of eastern Avalon peninsula, Newfoundland—a volcanic island complex: discussion. *Can. J. Earth Sci.* **9**, 1058–9.

Gale, N. H. (1982). Numerical dating of Caledonian times (Cambrian to Silurian). In *Numerical dating in stratigraphy* (ed. G. S. Odin), pp. 467–86. John Wiley, Chichester.

Glaessner, M. F. (1977). The Ediacara fauna and its place in the evolution of the Metazoa. In *Correlation of the Precambrian 1*, (ed. A. V. Sidorenko), pp. 257–68, Akad. Nauk.

Harland, W. B. (1983). More time scales. *Geol. Mag.*, **120**, (4), 393–400.

Harland, W. B., Smith, A. G., & Wilcock, B. (1964). The Phanerozoic time-scale. In *Q. J. geol. Soc. London.*, **120s**, (eds W. G. Harland *et al.*), pp. 260–2.

Harland, W. B., Cox, A. V., Llewellyn, P. G., Pickton, C. A. G., Smith, A. G., & Walters, R. (1982). *A geologic time scale*. Cambridge University Press.

Holmes, A. (1913). *The age of the earth*. Harper, London & New York.

Holmes, A. (1937). *The age of the earth*. Nelson, London, 263 pp.

Holmes, A. (1947). The construction of a geological time-scale. *Trans. geol. Soc. Glasgow*, **21**, 117–52.

Jenkins, R. J. F. (1981). The concept of an 'Ediacaran Period' and its stratigraphic significance in Australia. *Trans. R. Soc. Aust.* **105**: 179–94.

Jéury, A., Lancelot, J. R., Hamet, J., Proust, F., & Allègre, C. J. (1974). L'âge des rhyolites du Precambrien II du Haut-Atlas et le probleme de la limite Précambrien–Cambrien. 2e. *Réun. Ann. Sci. Terre, Nancy* **230**.

Kaye, C. A. & Zartman, R. F. (1980). A late Proterozoic Z to Cambrian age for the stratified rocks of the Boston Basin, Massachusetts, USA. In *Proceedings 'The Caledonides in the USA'*, (ed. D. R. Wones), pp. 257–61. Dept. Geol. Sci. Virginia Polytech. Inst. State Univ. Mem. 2.

Keller, B. M. & Krasnobaev, A. A. (1983). Late Precambrian geochronology of the European USSR. *Geol. Mag.*, **120**, 381–9.

King, L. H., Poole, W. H., & Wanless, R. K. (1985). Geological setting and age of the Flemish Cap granodiorite, East of the Grand Banks of Newfoundland. *Can. J. Earth Sci.*, 22, 1286–98.

Krogh, T. E., Strong, D. F., & Papezik, V. (1983). Precise U–Pb ages of zircons from volcanic and plutonic units in the Avalon Peninsula [sic]. *Geol. Soc. Am. Abstr. 18th Ann. Meeting, NE Section* **15**, (3), p. 135. [Ref. No. 13430].

Leech, G. B., Lowdon, J. A., Stockwell, C. H., & Wanless, R. K. (19863). Age determinations and geological studies (including isotopic ages—Report 4). *Geol. Surv. Can. Pap.*, **63–17**, 1–140.

McCartney, W. D., Poole, W. H., Wanless, R. K., Williams, H., & Loveridge, W. D. (1966). Rb/Sr age and geological setting of the Holyrood Granite, southeastern Newfoundland. *Can. J. Earth Sci.*, **3**, 947–57.

Ma Guogan, Lee Huaqin, & Xue Xiaofeng (1980). [Isotopic ages of the Sinian in the east Yangtze gorges with a discussion on the Sinian geochronological scale of China.] *Bull. Chinese Acad. geol. Sci.*, VIII **1**, (1), 39–55. [In Chinese with English summary.]

Ma Guogan, *et al.* (pers. comm.) (1982). A rediscussion of time limits of the Sinian System in South China [abstract circulated to IUGS–IGCP Project 29 Working Group on Precambrian–Cambrian Boundary during visit to China, November 1982.]

Mansuy, C. & Vidal, G. (1983). Late Proterozoic Brioverian microfossils from France: taxonomic affinity and implications of plankton productivity. *Nature*, **302**, 606–7.

Matthew, G. F. (1899). A Palaeozoic terrane beneath the Cambrian. *Ann. N.Y. Acad. Sci.*, **12**, 41–56.

Narbonne, G. M. (1988). Trace fossils, small shelly fossils and the Precambrian–Cambrian boundary. *Episodes*, in press.

Odin, G. S. (ed.) *et al.* (1982). *Numerical dating in stratigraphy*. Pt I and Pt II (2 volumes). John Wiley, Chichester.

Odin, G. S. *et al.* (1983). Numerical dating of the Precambrian–Cambrian boundary. *Nature*, **301**, 21–3.

Odin, G. S., Gale, N. H., & Doré, F. (1985). Radiometric dating of Late Precambrian, pp. 65–72. In *The chronology of the geological record*, (ed. N. J. Snelling), Geol. Soc. Lond. Mem., 10. Blackwell Scientific, Oxford.

O'Driscoll, C. F. & Gibbons, R. V. (eds) (1980). Geochronology report—Newfoundland and Labrador. *Current Res. Min. Dev. Div., Dep. Mines & Energy. St John's* [*Newfoundland*], *Rep*, **80–1**, 143–6.

Olszewski, W. J. & Gaudette, H. E. (1980). Musquash pluton. *Geol. Surv. Can. Pap.*, **80–1C**, 170–3.

Palmer, A. R. (1983). The Decade of North American Geology 1983. Geologic Time Scale. *Geology*, **11**, 503–4.

Pasteels, P. & Doré, F. (1982). Age of the Vire-Carolles granite. In *Numerical dating in stratigraphy*, (ed. Odin, G. S.), pp. 784–99. John Wiley, Chichester.

Patchett, P. J., Gale, N. H., Goodwin, R., & Humm, M. J. (1980). Rb–Sr whole-rock isochron ages of late Precambrian to Cambrian igneous rocks from southern Britain. *J. geol. Soc. Lond.*, **137**, 649–56.

Pelletier, B. R. (1971). A granodiorite drill-core from the Flemish Cap, eastern Canadian continental margin. *Can. J. Earth Sci.*, **8**, 1499–1503.

Peucat, J. J. (thesis) (1982). *Mem. Soc. géol. minéral, Bretagne.*

Poole, W. H. (1980*a*). Rb–Sr age of Shunacadie pluton, central Cape Breton Island, Nova Scotia. *Geol. Surv. Can., Pap.*, **80–1C**, 165–9.

Poole, W. H. (1980*b*). Rb–Sr age of some granitic rocks between Ludgate Lake and Negro Harbour, southwestern New Brunswick. *Geol. Surv. Can. Pap.*, **80–1C**, 170–3.

Preiss, W. V. & Forbes, B. G. (1981). Stratigraphy, correlation and sedimentary history of Adelaidean (Late Proterozoic) basins in Australia. *Precambr. Res.* **15**, 255–304.

Pringle, J. R. (1973). Rb–Sr age determinations on shales associated with the Varanger Ice Age. *Geol. Mag.*, **109**, 465–72.

Skehan, J. W. & Murray, D. P. (1980). A model for the evolution of the eastern margin (EM) of the northern Appalachians. In *Proceedings 'The Caledonides in the USA'*, (ed. D. R. Wones), pp. 67–72. Dep. geol. Sci. Virginia Polytech. Inst., State Univ. Mem., 2.

Snelling, N. J. (1985). An interim time scale. In *Chronology of the geological record*, (ed. N. J. Snelling), pp. 261–5. Geol. Soc. Lond. Mem., 10, Blackwells, Oxford.

Stockwell, C. H. (1964). Fourth report on structural provinces, orogenies and time classification of rocks of the Canadian Precambrian shield. *Geol. Surv. Can. Pap.*, **64–17.**

Williams, H. S. (1893). Elements of the geological time scale. *J. Geol.*, **1**, 283–95.

Xing Yusheng & Luo Huilin (1984). Precambrian–Cambrian boundary candidate, Meishucun, Jinning, Yunnan, China. *Geol. Mag.*, **121**, (3), 143–54.

Xing Yusheng, Ding Quxiu, Luo Huilin, He Ting-gui, & Chang Wentang (1984). The Sinian–Cambrian boundary of China and its related problems. *Geol. Mag.*, **121**, (3), 155–70.

Xue Xiaofeng (1984). Research on the isotopic age of the Sinian–Cambrian boundary at the Meishucun section in Jinning County, Yunnan Province, China. *Geol. Mag.*, **121**, (3), 171–3.

Zhang Zichao, Compston, W., & Page, R. W. (1984). The isotopic age of the Cambrian–Precambrian boundary from the Sinian–Lower Cambrian sequence in South China. *Abstract, 5th Inter. Meeting of Geochronology*, Nikko, Japan, July 1982, 2 pp.

10

Palaeoclimatology

W. B. Harland

10.1. INTRODUCTION

This chapter discusses the climatic record over the time spanning the Precambrian–Cambrian boundary.

At present, the stratigraphical record with its imprecise time-correlation is inadequte as a basis for the identification of a detailed short-term sequence of climatic fluctuations. Firstly, I will state my conclusions regarding the stratigraphical record, and then examine them in a little detail. At this stage I will restrict my attention to only two kinds of extreme climatic indicator in the stratigraphical record, namely, cold climate (i.e. glacial) indicators and hot climate indicators, as discussed below:

1. From the stratigraphical evidence we must consider seriously the possibility, (or even probability) of a glacial epoch at or near the Precambrian–Cambrian boundary, i.e. post-dating typical Ediacaran faunas. I have referred to it as the late Sinian Glacial Epoch (Harland 1983a) but it is still a doubtful event. This epoch would be entirely distinct from the very well-established pre-Ediacaran pair of Varangian glaciations and the still earlier Sturtian group of glacials.

2. It is significant that the glacial indications in Varangian strata, and possibly in Upper Sinian glacial strata, are in many instances associated more-or-less closely with hot climatic indicators such as reddened and dolomitic rocks, occasionally containing halite pseudomorphs. This suggests that the ambient climate which the cold spells interrupted was not typically polar. This view is largely supported by the many palaeomagnetic determinations which yield low or intermediate latitudes (e.g. Frakes 1979).

There is a weight of evidence for this opinion which cannot be ignored, accepting that many palaeolatitude determinations should not be taken at face value any more than all the other evidence discussed. The polar glacial proposition allows the assumption that tillites have little value in correlation, and this must be accepted as a possibility, but the weight of evidence is against it.

3. Assuming, for this argument, that some tillites were formed at low or intermediate latitudes, then we have a powerful correlation potential. In most successions, glacial deposits form a minor and intermittent part; they are exceptional rather than typical deposits, but they are easy to identify as tilloids (if not always to interpret as tillites). A glacial spell in an otherwise temperate or tropical climate is likely to be related to near-synchronous (coeval) effects over a large area, possibly a very large portion of the globe.

A degree of diachronism necessarily obtains for glaciation because it involves advance and retreat. The critical argument is, however, that diachronism is on a fine scale for a broadly synchronous, coeval climatic change. This is in opposition to the concept oi a coarse diachronism that would result from polar glaciation allied to rapid polar wandering. Glacial deposits in these polar-wander models would be coeval locally or even regionally, but not on a continental or global scale.

The evidence suggests that low or mid-latitude glaciation did occur, at least in the Varangian Epoch, from the palaeomagnetic evidence (e.g. Harland 1964). Added to this is the absence of evidence for rapid polar wander, or, in other words, the likelihood that a similar global climatic zonation occurred through, for example, late Vendian and early Cambrian times. This period was a time in which the Iapetus Ocean was spreading in an approximately equatorial position.

On this hypothesis there is a good chance that the Varangian glacial epochs and the possible late Sinian glaciation were separate global climatic cold periods, and that the longer intervening time-span was not cold enough to produce ice at seal-level in tropical or temperate latitudes.

It is not my purpose to argue the case in detail; it has been stated more than once (Harland & Herod 1975), and will be attempted elsewhere. For the purpose of this chapter, it is developed as just one of at least two conflicting hypotheses. The one that I favour involves infrequent glacial ages or epochs. Furthermore, tillites in particular can be mapped from rocks where sedimentary characters give good environmental discrimination, even into highly metamorphosed psephites. The correlation potential of tillites, however, depends on the possibility of distinguishing one tillite from another. At this point it is necessary to distinguish three kinds of historical events for use in correlation:

1. Progressive or evolutionary change allows unique or distinctive stages or steps to be identified. This applies supremely for biological evolution, and, in a sense, for the decay of a closed radioactive system. Although we can identify a point in time (an event), the uncertainty of a chronostratigraphical or chronometric age related respectively to either biological or radioactive change, may be very large in certain circumstances.

2. Distinct events may be identified by associated characters which, while not time dependent, may provide a potentially unique signature. Examples are of volcanic eruptions yielding tephra and bentonites or extraterrestrial bolides, each identified by a geochemical signature. These may be almost perfect time-correlation indicators, but they are so limited in duration that they are not easy to find and so of little practical use in routine correlation.

3. Environmental fluctuations provide a third correlation potential. The most obvious examples are of climatic and sea-level change and magnetic reversals. Nearly the whole sedimentary record is punctuated with evidence of such changes.

However, while the rate of change may be good for near-synchronous global correlation, the problem is that the evidence from one event looks much like that from another. This is even more troublesome in periodic phenomena related to sun-spot, annual, or other cycles (e.g. due to planetary perturbations). This whole group of phenomena will surely provide a future means of relatively precise correlation in favourable circumstances, provided that some method for distinguishing one cycle or fluctuation from another can be found. And so it is necessary to combine evidences of progressive phenomena or of unique signatures to calibrate the environmental fluctuations and cycles.

With respect to the distinction of the glacial episodes considered here, it seems probable that soon we will be able to calibrate, by biostratigraphical and radiometric means, the climatic, eustatic, and magnetic changes whose effects are ubiquitous. In other words, it may soon be possible, with confidence, to confirm or reject the supposed evidence for a late Sinian glacial episode.

4. A climatic fluctuation with a period of tens of

thousands of years is relevant to this argument. Cycles became fashionable again in about 1970, after a long period of scepticism. It is not necessary, for this argument, to decide between a 40 000-year period relating to the wobble of the Earth's tilt axis or a 90 000 to 100 000-year period for changes in the obliquity of the ecliptic; or combinations of these with precession of the equinox. It is sufficient to note that marked climatic fluctuation from this cause can be recognized, dated, and identified in later Quaternary time with glacials and interglacials. The cyclical mechanism is identifiable in oceanic deposits through much of Cenozoic time without temperatures reaching freezing point.

In other words the inevitability of the astronomical perturbations applying throughout later Earth history is confirmed in the gological record in some detail, back to Cenozoic time. It is not unexpected, therefore, that in appropriate Varangian glacial facies the same pattern of glacial and interglacial episodes may be distinguished within one glacial epoch.

This fact gives confidence in interpreting glacial and interglacial epochs—e.g. two Varangian and possibly one late Sinian when similar interglacial (tropical) environments (as indicated by reddening and dolomitization) may be observed on the much shorter planetary perturbation time-scale within at least some Varangian successions.

10.2. THE CHRONOSTRATIGRAPHICAL SCALE

The chronostratigraphical scale is the one relevant to our needs (e.g. Cowie *et al.* 1984). The naming and classification of proposed time divisions to accommodate the complex Precambrian rock units has been attempted in many different ways. A provisional scheme will be adopted for the purposes of this paper. It is essentially the same as the scheme used in Harland *et al.* 1982 and initiated for example in Harland & Herod (1975; see also Preiss & Forbes 1981); see table below. Vendian is accepted as the name for the latest Precambrian period, with an initial boundary probably near the beginning of the Varangian glacial epoch which is characterized throughout North-west Europe (e.g. Hambrey 1983).

The above scheme, however, is limited in that the time interval between Poundian Ediacaran faunas and Lower Tommotian faunas spans a great deal of history and perhaps deserves a name. We may look to the USSR, where good successions have been studied in detail.

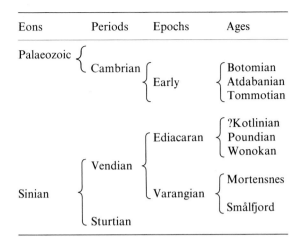

Eons	Periods	Epochs	Ages
Palaeozoic	Cambrian	Early	Botomian / Atdabanian / Tommotian
Sinian	Vendian	Ediacaran	?Kotlinian / Poundian / Wonokan
		Varangian	Mortensnes / Smålfjord
	Sturtian		

Sokolov & Fedonkin (1985) classified Vendian rock units of the East European platform into lower and upper divisions thus:

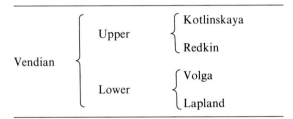

Vendian	Upper	Kotlinskaya / Redkin
	Lower	Volga / Lapland

In Sokolov & Fedonkin's volume 1, four divisions were listed based on biostratigraphical and climatic data:

Cambrian	
Vendian	Nemakit–Daldyn = Revenskiy / Kotlinian / Redkinian (Ediacarian) / Laplandian (Varangian)

We await with interest the detailed characterization of the Kotlinian and Nemakit–Daldyn, in case the hypothetical late Sinian glaciation may be fitted into that sequence.

10.3. THE CHRONOMETRIC SCALE

The initial Cambrian boundary has been calibrated many times and with widely divergent conclusions, e.g. Harland (1983b), Cowie & Johnson (1985). More recently, values ranging from 530 Ma to 620 Ma have been suggested. The lower estimates are based on good plutonic age determinations, but with equivocal structural relations (Cowie & Johnson 1985).

The higher estimates relate to whole-rock Rb–Sr determinations in which detrital clay minerals may have raised the age. An age of 570 Ma has long been used, but for an initial boundary which was later than the initial Tommotian. The age of 590 Ma for the initial Cambrian was used by Harland *et al.* 1982. Other time-scales give lower values.

The mid-Varangian age of 650 Ma was tentatively used in 1982. This suggested that an initial Vendian age might be *c.*650 to 670 Ma, so giving the Vendian a time span of 75 to 85 Ma.

Glauconite dates may be too young; whole-rock clay-mineral ages may be too old. Radiometric precision is still not quite adequate for the question in this contribution.

10.4. A TENTATIVE GLACIAL SEQUENCE

A tentative glacial sequence was proposed in 1983 to distinguish the hypothetical late Sinian from the Varangian and earlier events, because much, earlier literature had referred to 'a late Precambrian ice age', as though only one might be involved. The informal scheme prepared (Harland 1983) and elaborated (Hambrey & Harland 1985) was as follows:

Late Archaean to Early Proterozoic Glacial Era

Witwatersrand glacial period with four or more epochs
(*c.*2.65 Ga)
Huronian glacial period with three or more epochs
(*c.*2.3 Ga), possibly younger Early Proterozoic glaciations

Mid-Proterozoic Interglacial Era (2.0 to 1.0 Ga)

Late Proterozoic Glacial Era

Lower Congo glacial period with at least two epochs
(*c.*0.9 Ga)
Sturtian glacial period (with two main epochs)
(*c.*0.8 Ga)
Varangian glacial period (with two main epochs)
(*c.*0.65 Ga).
Late Sinian glacial epoch (*c.*0.60 Ga).

The above is a minimum history—many more glacial episodes may have filled out the story.

10.5. SOME POSSIBLE LATE SINIAN GLACIAL DEPOSITS

It is only fair to offer some details which form the basis of this hypothetical glaciation. These were listed in Harland 1983a, with occurrences in China, USSR, Poland, Sweden, Alaska, British Columbia, and southern Africa. It must also be added that, out of the hundreds of late Precambrian tillites, a large proportion cannot be dated, and so if a late Sinian glacial epoch were established, then many other occurrences might be candidates and many others might turn out to be Varangian—on the basis that many were formed at low or intermediate latitudes.

10.5.1. China

Two quite distinct provinces yield potential support from this hypothesis (Wang *et al.* 1981): a geosynclinal succession in Xinjiang and a platform succession in south-east China. The platform succession is nearly uniform over a large area but quite thin, so that the sequence must be incomplete—ranging as type Sinian through about 200 Ma. It is in this succession that the typical Nantuo tillite, possibly equivalent to the Sturtian glacials, was formed.

The Xinjiang geosynclinal sequence appears to be the most promising (Norin 1937; Wang *et al.* 1981), because three glacial epochs are represented in one succession. In the top unit of the Quruq Tagh Group is the Hankalchoug (Hangeerqiaoke) Formation (434 m thick). It occurs disconformably but always just beneath the phosphorus-rich and fossiliferous Lower Cambrian (e.g. Xitashan Formation). It is separated below from the Tereekan Formation (probably Varangian) by three formations. The uppermost of these, the Shuiquan Formation, contains microfossils similar to those in the Denying Formation of central and southern China. Elsewhere, *Vendotaenia* and Sabellitidae were found in strata correlated with those beneath.

Similar deposits, similarly placed, also occur in Qinghai, southern Shanxi, and western Henan, as well as on the northern slopes of the Quinling and Helan Shan Mountains in Henan and Shanxi. These are known as the *Luoquan Formation*, or in some areas as the *Zhengmuquan Formation*. They were described by Mu (1981) and in considerable sedimentological detail by Guan *et al.* (1981). In all these cases there is little doubt about the glacigenic origin of the sediments; only their age is in doubt. In each case there is a problem of their time-correlation, which it is difficult to assess even for those familiar with the rocks.

The Luoquan Formation which occurs in western Henan in a succession broken by disconformities, rests disconformably on the Dongjia (and older) to pre-Sinian rocks; striated pavements occur sporadically at the contact. It is overlain by the Dongpo Formation which disconformably underlies the trilobite-bearing Lower Cambrian Xinji Formation.

Abundant Dongjia and Luoquan microfloras range from Vendian through to Cambrian strata. Rocks in Anhui Province, possibly coeval with those immediately below Luoquan rocks, contain a rich fauna of Annelida and Pogonophora, and if correct a post-Varangian age is suggested. Radiometric data are numerous but not yet sufficiently constrained to be critical.

Guan *et al.* (1986) proposed the name 'Luoquan Glacial Period' for this suspected period of substantial glacial activity, in preference to the Late Sinian Glacial Epoch of Harland (1983a), because the new name can be referred to a particular stratigraphical section: as with the Varangian and Sturtian glacial periods. But this proposal could confuse the issue; the suspected period of glacial activity is late or post Ediacaran. The particular Luoquan stratigraphical sections may yet prove to be Varangian. There may be no such late Sinian Glacial Epoch, or it may be represented by the Hankalchoug Formation (Norin 1937), and not by Luoquan.

We still need an interim name for the hypothetical Late Sinian Glacial Epoch, to distinguish it from the Luoquan glacials which may or may not be of that age. The question remains: is the Luoquan Glacial Period late Sinian in age, in the sense of this chapter? It will be observed that the name 'Sinian' was selected because, in both regions of China, Sinian rocks provide the best chance to document this hypothetical event.

10.5.2. USSR

In the Tien Shan and other parts of Soviet Middle Asia, late Proterozoic tilloids are widely developed. The *Baykonor Formation* is one of at least two diamictite levels, and occurs close to the Precambrian–Cambrian boundary. However, it is neither of certain glacial origin nor certainly of late Sinian age.

10.5.3. North-west Africa

Tillites with apparent ages from 620 to 695 Ma occur at the base of the Bthaat Ergil Group in the Adrar Basin of Mauritania; at the base of the Serie Pourpree in Haggar, Algeria; and in the Pharusian Belt of Iforas in Mali.

10.5.4. Southern Africa

In the Schwarzrand Formation of Namibia, the Nomtas Formation overlies Ediacaran *Sabellidites*, but a glacial origin is not firmly established.

10.5.5. Other areas

Other possible contenders previously mentioned (e.g. Harland 1983) are probably no longer to be regarded as serious candidates. For example:

1. the Upper Tindit Group tillites of east-central Alaska were first thought to be latest Precambrian in age; but then significantly older;

2. the Siko tilloid and Vakkejokk breccia which overly Ediacaran fauna-bearing strata may not be glacial.

10.6. THE POSSIBLE CLIMATIC IMPACT ON THE BIOSTRATIGRAPHICAL RECORD

There has been a long-standing suspicion that the earliest Cambrian climates were cold. The Chinese name for the Cambrian Period, 'Hanwu Ji' (literally 'fiercely cold period') was given at a time when this notion was prevalent. Also, the difficulty in correlation at the Precambrian–Cambrian boundary (Harland 1974) may reflect a cold interruption.

The initial Cambrian transgression, which seems to be an eustatic phenomenon, could be related to melting of ice. With respect to sea-level, Brasier (1985) in summarizing events across the Cambrian threshold, noted two possible cold climatic episodes related to regressive conditions. The now well-established Varangian events were followed by the Ediacaran diversification of animal life, and the possible Kotlinian or late Sinian events were followed by the initial Cambrian 'explosion' of life. These two events were earlier confused and telescoped in proposing the argument for a climatic control over biological evolution (e.g. Harland & Wilson 1956; Harland & Rudwick 1964).

10.7. CONCLUSIONS

This chapter merely identifies an important area for research. Already the data available are probably sufficient to make significant advances if carefully addressed. It is predicted that these next five years will generate data sufficient to settle the question of a Late Sinian Glacial Epoch.

REFERENCES

Brasier, M. D. (1985). Evolutionary and geological events across the Precambrian–Cambrian boundary. *Geology Today*, Sept.–Oct. 1985, 141–46.

Cowie, J. W. *et al.* (1984). Chronostratigraphic definition of the Precambrian–Cambrian boundary. *Geol. Mag.* **121**, (6), 649.

Cowie, J. W. & Johnson, M. R. W. (1985). Late Precambrian and Cambrian geological time-scale. In *The chronology of the geological report*, (ed. N. J. Snelling), Geol. Soc. Mem., No. 10.

Frakes, L. A. (1979). *Climates throughout geologic time.* Elsevier, Amsterdam. 310 pp.

Guan Baode, Wu Ruitang, Hambrey, M. J., & Geng Wuchen (1986). Glacial sediments and erosional pavements near the Cambrian–Precambrian boundary in western Henan Province, China. *J. geol. Soc. Lond.*, **143**, 311–23.

Hambrey, M. J. (1983). Correlation of Late Proterozoic tillites in the North Atlantic region of Europe. *Geol. Mag.* **120**, (3), 209–32.

Hambrey, M. J. & Harland, W. B. (1985). The Late Proterozoic Glacial era. *Palaeogeogr., Palaeoclimatol., Palaeoecol.*, **51**, 255–72.

Harland, W. B. (1964). Critical evidence for a great Infra-Cambrian glaciation. *Geol. Rundsch.* **54**, 45–61.

Harland, W. B. (1974). The Pre-Cambrian–Cambrian boundary. In *Cambrian of the British Isles, Norden and Spitsbergen, Lower Palaeozoic rocks of the world*, Vol. 2, (Ed. C. H. Holland), pp. 15–42.

Harland, W. B. (1983*a*). The Proterozoic glacial record. *Geol. Soc. Am., Mem.*, **161**, 279–88.

Harland, W. B. (1983*b*). More time scales (essay review). *Geol. Mag.*, **120**, (4), 393–400.

Harland, W. B. & Herod, K. N. (1975). Glaciations through time. In *Ice ages, ancient and modern*, (eds A. E. Wright & F. Moseley), pp. 198–216. Geol. J. Spec. Iss., No. 6. Seel House Press, Liverpool.

Harland, W. B. & Rudwick, M. J. S. (1964). The great infra-Cambrian ice age. *Scient. Am.*, **211**, (2), 28–36.

Harland, W. B. & Wilson, C. B. (1956). The Hecla Hoek succession in Ny Friesland, Spitsbergen. *Geol. Mag.*, **93**, (4), 265–86.

Harland, W. B., Cox, A. V., Llewellyn, P. G., Pickton, C. A. G., Smith, A. G., & Walters, R. (1982). *A geologic time-scale*. Cambridge University Press.

Mu Yongji (1981). Luoquan Tillite of the Sinian System in China. In *Earth's pre-Pleistocene glacial record*, (eds M. J. Hambrey & W. B. Harland), pp. 402–13. Cambridge University Press.

Norin, E. (1937). Geology of the Western Qurug Tagh, Eastern Tien Shan. *Reports from the Scientific Expedition to the north western provinces of China under the leadership of Dr Sven Hedin. Sino-Swedish Expedition*, **3**, *Geology*, Thule, Stockholm.

Preiss, W. V. & Forbes, B. G. (1981). Stratigraphy, correlation and sedimentary history of Adelaidean (Late Proterozoic) basins in Australia. *Precambr. Res.* **15**, (3–4), 255–304.

Sokolov, B. S. & Fedonkin, M. A. (eds) Resp. Vendskaya sistema. Istoriko-geologicheskoye i paleontologicheskoye obsnovaniye. [The Vendian System. Historical–geological and palaeontological basis.] Tom 2, Stratigrafiya i geologicheskiye protsessy. [Vol. 2, Stratigraphy and geological processes.] An SSSR, Otd. Geol., Geofiz. in Geokhim., Nauka, Moscow.

Wang Yuelun, Lu Songnian, Gao Zhenjia, Lin Weixing, & Ma Guogan (1981). Sinian tillites of China. In *Earth's pre-Pleistocene glacial record*, (eds M. J. Hambrey & W. B Harland), pp. 386–401, Cambridge University Press.

11

Concluding remarks

M. D. Brasier & J. W. Cowie

11.1. INTRODUCTION

The problem of the stratigraphical and geographical placing of the Precambrian-Cambrian Global Stratotype Section and Point (GSSP) is not solved at present. Fortunately, the interest in this part of the geological column is active and lively in most parts of the world, so that progress is rapid and success is probable. However, it is not possible to claim a consensus for any agreed solution yet. Therefore this chapter is not headed 'Conclusions' because any conclusion could be swept aside by the rapid rising tide of research results and hypotheses.

The remaining three parts of the chapter are:

11.2. the turning point—palaeoceanographic comments (M.D.B.);

11.3. a voting result from a recent field trip in Newfoundland (J.W.C.);

11.4. a compilation of opinion, unavoidably subjective and selective, on the criteria (mainly from Cowie *et al.* 1986) applied to stratotype candidates in China, Newfoundland (Canada), and Siberia (USSR) (J.W.C.)

11.2. THE TURNING POINT [M.D.B.]

Previous chapters reveal geological and biological changes of great magnitude across the Precambrian–Cambrian boundary interval. Shelf successions, for example, display profound changes in sedimentary rate and style, probably related to transgressions (interspersed with regressions) and block movements, as the ancient Precambrian continents began to move apart and subside. A general pattern also emerges of increasing biotic diversity over this time-span, embracing invertebrate traces, skeletal and sclerotized invertebrates, calcareous algae, cyanobacteria, and phytoplankton. The variety of forms involved in this evolutionary explosion was uniquely great, raising many questions about cause and effect (e.g. why did cyanobacteria biomineralize at this time? see Riding 1985) and patterns of evolution (e.g. did Darwinian selection operate? see Conway Morris & Fritz 1984). This story, from the appearance of the first biomineralized worm tubes in the late Precambrian to the progressive biomineralization of trilobite cuticle and echinoderm tissue through approximately Atdabanian times (and approximately China Zone III to V times), probably spans a considerable period. Since there is no consensus on the absolute age span of this interval (Cowie & Johnson 1985; Odin *et al.* 1985), some obvious questions remain unanswered: how rapid was the whole 'Cambrian Explosion'? Does this broad succession contain an interval (i.e. a turning point) of highly accelerated evolution?

The effects of external factors on evolution at this time, e.g. sea-level rise (Brasier 1979, 1980, 1982); of associated igneous–tectonic events (Brasier 1985); and even of related marine geochemical environments (Brasier 1986a,b), have been considered by the writer. Do any of these show a burst of activity, suggestive of a 'geological turning point'?

Although there is clear evidence for sea-level rise, with widespread transgressive-regressive cycles on most platforms (e.g. Brasier 1980; Mount & Rowland 1981; Palmer 1981), globally synchronous sea-level events have not yet been proven. This may be partly because of the primitive state of biostratigraphy and chronostratigraphy at this level. It may also reflect the possibility that eustatic rise was essentially gradual, driven by oceanic ridge production, whereas smaller cycles responded to crustal buoyancy in differing tectonic settings (e.g. Donovan & Jones 1979; Wyatt 1986). Even so, events can be traced across cratons and into their adjacent terranes, so that provisional sea-level curves can be drawn (e.g. Notholt & Brasier 1986). Although such curves may prove useful for local correlation, the ?epeirogenic discontinuities at cycle tops provide the sharpest data points, and more effort should be spent in obtaining chronostratigraphical control on these. Ultimately, it may be possible to tie-in the sea-level curves, discontinuities, or unconformities, and more deep-seated tectonic and igneous events. Probable relationships between sea-level, habitat, and stages in the Cambrian radiations have already been discussed by the writer at some length (Brasier 1979, 1982).

Of the various geological changes at this time, igneous and tectonic events seem likely to have been the broadest in time-scale. Breakup of the Precambrian 'Pangaea', followed by rapid polar wandering, was probably initiated by late Precambrian times (e.g. Piper 1982; Thorpe *et al.* 1984), and the increase in the number of smaller crustal blocks with thinned margins may arguably have led to lowered elevation and transgressions (cf. Wyatt 1986). The latter may also have been enhanced by displacement of water from increased production of oceanic ridge material (cf. Pitman 1978). The 'Pan-African' thermal events of this time, however, suggest that something more than simple plate-tectonic processes are implicated: more than half the African craton and equally extensive areas in surrounding continental regions were involved in thermal and metamorphic events, with dates of African syntectonic magmatism tending to fall around 600 Ma, and modal dates (including cooling effects) of *c.*510 Ma (e.g. Shackleton 1981). Late Cadomian dates of around 540 Ma are also widespread in Western Europe and maritime Canada (see Odin *et al.* 1983, 1985; Thorpe *et al.* 1984; Cowie & Johnson 1985). These Pan-African events are generally mapped as pre-Lower Cambrian and can be seen below fossiliferous strata in Morocco (pre-?Tommotian algae and pre-*Eofallotaspis*), northern France (pre-*Allonnia*), England (post-*Charnia* and pre-*Sunnaginia neoimbricata* at Nuneaton; post-*Arumberia* and pre-*Mobergella* in Shropshire), and Massachusetts (pre-*Aldanella attleborensis*). It is therefore arguable that the late phase of plutonic intrusion in these regions is pre-*Aldanella* and post-*Arumberia* (e.g. Brasier, this volume, p. 99). If the *Aldanella attleborensis* fauna is simplistically correlated with China Zone II, and *Arumberia* with parts of China Zone 0, these Pan-African thermal events may have taken place during later Zone 0 or Zone I times. This would be remarkably close to the appearance of the earliest skeletal faunas. Is it possible that these profound thermal events helped to bring about corresponding changes in crustal buoyancy, palaeogeography, ocean chemistry, and atmosphere, thereby stimulating the evolutionary explosion?

A major phosphogenic episode at the Precambrian–Cambrian boundary (Cook & Shergold 1986) should be seen in this context of thermal and tectonic events. Phosphorites are widespread and episodic throughout the latest Precambrian and Cambrian (e.g. Notholt & Brasier 1986), but there is a particularly well-defined peak of phosphogenesis along the Palaeotethyan margin from China and India to Kazakhstan and Iran in strata of China Zone I times (Shergold & Brasier 1986; Brasier, this volume, pp. 66–7).

A scenario may be envisaged for this, involving an anoxic oceanic water mass that was overturned and flushed into shallow, formerly hypersaline, bays and lagoons, by a major pulse of sea-level and/or a phase of climatic cooling (e.g. Cook & Shergold 1986). The increasing evidence for a heavy $\delta^{13}C$ excursion in the pre-Tommotian Yudoma dolomites (Magaritz *et al.* 1986), correlated Zone 0 Dengying dolomites (Lambert *et al.* 1987), Krol E dolomites (Aharon *et al.* 1987) and the *Dolomie Inférieur* of Morocco (Tucker 1986) lend support to the hypothesis of an anoxic event causing loss of ^{12}C into black shales. The succeeding phosphorites, and the *sunnaginicus* Zone of the Lower Tommotian, have very light $\delta^{13}C$ values (op. cit.; Banerjee 1986), which have, rather implausibly, been explained as being due to mass extinction (Hsü *et al.* 1985) or a reduction in biomass (Margaritz *et al.* 1986). It seems more likely that negative $\delta^{13}C$ values resulted from several natural phenomena associated with the phosphogenesis. First and foremost, light organic carbon may have 'upwelled' into more extensive neritic environments, where bacterial oxidation led to liberation of phosphorus (e.g. Southgate 1986) and ^{12}C. Secondly, the associated explosive evolution of scavenging, heterotrophic invertebrates (signalled by increasing numbers of shells and trace fossils) must have dramatically increased the respiration of organic carbon, with similar effects. This increased heterotrophic activity is likely to have been stimulated by increased availability of upwelled phosphorus (Cook & Shergold 1986).

It does seem that there was a natural turning point, marked by the turnover from heavy to light $\delta^{13}C$ and the widespread appearance of phosphorites. This point lies at the base of the Tommotian, *sunnaginicus* Zone in Siberia and in the upper Krol dolomites of India (Aharon *et al.* 1987). A major transgression is indicated at about this level by a change from dolomites to phosphorites on the Yangtze Platform, and from dolomites to argillaceous limestones on the Siberian Platform. A second decline in $\delta^{13}C$ is recorded above China Zone II, around the undated base of the Badaowan Member in the Yangtze Gorges and Meishucun (Hsü *et al.* 1985; Xu *et al.* 1985; Lambert *et al.* 1987).

The next stage of exploration requires integrated analysis of early skeletal fossils and isotopes from sections with Manykayan/Nemakit–Daldynian to Meishucunian and Tommotian assemblages. The most promising sections for this are in the Anabar region of Siberia, the Elburz Mountains of Iran, and the western Yangtze Platform of China. It would be ideal to have this information available when making a decision on the placing of the Precambrian–Cambrian boundary. But whatever the decision, integrated information of this kind will provide for better stratigraphy and some more-informed opinions on these highly interesting events.

11.3. VOTE OF FIELD-TRIP PARTICIPANTS ON THE SUITABILITY OF THE BURIN PENINSULA SECTIONS FOR THE GLOBAL STRATOTYPE POINT FOR THE PRECAMBRIAN–CAMBRIAN BOUNDARY, 17 AUGUST 1987 (J.W.C.)

The following three questions were put to the 31 participants (including 10 Voting members of the IUGS Working Group on the Precambrian–Cambrian Boundary) during the field excursion and the returned secret ballot papers were collected at the end of the excursion and are tabulated as follows:

1. The stratigraphic succession in the Chapel Island Formation on the southern Burin Peninsula, Newfoundland, illustrated by exposures at Grand Bank Head, Fortune Head, and Little Dantzic Cove, is (good, fair, poor) with regard to each of the following criteria proposed by Cowie et al. (1986) for the placement of a global boundary stratotype.

Criterion		Good	Fair	Poor	Abstain
a.	Correlation on a global scale	17	12	1	1
b.	Completeness of exposure	26	4	0	1
c.	Adequate completeness of deposition	21	8	0	2
d.	Abundance & diversity of well preserved fossils	8	16	6	1
e.	Favourable facies for widespread correlation	18	11	1	1

Criterion		Good	Fair	Poor	Abstain
f.	Freedom from structural complication	25	5	0	1
g.	Freedom from metamorphism	8	18	4	1
h.	Amenability to magnetostratigraphy	1	5	7	18
i.	Amenability to geochronometry	1	2	11	17
j.	Accessibility and conservation	27	3	0	1
k.	Adequate thickness of sediments	25	5	0	1

2. Based on the above criteria, the stratigraphical succession in the Chapel Island Formation on the southern Burin Peninsula is suitable for placement of a Precambrian–Cambrian boundary stratotype point.

Decision	Voting members	All participants
Yes	7	25
No	3	5
Abstain	0	1

3. The base of the *Phycodes pedum* ichnofossil Zone, within the *Sabellidites cambriensis* interval in the section at Fortune Head, provides an acceptable level for the Precambrian–Cambrian boundary.

Decision	Voting members	All participants
Yes	7	20
No	3	8
Abstain	0	3

Vote invigilators: A. R. Palmer &
                                 M. D. Brasier
J. W. Cowie, Chairman Precambrian–Cambrian Boundary Working Group

11.4. SUBJECTIVE ASSESSMENT OF THREE GSSP CANDIDATES (J.W.C.)
(Estimated at present state of research)

Criterion	Meishucun, China	Burin Peninsula, Canada	Ulakhan–Sulugur, USSR
1. Correlation on a global scale	Fair	Fair	Fair
2. Completeness of exposure	Good	Good	Good
3. Adequate thickness of sediments	Poor	Good	Fair
4. Adequate completeness of deposition	Fair	Good	Fair
5. Abundance and diversity of well-preserved body fossils	Good	Fair	Good
6. Favourable facies for widespread correlation	Fair	Good	Fair
7. Freedom from structural complication	Good	Good	Good
8. Freedom from metamorphism	Good	Fair	Good
9. Amenability to magnetostratigraphy	Good	Poor	Good
10. Amenability to geochronometry	Good	Poor	Fair
11. Accessibility—all aspects	Good	Good	Fair
12. Conservation	Good	Good	Good
13. Abundance and diversity of trace-fossils	Good	Good	Poor

REFERENCES

Aharon, P., Schidlowski, M., & Singh, I. B. (1987). Chronostratigraphic markers in the end-Precambrian carbon isotope record of the Lesser Himalaya. *Nature*, **327**, 699–702.

Banerjee, D. M. (1986). Proterozoic and Cambrian phosphorites—Regional review: Indian subcontinent. In *Proterozoic and Cambrian phosphorites*, (eds P. J. Cook & J. H. Shergold), pp. 70–90. Cambridge University Press.

Brasier, M. D. (1979). The Cambrian radiation event. In *The origin of major invertebrate groups*, (ed. M. R. House), pp. 103–59. Syst. Assoc. Spec. Vol., 12. Academic Press, London.

Brasier, M. D. (1980). The Lower Cambrian transgression and glauconite–phosphate facies in western Europe. *J. geol. Soc. Lond.*, **137**, 695–703.

Brasier, M. D. (1981). Sea-level changes, facies changes and the late Precambrian–early Cambrian evolutionary explosion. *Precambr. Res.*, **17**, 105–23.

Brasier, M. D. (1985). Evolutionary and geological events across the Precambrian–Cambrian boundary. *Geol. Today*, **1**, 141–6.

Brasier, M. D. (1986a). Precambrian–Cambrian boundary biotas and events. In *Global Bio-Events*, (ed. O. Walliser), Lecture Notes in Earth Sciences, **8**, Springer-Verlag, Berlin.

Brasier, M. D. (1986b). Why do lower plants and animals biomineralise? *Paleobiology*, **12**, 241–50.

Conway Morris, S. & Fritz, W. H. (1984). *Lapworthella filigrana* n. sp. (*incertae sedis*) from the Lower Cambrian of the Cassiar Mountains, northern British Columbia, Canada, with comments on the possible levels of competition in the early Cambrian. *Paläontol. Z.*, **58**, 197–209.

Cook, P. J. & Shergold, J. H. (1986). Proterozoic and Cambrian phosphorites—nature and origin. In *Proterozoic and Cambrian phosphorites*, (eds P. J. Cook and J. H. Shergold), pp. 369–86. Cambridge University Press.

Cowie, J. W. & Johnson, M. R. W. (1985). Late Precambrian and Cambrian geological time scale. In *The chronology of the geological record*, (ed. N. J. Snelling), pp. 47–64. Mem. geol. Soc. Lond., 10.

Cowie, J. W., Ziegler, W., Boucot, A. J., Bassett, M. G., & Remane, J. (1986). Guidelines and Statutes of the International Commission on Stratigraphy (ICS). *Cour. Forsch.-Inst. Senckenberg*, **83**, 1–14.

Donovan, D. T. & Jones, E. J. W. (1979). Causes of world-wide changes in sea level. *J. geol. Soc. London.*, **136**, 187–92.

Hsü, H. J., Oberhansli, H., Gao, J. Y., Sun Shu, Chen Haihong, & Krahenbühl, U. (1985). 'Strangelove ocean' before the Cambrian explosion. *Nature*, **316**, 809–11.

Lambert, I. B., Walter, M. R., Zang Wenlong, Lu Songnian, & Ma Guogan (1987). Palaeoenvironment and carbon isotope stratigraphy of Upper Proterozoic carbonates of the Yangtze Platform. *Nature*, **325**, 140–2.

Magaritz, M., Holser, W. T., & Kirschvink, J. L. (1986). Carbon-isotope events across the Precambrian–Cambrian boundary on the Siberian Platform. *Nature*, **320**, 258–9.

Mount, J. F. & Rowland, S. M. (1981). Grand Cycle A (Lower Cambrian) of the southern Great Basin: a product of differential rates of relative sea-level rise. *US geol. Surv. Open-File Rep.*, **81–743**, 143–6.

Notholt, A. G. & Brasier, M. D. (1986). Proterozoic and Cambrian phosphorites—Regional review: Europe. In *Proterozoic and Cambrian phosphorites*, eds P. J. Cook and J. H. Shergold). Cambridge University Press.

Odin, G. S. *et al.* (1983). Numerical dating of Precambrian–Cambrian boundary. *Nature*, **301**, 21–3.

Odin, G. S., Gale, N. H., & Doré, F. (1985). Radiometric dating of Late Precambrian times. In *The chronology of the geological record*, (ed. N. J. Snelling), pp. 65–72. Mem. geol. Soc. Lond., 10.

Palmer, A. R. (1981). On the correlatability of Grand Cycle tops. *US geol. Surv. Open-File Rep.*, **81–743**, 156–9.

Piper, J. D. (1982). The Precambrian palaeomagnetic record: the case for the Proterozoic supercontinent. *Earth planet. Sci. Lett.*, **59**, 61–89.

Pitman, W. C. (1978). Relationships between eustasy and stratigraphic sequences of passive margins. *Bull. geol. Soc. Am.*, **89**, 1389–1403.

Riding, R. (1985). Cyanophyte calcification and changes in ocean chemistry. *NERC News J.*, **3**, 11–12.

Shackleton, R. M. (1981). Introduction. In *Lower Palaeozoic of the Middle East, Eastern and Southern Africa, and Antarctica*, (ed. C. H. Holland), pp. 1–4, Wiley-Interscience, Chichester.

Shergold, J. H. & Brasier, M. D. (1986). Proterozoic and Cambrian phosphorites—specialist studies. Biochronology of Proterozoic and Cambrian phosphorites. In *Proterozoic and Cambrian Phosphorites*, (eds P. J. Cook & J. H. Shergold). Cambridge University Press.

Southgate, P. N. (1986). Proterozoic and Cambrian phosphorites—specialist studies: Middle Cambrian phosphatic hardgrounds, phoscrete profiles and stromatolites and their implications for phosphogenesis. In *Proterozoic and Cambrian phosphorites*, (eds P. J. Cook and J. H. Shergold), pp. 327–51. Cambridge University Press.

Thorpe, R. S., Beckinsale, R. D., Patchett, P. J., Piper, J. D. A., Davies, G. R., & Evans, J. A. (1984). Crustal growth and late Precambrian–early Palaeozoic plate tectonic evolution of England and Wales. *J. geol. Soc. Lond.*, **141**, 521–36.

Tucker, M. E. (1986). Carbon isotope excursions in Precambrian/Cambrian boundary beds, Morocco. *Nature*, **319**, 48–50.

Xu Dao-Yi, Zhang Qin-Wen, Sun Yi-Yang, & Yan Zheng (1985). Three main mass extinctions—significant indicators of major natural divisions of geological history in the Phanerozic. *Modern Geol.*, **9**, 1–11.

Wyatt, A. R. (1986). Post-Triassic continental hypsometry and sea level. *J. geol. Soc. Lond.*, **143**, 907–10.

Index

The symbol (R) after a term indicates that it refers to a rock suite or lithostratigraphic or chronostratigraphic subdivision: groups, formations, series, plutonic bodies, volcanics etc.